RESONANCE IONIZATION SPECTROSCOPY 2000

Previous Proceedings in the Series of
Symposia on Resonance Ionization Spectroscopy

Year	Symposium	Publisher	ISBN
1998	Ninth	AIP Conf. Proceedings Vol. 454	1-56396-810-X
1996	Eighth	AIP Conf. Proceedings Vol. 388	1-56396-611-5
1994	Seventh	AIP Conf. Proceedings Vol. 329	1-56396-437-6

Other Related Titles from AIP Conference Proceedings

559   Spectral Line Shapes, Volume 11, 15th ICSLS
Edited by Joachim Seidel, April 2001, 1-56396-991-2

551   Atomic Physics 17: XVII International Conference on Atomic Physics; ICAP 2000
Edited by Ennio Arimondo, Paolo De Natale, and Massimo Inguscio, February 2001, 1-56396-982-3

543   Atomic and Molecular Data and Their Applications: ICAMDATA—Second International Conference
Edited by Keith A. Berrington and Kenneth L. Bell, November 2000, 1-56396-971-8

477   Atomic Physics 16: Sixteenth International Conference on Atomic Physics
Edited by William E. Baylis and Gordon W. F. Drake, May 1999, 1-56396-752-9

467   Spectral Line Shapes: Volume 10, 14th ICSLS
Edited by Roger M. Herman, March 1999, 1-56396-754-5

430   Fourier Transform Spectroscopy: Eleventh International Conference
Edited by James A. de Haseth, May 1998, 1-56396-746-4

364   Fast Elementary Processes in Chemical and Biological Systems: 54th International Meeting of Physical Chemistry
Edited by Andre Tramer, May 1996, 1-56396-564-X

To learn more about these titles, or the AIP Conference Proceedings Series, please visit the webpage **http://www.aip.org/catalog/aboutconf.html**

RESONANCE IONIZATION SPECTROSCOPY 2000

Laser Ionization and Applications Incorporating RIS
10th International Symposium

Knoxville, Tennessee   8–12 October 2000

EDITORS
James E. Parks
The University of Tennessee, Knoxville, Tennessee

Jack P. Young
Oak Ridge National Laboratory, Oak Ridge, Tennessee

Melville, New York, 2001
AIP CONFERENCE PROCEEDINGS ■ VOLUME 584

Editors:

James E. Parks
Department of Physics and Astronomy
The University of Tennessee
401 Nielsen Physics Building
Knoxville, TN 37996-1200
USA

E-mail: jeparks@utk.edu

Jack P. Young
Chemical and Analytical Sciences Department
Oak Ridge National Laboratory
P.O. Box 2008, MS 6142
Oak Ridge, TN 37831-6142
USA

E-mail: youngjp@ornl.gov

The article on pp. 112-122 was authored by U.S. Government employees and is not covered by the below mentioned copyright.

L.C. Catalog Card No. 2001093539
ISBN 0-7354-0024-5
ISSN 0094-243X
Printed in the United States of America

CONTENTS

SESSION I

SESSION II

SESSION VI

SESSION VII

SESSION VIII

POSTERS

PREFACE

Laser Ionization and Applications Incorporating RIS: The Tenth International Symposium on Resonance Ionization Spectroscopy and Its Applications (RIS-2000), was hosted by The University of Tennessee's Institute of Resonance Ionization Spectroscopy (IRIS) in Knoxville, Tennessee, October 8-12, 2000. This tenth anniversary meeting brought the symposium back to East Tennessee, where it originated in 1981. IRIS has served as the "academic home" of RIS since 1985. RIS-2000 paid homage to the past with an historical perspective on RIS presented by Dr. G. Samuel Hurst, an inventor of the technology and the first IRIS director. A shifting research emphasis in the RIS community over the years has gradually changed the tenor of the symposium and has broadened its appeal to other scientific fields. A future 2004 symposium is planned and will be held in Leuven, Belgium where this new direction will be more apparent.

The outstanding work presented at RIS-2000 and included in these proceedings clearly illustrates that excellent research based on RIS technology is alive and well. The strong participation of graduate students was another encouraging sign. RIS-2000 opened with a well-attended short course on the basics of RIS. This popular program offers a solid overview of the field for researchers new to RIS, such as graduate students, as well as those looking for new ways to apply the technology. The short course and a welcoming reception were followed by a week's worth of presentations, both oral and poster. The meeting concluded with a banquet on October 12, where awards were presented for the best work presented by a student and in a poster presentation.

In 1992, the RIS International Advisory Committee set out to endow an award that would recognize exceptional work contributed by a graduate student at each RIS symposium. With the support of several RIS supporters, the fund has met its endowment mark and now provides a cash award for each conference. The RIS-2000 award was split between two students, Iain Spencer of the University of Glasgow and Klaus Blaum of the Universität Mainz. Iain's work was on "High Intensity Laser Generation of Proton Beams for the Production of ß^+ Sources in Positron Emission Tomography." Klaus' presentation was on "High-Resolution, Three-Step Resonance Ionization Mass Spectrometry of Gadolinium." Both students won $500 for their work, which is presented in these proceedings.

The poster session is another important aspect of the RIS symposium tradition. Designed to encourage discussion and share results, this session truly allows for a meaningful exchange of ideas among scientists. One outstanding poster presentation also wins an award. For RIS-2000, the award went to Lynne Robson, a graduate student from the University of Glasgow, for her work entitled, "Ultrafast Laser Time-of-Flight Mass Analysis of Laser-Desorbed Atoms and Molecules." Lynne's work is included in these proceedings as well. The participation of graduate students was a key factor in the success of RIS-2000, as all presented quality work and demonstrated an excellent capability to move RIS forward into new frontiers of scientific discovery.

The University of Tennessee was pleased to host RIS-2000, with the Hilton Knoxville, very near the University campus, serving as the conference venue.

Accommodations and meals were provided at the site, allowing participants to give maximum attention to the details of the conference. The meeting's success can be attributed largely to the hard work and sound advice of the RIS International Advisory Committee, as well as the RIS-2000 local organizing committee. Financial support from the University of Tennessee, Atom Sciences, Inc., and UT-Battelle LLC., was also crucial to bringing the symposium to fruition. Catherine Longmire, publications coordinator for the University of Tennessee Department of Physics and Astronomy, is appreciated for her work on all aspects of the conference planning and operations, including the final coordination of these proceedings.

James E. Parks
Jack P. Young

SPONSORS

Financial support from the following organizations is gratefully acknowledged:

Atom Sciences, Inc.

The University of Tennessee Office of Research

The University of Tennessee Department of Physics and Astronomy

The University of Tennessee Science Alliance

UT-Battelle, LLC

HOST

The University of Tennessee hosted RIS-2000, which was held at the Hilton Knoxville, 501 W. Church Ave., Knoxville, Tennessee, USA.

Laser Ionization and Applications Incorporating RIS: The Tenth International Symposium on Resonance Ionization Spectroscopy and Its Applications, was coordinated through the Institute of Resonance Ionization Spectroscopy of The University of Tennessee.

CONFERENCE ORGANIZATION

CONFERENCE CHAIRMAN

James E. Parks

LOCAL ORGANIZING COMMITTEE

C. H. (Winston) Chen
Catherine Longmire
John Miller
James E. Parks
Tom Whitaker
Jack Young

INTERNATIONAL ADVISORY COMMITTEE

Bruce Bushaw	USA	John Miller	USA
John Crawford	Canada	V.I. Mishin	Russia
Costas Fotakis	Greece	Nicolo Omenetto	Italy
Juergen Grotemeyer	Germany	James E. Parks	USA
Dag Hanstorp	Sweden	John Vickerman	UK
Wim Hogervorst	The Netherlands	Klaus Wendt	Germany
Sam Hurst	USA	Nicholas Winograd	USA
Kenneth Ledingham	UK	Xiangyuan Xu	China
Bernhard Lehmann	Switzerland		
Ian Lyon	UK		
Naohiko Mikami	Japan		

SESSION I

Photoionization of Dipeptides with Femtosecond Laser Pulses

Shixin Sun, Vasil Vorsa,[1] and Nicholas Winograd

184 Materials Research Institute Building, Department of Chemistry, The Pennsylvania State University, University Park, Pennsylvania 16802

Abstract. Imaging time-of-flight secondary ion mass spectrometry is a very important technique for characterizing organic and biological systems. However, the signal is very low due to small amounts of desorbed molecules. During the ion bombardment process, neutral species as well as positive and negative ions are desorbed, with neutral molecules being the main species. Therefore, ionization of the neutral species should improve sensitivity. Here, we report the use of femtosecond laser photoionization of ion beam desorbed dipeptides to enhance the detection sensitivity. Ionization using 200 nm laser pulses is found to break the C-C bond in the C-C=O group or the C-C bond between the functional group and the backbone, which agrees with the α-cleavage mechanism proposed previously. Photoionization produces 5-10 fold higher dipeptide ion yields than the secondary ion yields achieved with the use of the incident ion beam only.

INTRODUCTON

Imaging time-of-flight secondary ion mass spectrometry (SIMS) is an important technique used to characterize organic, inorganic and biological systems [1,2]. During the ion bombardment process, neutral species as well as positive and negative ions are desorbed, and the yield of neutral species is a few orders of magnitude higher (typically 10^4:1) than that of the ions. Therefore, efficient analysis of the neutral species desorbed would potentially produce more signal compared to desorbed ion detection. Earlier efforts focused on using nanosecond lasers to photoionize neutral species desorbed from surfaces [3,4]. This has been a success with atomic species, however, this results in extensive photo-induced fragmentation in molecular systems.

Ultrashort pulsed lasers are promising new ionization sources for the detection of molecules desorbed from surfaces. The enhanced absorption rates generated by high-power ultrashort pulses make it possible to "outrun" the neutral fragmentation channels, which are prevalent when employing nanosecond excitation of molecular

---

[1] *Air Products and Chemicals, Inc., 7201 Hamilton Blvd., Allentown, PA 18195*

CP584, Resonance Ionization Spectroscopy 2000: 10th Int'l. Symp., edited by J. E. Parks and J. P. Young
© 2001 American Institute of Physics 0-7354-0024-5/01/$18.00

species. Research [5,6] shows that the degree of fragmentation can be reduced considerably by using ultrashort laser pulses for ionization.

Here, we report the use of femtosecond laser photoionization of ion beam desorbed dipeptides to enhance the detection sensitivity. The ionization mechanism is examined and found to agree with the α-cleavage mechanism proposed previously [7].

EXPERIMENTAL

The dipeptides were purchased from Sigma and used without further purification. The analytes were pressed into In foil directly and transferred into the vacuum chamber for analysis. The details of the time-of-flight secondary ion mass spectrometer (TOF-SIMS) apparatus have been described previously [3]. Briefly, molecules are desorbed from the surface by a 25 KeV Ga^+ ion beam (Ionoptika) in forms of positive ions, negative ions or neutral species. Ions formed at the sample surface are pulse extracted into a dual-field reflectron TOF mass analyzer by applying a positive or negative potential directly to the sample stage. Ions are separated from each other in the flight tube according to their masses and detected by a micro-channel-plate (MCP) detector. The signal from the MCP detector is sent to a analog-to-digital converter (ADC) whose output is transferred to a PC for storage and later analysis. The sample block is cooled to -125°C with liquid nitrogen to minimize the background signal from sample sublimation.

The Ti:sapphire laser system (Clark-MXR, Inc.) employed in the experiments has been reported elsewhere [8]. Briefly, an Argon ion laser pumps the mode locked Ti:sapphire oscillator which produces 800 nm pulses with ~50 fs pulse width and 3 nJ per pulse energy. The 800 nm pulses are stretched to ~300 ps by multiple passes on a single grating and then amplified by a Ti:sapphire regenerative-amplifier and a Ti:sapphire post-amplifier. Both amplifiers are pumped by the second harmonic of Nd: YAG lasers. The amplified beam has 5.5 mJ per pulse energy. Finally, the beam is compressed to ~100 fs with 3.5 mJ per pulse energy. The output of this laser system is frequency doubled, tripled or quadrupled to produce 400, 267 and 200 nm respectively. The beam is coupled into the analysis chamber by a 25-cm focal length CaF_2 lens. The laser spot size and position over the sample is controlled by moving the lens mounted on an x, y, z manipulator outside the chamber.

RESULTS AND DISCUSSION

The 200 nm photoionization mass spectra of Val-Val and Leu-Trp are shown in Figure 1 and Figure 2 respectively. Neither of these two dipeptides produces significant molecular ions, although molecular ions can be easily seen from SIMS spectra. The base peak in the Val-Val spectrum results from α-cleavage (Figure 3), which was proposed previously [8].

A previous study shows that for aliphatic amino acids, the lowest ionization energies correspond to the removal of one lone pair electron from the nitrogen atom and form a radical site on the nitrogen atom [7]. The radical site on the nitrogen atom initiates the reaction by donating an electron to form a double bond between the carbon and nitrogen atoms. The amido group donates another electron, and the C-C bond is broken and a carboxyl group radical is formed. In Val-Val, there are two nitrogen atoms. Due to the electron withdrawing ability of the C=O group, the lone pair ionization energy of the amido group nitrogen is a little bit higher than that of the other nitrogen, so only the C-C bond shown in Figure 1 is broken.

For tryptophan and tyrosine, the lowest ionization energy corresponds to the removal of an electron from the functional group. The ionization potential for tryptophan is 7.2 eV [9], corresponding to the energy required to remove one electron from the functional group. The ionization potential for leucine is 8.5 eV [9], corresponding to the energy required to remove one electron from the N atom. In Leu-Trp, due to the electron withdrawing capability, the energy required to remove one electron from the functional group is increased. Therefore, the energies required to remove one electron from the functional group or from the N atom are similar, so the fragmentation from both ends can be seen in the spectrum, with the functional group signal intensity (130 amu) a little bit higher.

Comparing the spectra of SIMS (data not shown) and photoionization, it is very obvious that molecular ion signal can be easily detected in SIMS experiments, which means that some of the molecules are desorbed and ionized intact. However, there are no molecular ions detected in photoionization experiments. The reason for excessive fragmentation has to be further investigated. In addition to comparing mass spectra, it is useful to compare total ion yields formed by SIMS and photoionization. Comparing the spectra of SIMS (data not shown) and photoionization, it is very obvious that the total ion yield resulting from photoionization is much higher than that from SIMS. For Leu-Trp, it is 25 fold higher.

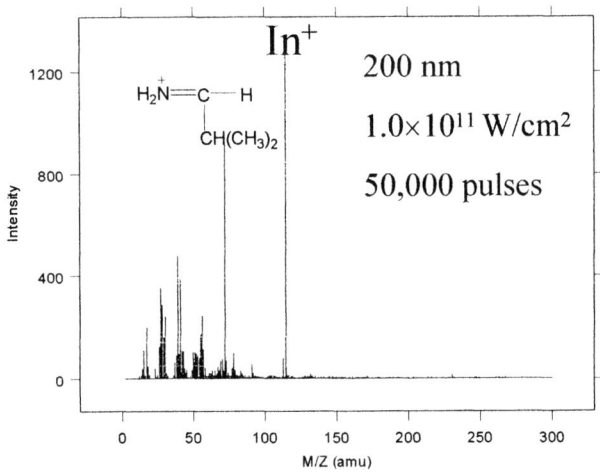

Figure 1. Laser photoionization spectrum of Val-Val

5

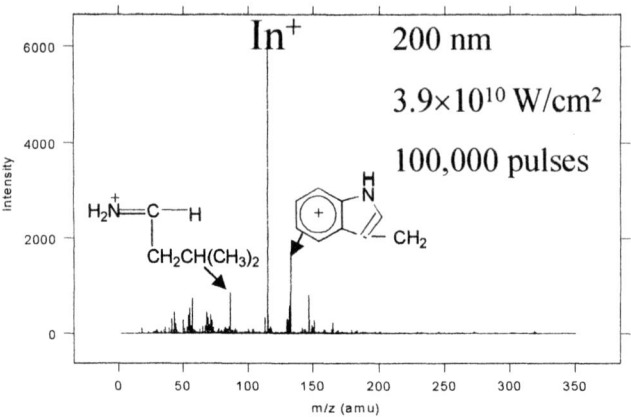

Figure 2. Laser photoionization spectrum of Leu-Trp

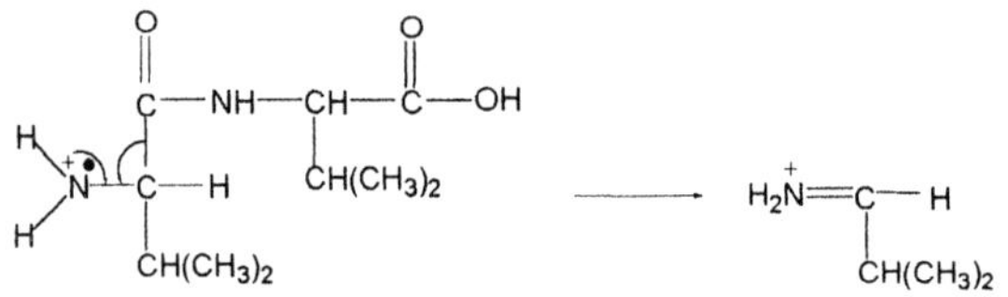

Figure 3. α-cleavage mechanism

CONCLUSIONS

For photoionization of dipeptides with 200 nm femtosecond laser pulses, there is no molecular ion signal detected and the α-cleavage is the main fragmentation mechanism. When comparing the total signal intensity produced with photoionization and SIMS, photoionization produces more signal.

ACKNOWLEDGEMENTS

The authors want to thank the National Institute of Health and the National Science Foundation for financial support.

REFERENCES

1. Winograd, N., *Anal. Chem.*, **65**, 622A (1993)

2. Benninghoven, A., Hagenhoff, B., Niehius, E., *Anal. Chem.*, **65**, 630A (1993)

3. Wood, M., Zhou, Y., Brummel, C. L., Winograd, N., *Anal. Chem.*, **66**, 2425 (1994)

4. Grotemeyer, J., Schlag, E. W., *Org. Mass Spectrum.* **23**, 388 (1988)

5. Kilic, H. S., Ledingham, K. W. D., Kosmidis, C., McCanny, T., Singhal, R. P., Wang, S. L., Smith, D. J., Langley, A. J., Shaikh, W., *J. Phys. Chem. A*, **101**, 817 (1997)

6. Szaflarski, D., El-Sayed, M. A., *J. Phys. Chem.*, **92**, 2234 (1988)

7. Vorsa, V., Kono, T., Willey, K. F., Winograd, N., *J. Phys. Chem. B*, **103**, 7889 (1999)

8. Willey, K. F., Vorsa, V., Braun, R. M., Winograd, N., *Rapid Commun. Mass Spectrom.*, **12**, 1253 (1998)

9. Campbell, S., Beauchamp, J. L., Rempe, M., Lichtenberger, D. L., *Int. J. Mass Spectrom. Ion Processes*, **117**, 83 (1992)

Laser Ion Kinetics: Application of a Single Shot Femtosecond Pump-Probe Technique

Ralf Heinicke, Carsten Grun and Jürgen Grotemeyer

Institut für Physikalische Chemie, Christian-Albrechts Universität zu Kiel,
Olshausenstraße 40, D-24098 Kiel, Germany

Abstract : In this contribution the measurements of a single shot femtosecond laser pump-probe technique on substituted benzalacetones are reported. The technique is based on counter propagating femtosecond laser pulses in a supersonic beam of low density of sample molecules and simultaneous probe detection by ion or fragment ion formation through a reflectron time-of-flight mass spectrometer. The application of this technique to medium sized organic molecules reveals some insight into the electron transfer process during ionization through a 1+1 multi photon absorption procedure.

INTRODUCTION

Isomerization and fragmentation reactions of ion from organic molecules have been a major subject of mass spectrometric investigations throughout the last five decades. On one hand the understanding of the details of a certain fragmentation reaction in one molecule allows a globalization for analytic purposes. On the other hand investigations in the elementary steps of a reaction in a mass spectrometer yield in information on the dynamics and energetics of the process.

Using lasers for ion formation has opened new ways to investigate the dynamics of elementary reactions. Pioneered by the work of A. Zewail [1,2] , laser pump-probe experiments together with mass spectrometric detection allow a direct measurement of the intramolecular dynamics.

Usually a pump-probe experiment consists of a pump laser pulse which induces in case of femtosecond laser pulses a coherent population in an excited electronic state of an atomic or molecular species. This state is interrogated by a time delayed probe laser pulse. In widening the pump-probe techniques to larger molecular systems we have introduced a new technique, the laser ion kinetics [3], to different organic molecules. The concept behind this technique is the use of counterpropagating laser pulses in a special mannerwhich allows the storage of the complete pump-probe spectrum of a molecule or ion as a one dimensional spatial map of ion densities. By use of a time-of-flight mass spectrometer these ion densities can then be transferred into flight time differences which allow the unfolding of this ion density map into discrete values of intensities at different

CP584, *Resonance Ionization Spectroscopy 2000: 10th Int'l. Symp.*, edited by J. E. Parks and J. P. Young

delays between pump and probe laser pulses. As a result the direct measurement of the dynamics of intramolecular processes in isolated molecular systems is possible.

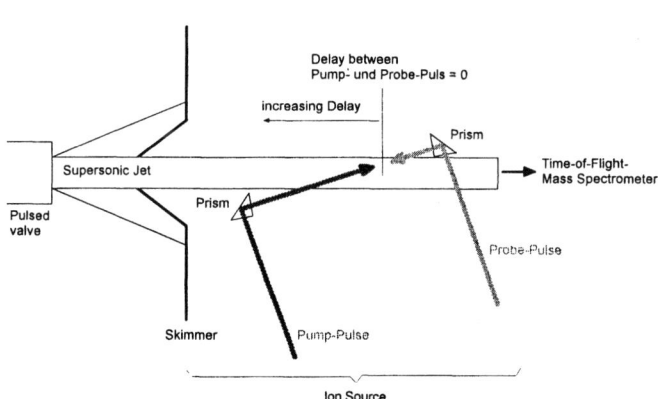

Figure 1. General fragmentation of substituted benzalacetones

In this contribution we are reporting on investigations of substituted benzalacetones, which are undergoing a intramolecular electrophilic aromatic substitution reaction [4, 5] as shown in figure 1.

EXPERIMENTAL SETUP

The mass-spectrometric system used for the investigations presented here is described in detail elsewhere [6]. It consists of three main parts, a supersonic beam together with a laser desorption setup, a modified laser ionization source and a reflectron time-of-flight system. A supersonic molecular beam of Argon and the analyte is produced through a pulsed nozzle. In case of molecules with low volatility, a laser desorption setup with an infra red laser is used. The laser is focused onto a probe tip mounted directly in front of the nozzle in the vacuum thus desorbing the sample into the supersonic beam.

For the 1+1 multi photon ionization a femtosecond dye laser pump by an excimer laser is used. This laser delivers energies up to 80 µJ at pulse durations ranging from 100 to 500 fs. The setup for the laser ion kinetics method is shown in figure 2. The femtosecond laser pulse is separated by using a 50% mirror which splits the laser beam into two equal parts with nearly the same intensity. While the first pulse is focused into the ion source of the mass spectrometer the second part is delayed via an optical delay line. In the ion source both pulses are introduced through a window into a prism setup which allows the adjustment as counter propagating pulses in a slanting way. The adjustment of both pulses can be done in several ways.

Figure 2. Setup for the laser ion kinetics experiments. The overlap of the counter propagating laser pulses with the supersonic molecular beam is approximately 250 µm.

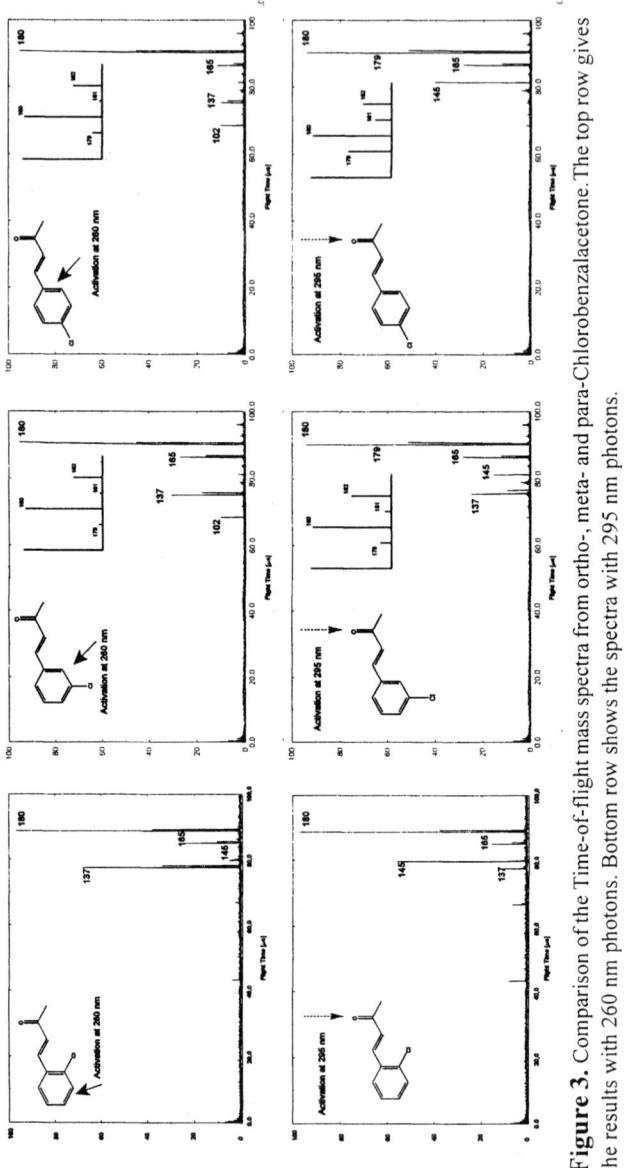

Figure 3. Comparison of the Time-of-flight mass spectra from ortho-, meta- and para-Chlorobenzalacetone. The top row gives the results with 260 nm photons. Bottom row shows the spectra with 295 nm photons.

In the normal setup the probe pulse is illuminating the complete neutral beam before the probe pulse enters the part of the neutral beam from the opposite side. By moving the optical delay line the first point of the intercept of both pulses can be moved toward the center of the neutral beam. As a result the onset of the pump-probe experiment can be investigated in greater detail, while the first experiment guarantees that the most of the pump pulse can be probed with a second pulse.

RESULTS AND DISCUSSION

Wavelength dependence of the fragmentation in substituted Benzalacetones.

In figure 3 the multi photon ionization mass spectra of the three isomeric chloro-benzalacetones are displayed taken at two different wavelength (260 & 295 nm) with a 250 femtosecond laser pulse. Comparing the mass spectra strong differences in the fragmentation pattern are observed not only by shifting the wavelength from 295 nm to 260 nm but also by changing the position of the substituent attached to the aromatic ring.

The striking difference between the mass spectra at the various wavelength is due to the formation of the charge at either the oxygen or the aromatic ring system. Using 295 nm photons for ionization the most prominent fragment ion is the chromylium moiety. This ion is formed by the intramolecular substitution reaction. This ion can only be produced if the charge is established at the oxygen. The differences in intensity is easily explained by standard rules of organic synthesis. The chloro substituent has an inductive and meso-meric influence which slows down the ring closure reaction and the subsequent fragmentation reaction. Therefore other reactions are found with higher intensities (Scheme 1).

Shifting the wavelength to 260 nm changes the fragmentation behavior. Here the loss of an acetyl group resulting in the formation of the ion at m/z 137 is strongly favored over the intramolecular aromatic substitution reaction. Therefore the charge must be formed at the aromatic moiety. The charge has to be transported to the oxygen via the side chain. At that point a change in the overlap of the electronic states from a molecular orbital to an atomic orbital has to take place, which is known to need more time than the direct cleavage reaction leading to the ion at mass 137. The differences in the mass spectrometric observable fragmentation reactions are a clear prove for the concept of localized charge states within an organic molecule [7].

Investigation of the dynamics of the ion formation processes

As demonstrated in the previous part the differences in the ionization and fragmentation pattern can be accounted to initial position of the charge formation and to its transport through the molecule. Since this process must show clearly a dynamic behavior, our laser ion kinetic pump-probe setup should yield in a measurement of the velocity of the charge distribution within the molecule. Figure 4 displays the results of this pump-probe measurement. Clearly the mass signals for the formation of the molecular ion show a different behavior for the various isomers and wavelength applied. Since the technique of counter propagating laser pulses is used, the each point of the falling flank of the mass signal is due to a different spatial point in the ion source, thus representing a different delay between pump and probe beam. Starting with an activation with 260 nm photons the transients for the formation of the molecular ion from the three chloro-isomers follow nearly a Gaussian distribution function. As shown in figure 4 the formation of the molecular ion is observed up to a time limit of 1.5 ps. Fitting these time multiplexed transient mass signals result in an observation of a bi-exponential decay with two different transients as shown in table 1. The shape of the time multiplexed signal is completely different at 295 nm. The most pronounced difference in the transient signal at 295 nm, compared to the activation at 260 nm, is the clear and intense bi-exponential behavior. As shown in table 1 the first transient has decay times similar to these observed at 260 nm activation.

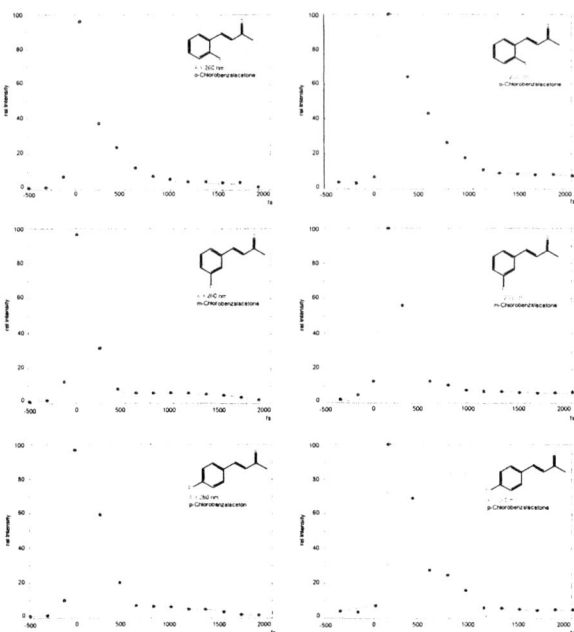

From theoretical considerations these results can be interpreted in the following way. Inducing a charge through the $n-\pi^*$ transition in the oxygen of the carbonyl function should have no immediate effect to the electronic structure of the aromatic ring system. Therefore it can be assumed that the ionization in this moiety should yield the same decay times for each isomeric compound. Comparing the values of table 1 the fast transient has nearly the same decay for the different isomers as well as for the investigated wave length. As a result this transient must be due to the charge formation at the carbonyl group.

Figure 4. Femtosecond transients ionization spectra of the molecular ions from the three different isomeric Chlorobenzalacetones. (Laser pulse duration ~ 120 fs). The transients are accumulated over 50 laser shots. The transients of the left side are acquired with 260 nm, on the right side with 295 nm

Tabelle 1 Decay times from the fits of the transients for the parent ions of the three isomeric Chloro- and Methoxy-benzalacetones at 260 and 295 nm wavelength

Compound	($\lambda = 260$ nm)		($\lambda = 295$ nm)	
	τ_{decay} (fs); 1. transient	τ_{decay} (fs); 2. transient	τ_{decay} (fs); 1. transient	τ_{decay} (fs); 2. transient
o-Chloro-benzalacetone	210 ± 15	455 ± 60	$230 \pm (20$	490 ± 75
m-Chloro-benzalacetone	190 ± 10	420 ± 70	215 ± 20	430 ± 60
p-Chloro-benzalacetone	220 ± 15	470 ± 60	235 ± 20	510 ± 75

Taking these results together a clear picture for the different charge movements through the isomeric molecules arises. Starting with 260 nm photons the charge is induced in the aromatic ring system of the organic molecule. Shifting the charge over the molecule leads to the fragmentation reaction producing the ion at mass 137. The intensity of this reaction in the mass spectrum as well as the decay of the transient in the pump-probe experiment supports also the influence of the substituent placed at the aromatic ring system. It is well known from basic organic chemistry that a halogene attached to an aromatic ring increases the basicity of the aromatic ring system and has simultaneous supports mesomeric effects. Therefore the differences in the decay of the molecular ions are understandable.

Conclusions

The ultra fast photo dissociation dynamics the isomeric benzalacetones have been investigated in the gas phase using a combination of femtosecond laser pulses and reflectron time-of-flight mass spectrometry. By using the method of laser ion kinetics or time multiplexing is shown for the first time that a complete set of dynamic data can be recorded from one single shot femtosecond pump-probe experiment. The combination of normal mass spectra as well as pump-probe data allows a detailed insight into the dynamics of molecular processes.

Acknowledgment

This work is supported by the Deutsche Forschungsgemeinschaft (Project Gr 917/19-1,2) and in part by the Volkswagen Stiftung. Support from the Fonds der Chemischen Industrie is gratefully acknowledged. The authors thank Dr. C. Weickhardt for his help throughout the measurements.

References

1. A.H. Zewail: *Ferntochemistry - ultrafast dynamics of the chemical bond,* World Scientific Publishing Vol. I & 11, Singapore (1994).
2. J. Manz, L. Wöste: *Femtosecond Chemistry,* VCH VerlagsgesellschftmbH, Weinheim, Germany (1995).
3. a) R. Heinicke, J. Grotemeyer; Appl. Phys. B, **71**, 419 (2000).
    b) R. Heinicke, C. Grun, J. Grotemeyer, Eur..J. Mass Spectrom. **6**, 143 (2000).
4. J. March, *Advanced Organic Chemistry*, McGraw-Hill 1977
5. B. Schaldach, B. Grotemeyer, J. Grotemeyer, H.Fr. Grützmacher; Org.Mass Spectrom. 16, 410 (1981).
6. U. Boesl, K. Walter, J. Grotemeyer, E.W. Schlag; Anal. Instrum. **16**, 151 (1987).
7. R. Weinkauf, P. Schanen, D. Yang, S. Soukara, E.W. Schlag; J. Phys.Chem. 99, 11255 (1995).

Ultrafast Laser Time-of-Flight Mass Analysis of Laser-Desorbed Atoms and Molecules

S.M. Hankin[1], L. Robson[1], A.D. Tasker[1], K.W.D. Ledingham[1],
T. McCanny[1], R.P. Singhal[1], C. Kosmidis[2], P. Tzallas[2], A.J. Langley[3],
P.F. Taday[3], E.J. Divall[3]

[1]Department of Physics & Astronomy, University of Glasgow, G12 8QQ, UK
[2]Department of Physics, University of Ioannina, 45110 Ioannina, Greece
[3]ASTRA Laser Facility, CLRC Rutherford Appleton Laboratory, Chilton, OX11 0QX, UK

Abstract. Femtosecond time-of-fight mass spectra of laser-desorbed gallium arsenide (GaAs) and trinitrobenzene (TNT) reveal the characteristic features and differences of femtosecond atomic and molecular ionisation. Significant yields of multiply-charged atomic fragments, parent molecular ions and structure-specific fragment ions of labile molecules have been obtained. An optimum desorption laser intensity for the transfer of neutral intact parent molecules in to the gas phase has been observed to be molecule-dependent. This work demonstrates the potential of ultrafast laser ionisation and has significant implications for analytical and environmental studies of solid materials.

INTRODUCTION

The emerging field of ultrafast laser-molecule interactions has highlighted new areas of physics and chemistry and is becoming an important tool for trace molecular analysis [1-4]. From a mass spectrometry perspective, the generation of molecular or structurally-characteristic ions is a key feature in the technique's analytical ability. Short-pulse laser radiation offers the unique attributes of very high intensities (10^{12-17} W cm^{-2}) with a pulse duration shorter than the vibrational, rotational and dissociative timescales of molecules. These attributes have led to suggestions that femtosecond laser radiation can be regarded as an efficient, universal and soft ionisation source [5-7].

Comparative studies [8-11] using short-pulse (ps and fs) and nanosecond lasers have been carried out for atoms and molecules in the gas phase to show the advantages of ultrafast ionisation. However, the evolution of FLMS as an analytical technique requires the use of methods to introduce solid-phase species to the gas-phase for ionisation. In addition to studies of elemental and inorganic materials [12,13], the potential of FLMS for the analysis of biomolecules [14] and the detection of environmentally-hazardous materials is considerable [7]. Current research by the Glasgow group focuses on coupling femtosecond laser ionisation with a laser desorption source for the analysis of solid-phase labile molecules.

CP584, *Resonance Ionization Spectroscopy 2000: 10th Int'l. Symp.*, edited by J. E. Parks and J. P. Young
© 2001 American Institute of Physics 0-7354-0024-5/01/$18.00

EXPERIMENTAL

i. Time-of-Flight Mass Spectrometer

The instrument used in this work, schematically shown in Figure 1, was a reflectron time-of-flight (ToF) mass spectrometer comprising a sample load-lock, source chamber and flight tube. The core of the system is a spherical stainless steel chamber of 30 cm diameter. Ports on the chamber are fitted with fused quartz windows allowing laser irradiation and direct viewing of the sample. The 1.5 m long flight tube houses the reflectron and MCP detector. The source chamber and flight tube are pumped to a base pressure of ~10^{-9} Torr using two rotary-backed turbomolecular pumps.

Macroscopic samples are glued to a stainless steel stub using conducting silver-epoxy adhesive. Powder samples for analysis are dissolved in a suitable volatile solvent, deposited on to the stub and allowed to dry. Following evaporation of the solvent, the stub is admitted to the source chamber by means of a rotary-pumped load-lock and transfer arm. The stub is positioned on a manipulator arm in the centre of the chamber which allows for the movement in three dimensions and rotation about the vertical axis.

Ion extraction and acceleration is achieved using the sample stage and a two-plate ion optic designed specifically for this instrument [15]. An electrostatic potential of +2.7 kV is applied to the sample stage from a HT power supply. Potentials of +2 kV and +520 V are applied to the first and second plates respectively. The ions are guided into a Reflectron electrostatic mirror using XY deflection plates mounted after the ion extraction optics. The ions are detected by a multi-channel plate detector (Galileo) maintained at +2kV. Signal output from the detector is coupled to a digital oscilloscope (LeCroy, 9344C) for single-shot and averaged data collection. A PC installed with GRAMS/32 software (Galactic), connected to the oscilloscope through a GPIB interface, is used for data acquisition, mass calibration and analysis.

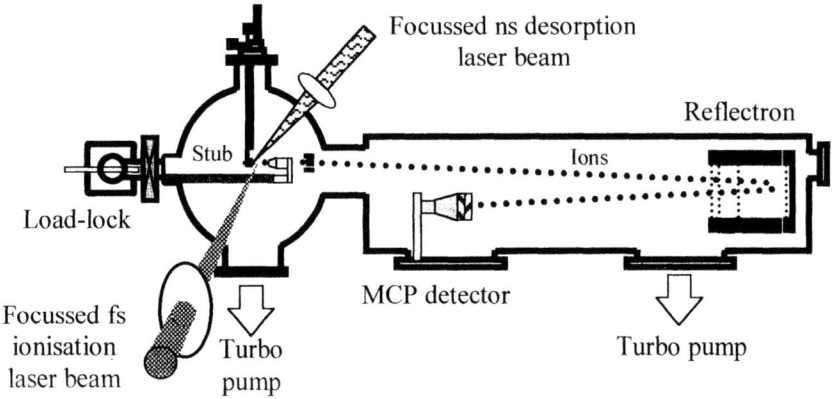

FIGURE 1. Reflectron Time-of-Flight Mass Spectrometer.

ii. Laser Systems

Laser desorption of the solid samples was achieved using the fourth harmonic output (266 nm, 5 ns) from a Nd:YAG laser (Minilite I, Continuum) focussed onto the sample stub. The energy of the desorption laser beam was controlled using an optical attenuator (Newport, 935-5) and monitored using a joulemeter (Molectron, J4-09). Focussing the ~7 mm diameter beam using a 27 cm focal length lens generated intensities up to 1×10^{10} W cm^{-2}.

Post-ionisation was achieved using the ASTRA femtosecond laser system [16] of the Rutherford Appleton Laboratory. The Ar-ion pumped mode-locked titanium-sapphire oscillator (Spectra-Physics) produced pulses of ~9 nJ pulse^{-1} at a wavelength of ~790 nm and pulse duration of 50 fs. The low-energy pulses were stretched to ~300 ps, amplified in a second Ti:Sapphire rod pumped by 140 mJ in 20 ns of the second harmonic from a Nd:YAG laser (Spectra Physics). Variable attenuation of the beam was achieved using a $\lambda/4$ waveplate & polariser positioned before the pulses were re-compressed to eliminate any non-linear effects and optical damage in the attenuator at high intensity. The amplified pulses (~5 mJ pulse^{-1}) were re-compressed to 50 fs in a grating pair and directed into the source of the mass spectrometer for ionisation of the laser-desorbed sample. Focussing the ~1 cm diameter beam using a 30 cm focal-length lens yielded intensities as high as 5×10^{15} W cm^{-2}, determined by calculation and from the thresholds for Ar^{n+} ionisation.

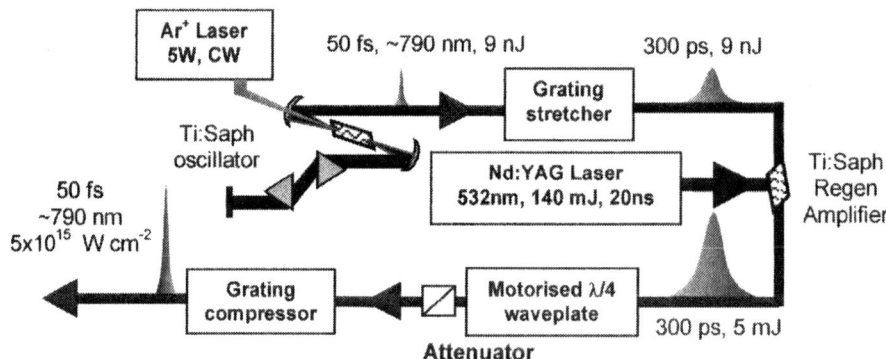

FIGURE 2. Ti:Sapphire Regenerative Amplifier Femtosecond Laser System.

Accurate spatial and temporal alignment of the desorption and ionisation laser pulses is critical for achieving maximum ion signal or studying the dynamics of laser desorption. The timing between the desorption and ionisation laser pulses was accurately controlled using a pulse/delay generator (SRS, DG535). A master trigger (T$_0$) from the femtosecond oscillator was used to trigger the delay generator. A variable 37 μs delay TTL pulse fired the flashlamps of the desorption laser. The Q-switch was fired at a fixed delay of 150 μs with respect to the flashlamps for optimum energy. The delay between desorption and post-ionisation was variable from 0-37 μs.

RESULTS AND DISCUSSION

With atoms, ionisation is the only channel. However, with molecules this is not the case: dissociation may compete with ionisation and fragmentation may play an important role. In addition, multiply-charged ions formed at higher intensities may be highly unstable due to Coulombic repulsion. These initial results show the qualitative differences between atomic and molecular ionisation under strong-field laser irradiation.

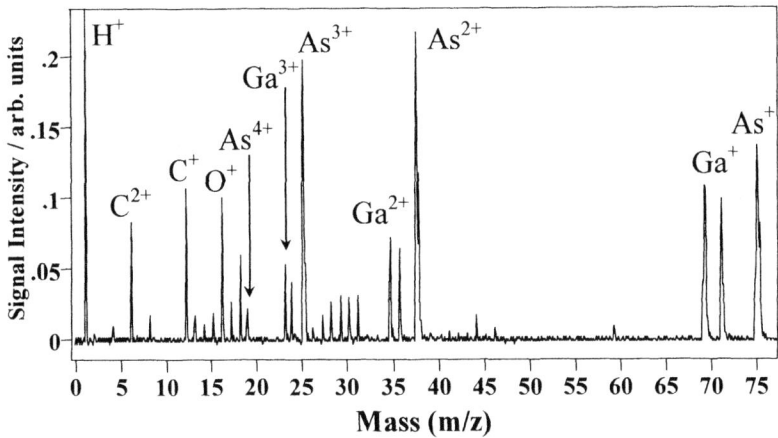

FIGURE 3. Femtosecond Ionisation ToFMS of Laser-Desorbed Gallium Arsenide (GaAs).

Figure 3 shows the femtosecond ionisation mass spectrum of laser-desorbed Gallium Arsenide recorded at intensities of 1×10^9 W cm^{-2} and 1.8×10^{15} W cm^{-2} for desorption and ionisation respectively. The spectrum exhibits singly-charged atomic ions of As (m/z=75) and the isotopes of Ga (m/z=69 & 71) in agreement with the natural abundances. Multiply-charged atomic ions were observed unambiguously up to the +3 charge-state for Ga and +4 for As. The isotopic abundances of ^{69}Ga and ^{71}Ga were resolved for all charge-states. The As^{2+} and As^{3+} signals dominate the spectra at intensities greater than 5×10^{14} W cm^{-2}. This is a reflection of the relative differences in the ionisation potentials for the two atoms as shown in Table 1.

TABLE 1. Ionisation Potentials [17] for Ga^{n+} and As^{n+} (n=1-3).

Atom	1st IP / eV	2nd IP / eV	3rd IP / eV	4th IP / eV
Ga	5.99	20.51	30.71	64
As	9.81	18.63	28.35	50.13

The low mass signals assigned as H$^+$, C^{n+} and O$^+$ arise from ionisation of hydrocarbon and water surface-contamination co-desorbed with the GaAs. These ion signals decreased as the surface layers were ablated, exposing material in the bulk to desorption. At laser intensities greater than 5×10^{14} W cm^{-2} minor higher mass ion peaks were observed at m/z=150 and 225 and were attributed to As$_2^+$ and As$_3^+$ clusters.

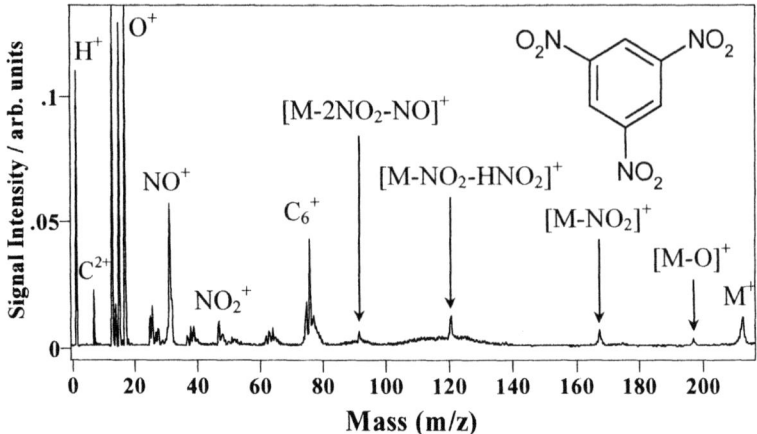

FIGURE 4. Femtosecond ToFMS of Laser-Desorbed Trinitrobenzene.

Figure 4 shows the mass spectrum of trinitrobenzene (TNB) recorded at intensities of 3.3×10^8 W cm^{-2} and 4×10^{15} W cm^{-2} for desorption and ionisation respectively. The spectrum exhibits a significant degree of molecular fragmentation yielding structurally-specific and non-specific fragments. In addition to the small parent ion at m/z=213, the mass spectrum shows ion signals characteristic of the TNB molecule: $[C_6H_3N_3O_5]^+$ (m/z=197), $[C_6H_3N_2O_4]^+$ (m/z=167), $[C_6H_2NO_2]^+$ (m/z=120) and $[C_6H_3O]^+$ (m/z=91). These dissociative channels arise from the loss of H, O, NO and NO_2 from the parent. Low mass fragment ions were identified as H^+, C^{2+}, N^+, O^+, C_n^+ (n=1-6), NO^+ and NO_2^+. The effect of the desorption and ionisation laser intensities on the ion yields for TNB and other nitro-aromatic molecules has been studied and will be reported elsewhere.

The laser desorption step plays as important a role as the ionisation step in terms of inducing molecular fragmentation. Molecular dissociation prior to ionisation may occur under intense desorption conditions yielding small neutral fragments. Preliminary investigations have revealed that higher ion yields of the parent ions and structure-specific fragments are accompanied by lower yields of the small fragment ions.

An optimum laser desorption intensity is desirable for desorption of intact neutral species representative of the sample, prior to direct ionisation during the nanosecond laser pulse. Furthermore, the optimum laser intensity was also observed to be dependent on the molecule being analysed [18]; in these experiments the optimum intensity for TNB was found to be $\sim 3.3 \times 10^8$ W cm^{-2}. Notably, the peak shapes for laser-desorbed analytes exhibit a significantly different appearance to that of gas phase analytes. This is due to the large kinetic energy distribution in the ions as a result of the desorption event and energetic fragmentation [19,20].

CONCLUSIONS

The coupling of a nanosecond laser for molecular desorption with the ASTRA femtosecond laser has extended the applicability of femtosecond laser mass spectrometry to the analysis of involatile solid materials. The time-of-fight mass spectra of GaAs and TNB have revealed the characteristic features and differences of femtosecond atomic and molecular ionisation of laser-desorbed species. Significant yields of multiply-charged atomic fragments, parent molecular ions and structure-specific fragment ions of labile molecules have been obtained. An optimum desorption laser intensity for the transfer of neutral intact parent molecules in to the gas phase has been observed to be molecule-dependent. This work demonstrates the attributes of femtosecond ionisation of laser-desorbed species and has significant implications for analytical studies of semi-conductors, nitro-aromatic explosive molecules and photo-labile environmental pollutants.

ACKNOWLEDGMENTS

The authors acknowledge NERC for a Postdoctoral Fellowship (SMH) and EPSRC for funding and Studentships (LR and ADT).

REFERENCES

1. Corkum, P.B., MY Ivanov, M.Y., Wright, J.S., *Annu. Rev. Phys. Chem.* **48**, 387, (1997).
2. Codling, K., Frasinski, L.J., *J. Phys. B: At. Mol. Opt. Phys.* **26**, 783, (1993).
3. Burlingame, A.L., Boyd, R.K., Gaskell, S.J., *Anal. Chem.* **70**, R647, (1998).
4. Ledingham, K.W.D., Singhal, R.P., *Int. J. Mass Spectrom. Ion Proc.* **163**, 149, (1997).
5. Hankin, S.M., Villeneuve, D.M., Corkum, P.B., Rayner, D.M., *Phys. Rev. Lett.* **84**, 5082, (2000).
6. Matsumoto, J., Lin, C-H., Imasaka, T., *Anal. Chem.* **69**, 4524, (1997).
7. Grun, C., Heinicke, R., Weickhardt, C., Grotemeyer, J., *Int. J. Mass Spectrom.* **185/186/187**, 307, (1999).
8. Weinkauf, R., Aicher, P., Wesley, G., Grotemeyer, J., Schlag, E.W., *J. Phys. Chem.* **98**, 8381, (1994).
9. Ledingham, K.W.D., Kosmidis, C., Georgiou, S., Couris, S., Singhal, R.P., *Chem. Phys. Lett.* **247**, 555, (1995).
10. Matsumoto, J., Lin, C-H., Imasaka, T., *Anal. Chim. Acta* **343**, 129, (1997).
11. Lockyer, N.P. and Vickerman, J.C., *Int. J. Mass Spectrom.* **176**, 77, (1998).
12. He, C., Basler, J.N., Becker, C.H., *Nature* **385**, 797 (1997).
13. Nicolussi, G.K., Pellin, M.J., Lykke, K.R., Trevor, J.L., Mencer, D.E., Davies, A.M., *Surf. Interface Anal.* **24**, 363, (1996).
14. Vorsa, V., Willey, K.F., Winograd, N., *Anal. Chem.* **71**, 574, (1999).
15. McLean, C.J., McCombes, P.T., Jennings, R., Ledingham, K.W.D., Singhal, R.P., *Nuc. Instrum. Meth. Phys. Res. B* **62**, 285, (1991).
16. Langley, A.J., Girard, N., Mohammed, I., Ross, I.N., Taday, P.F., *1998-99 CLF RAL Annual Report* (RAL-TR-1999-062), 187.
17. Weast, R.C. (Ed), *Handbook of Chemistry and Physics*, CRC Press: Boca Raton, 1988.
18. Kinsel, G.R., Lindner, J., Grotemeyer, J., Schlag, E.W., *J. Phys. Chem.* **95**, 7824, (1991).
19. Reid, S.A. *Chem. Phys. Lett.* **301**, 517 (1999).
20. Miller, J.C. and Haglund, R.F. (Eds.), *Laser Ablation: Mechanisms and Applications*, Springer: Berlin, 1991.

Improvement of the detection limit of hydrogen by Lyman-α RIS

Y. Miyake[1], S. Makimura[1], K. Shimomura[1], Y. Matsuda [2],
P. Bakule[2], P. Strasser [2], R.J. Scheuermann[2], K. Nagamine[1,2]

[1] *Meson Science Laboratory, High Energy Accelerator Research Organization (KEK-MSL)
Oho, Ibaraki-305-0801, Japan*
[2] *Muon Science Laboratory, Institute of Physics and Chemical Research (RIKEN),
Saitama 351-01, Japan*

Abstract. We have been pursuing an ultra-sensitive method for detecting and extracting hydrogen isotopes by utilizing a resonant ionization scheme via the 1S-2P-unbound transition. A time-of-flight measurement by the use of pulsed lasers coupled with mass separation enables us to distinguish hydrogen isotopes with a very low background. By preparing a new "Slow Ion Optics" to extract hydrogen isotope ions, the detection limit of the present method is improved to be much better than 3×10^3 atoms/cm^3.

INTRODUCTION

At KEK -MSL we have been pursuing the "Ultra Slow Muon Project", in which thermal Muonium (designated as Mu; consisting of a μ^+ and an e^-, can be considered to be a light isotope of hydrogen) atoms are generated from the surface of a hot tungsten foil [1] placed at the primary 500 MeV proton beam line and ionized by intense lasers synchronized with the emission of Mu.

In the course of the development of the laser resonant ionization method to ionize Mu utilizing the Lyman-α generation, it was found that Lyman-α RIS is also powerful tool for the detection and extraction of hydrogen isotopes [2]. In this paper, we report results obtained by an apparatus recently constructed in order to extract hydrogen isotopes with a better transmission.

EXPERIMENT

Experiments are performed using Slow Ion Optics which consists of an immersion lens (SOA lens [3]), a magnetic bend for mass separation, an electrostatic mirror, and five sets of electrostatic quadrupoles. A high purity (99.9999%) tungsten (W) foil with 50 μm thickness (obtained from Metallwerk Plansee GmbH) is placed inside the SOA target chamber . The W foil is capable of being heated up to 3000

CP584, *Resonance Ionization Spectroscopy 2000: 10th Int'l. Symp.*, edited by J. E. Parks and J. P. Young
© 2001 American Institute of Physics 0-7354-0024-5/01/$18.00

K by a pulsed DC current which is turned off for 1 ms, in order to avoid interaction of the magnetic field produced by the DC current with the resonantly ionized ions. The target chamber is evacuated down to 2.0×10^{-9} mbar at 2300 K. The tungsten target is cleaned by surface treatment in 3.0×10^{-7} mbar of oxygen at 1800 K for about 10 hours, and then heated up to 2300 K for the experiments. The heated W foil is used as a catalyzer to dissociate H_2 molecules into H atoms through the interaction between the molecules and the hot W surface. The experimental set-up is shown in Fig.1(left).

In order to efficiently ionize the hydrogen atoms dissociated from the H_2 molecules near the W surface, we are adopting a resonant ionization scheme via the 1s → 2p → unbound transition. The generation of the Lyman-α light is achieved by a sum-difference frequency mixing method using two 212.55 nm photons for two-photon resonant excitation of the $4p^5 5p[1/2]_0$ state in krypton, subtracted by a photon with a tunable difference wavelength [4]. Fig. 1 (right) shows the resonant ionization scheme for the hydrogen atom isotopes and the frequency wave mixing scheme for the Lyman-α wavelength generation. The 212.55 nm beam is obtained by quadrupling a single mode OPO (850 nm) laser (Continuum Mirage 800) output using β-Ba_2BO_4 crystals. The output energy of the 212.55 nm beam used for this experiment is 1-2 mJ/pulse, since further amplification of the OPO output is not performed. The difference wavelength between the $4p^5 5p[1/2,0]$ state of krypton and the desired Lyman-α wavelength is generated by a broad-band TiS laser system (STI LRL). In order to ionize hydrogen, the difference wavelength is adjusted to 845.0 nm in order to obtain the hydrogen Lyman-α of 121.57 nm. For ionizing

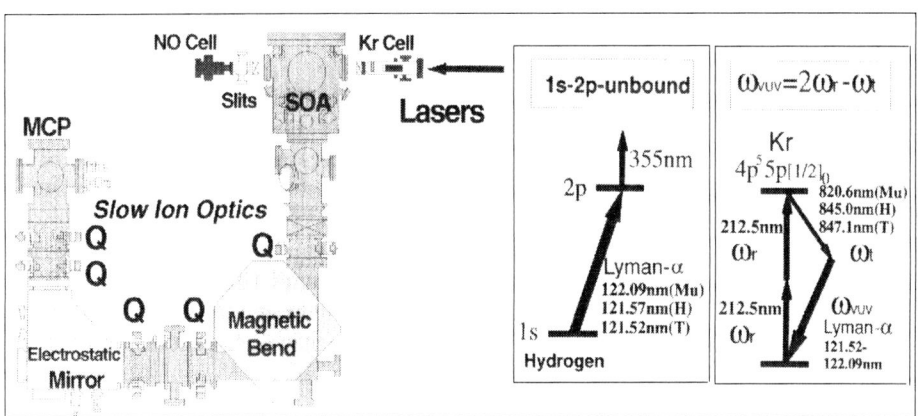

FIGURE 1. Layout of the slow ion optics consisting of an immersion lens (SOA), a magnetic bend, an electrostatics mirror and five sets of electrostatic quadrupoles (left figure). Resonant ionization scheme via the 1s→ 2p → unbound transition for the hydrogen atom isotopes and the scheme for the Lyman-α generation via the sum-difference frequency mixing method using two 212.55 nm photons for two-photon resonant excitation of the $4p^5 5p[1/2]_0$ state in krypton, subtracted by a photon with a tunable difference wavelength (right figure).

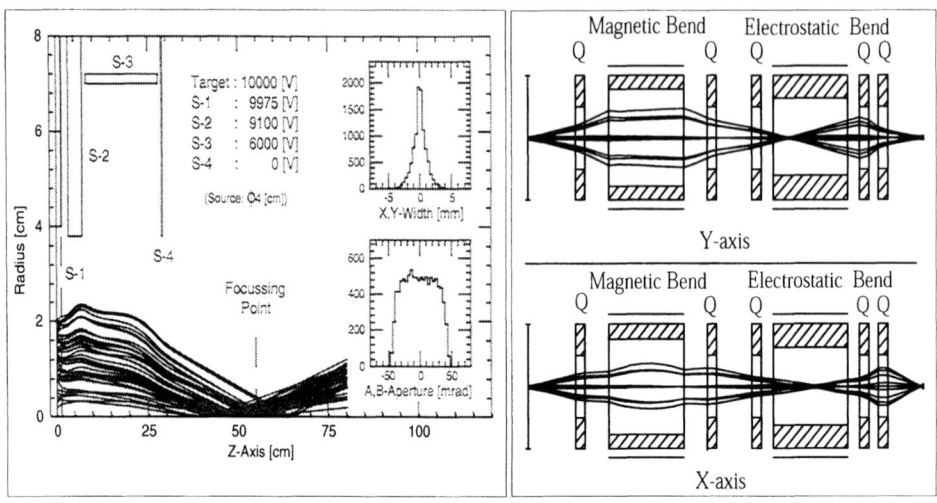

FIGURE 2. Monte Carlo simulation of the slow muon extraction by the SOA Lens (left figure). Simulations of the beam transport optics by the GIOS code (right figure).

the 2p state of H, frequency-tripled light from a Nd:YAG laser (Spectra-Physics GCR130) with a wavelength of 355 nm is used.

Ions accelerated by the SOA lens are transported to a micro-channel plate (MCP) detector located 3.8 m from the target. Any ion can be distinguished by setting the bending magnet in the slow optics and identified through the mass/charge(Q) value by the time-of-flight (TOF) spectrum with reference to the laser pulse (two-dimensional measurement) [5] . Fig. 2 shows the results of a Monte Carlo simulation of the charged particles by laser extraction with the SOA Lens, and simulations of the beam transport optics calculated by the GIOS [6] code.

LYMAN-α BEAM PROFILE

The 212.55 nm and the difference wavelength of 845.0 nm beams are focussed by a lens with a focal length of $f = 500$ mm into an overlapping region in a Kr (mixed with Ar gas for phase matching) chamber for the wave-mixing. Then resulting the Lyman-α beam generated is focused only in the horizontal plane by a MgF_2 cylindrical lens with $f = 120$ mm to form an elliptical beam shape in the vicinity of the W target which matches the spatial distribution of the evaporated H atoms. The intensity of the Lyman-α beam is measured by an NO cell containing 500 Pa of pure NO gas by observing the ionization signal of the NO gas after passing through the W target region, as is shown in Fig.1.

The beam profile and the alignment of the invisible Lyman-α beam are obtained by moving the slits in the Up-Down and Left-Right directions. Fig.3 shows the NO signal as a function of the slit position. Since the resulting signal is a partial section

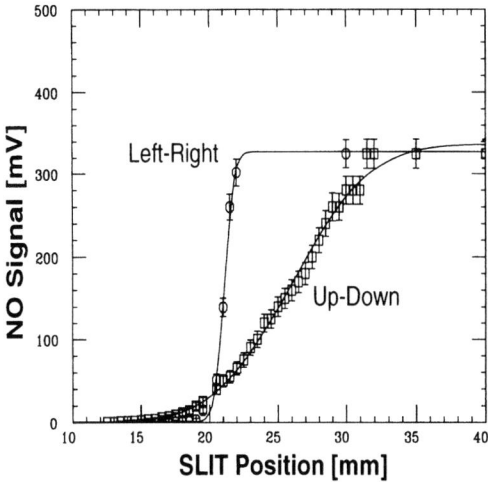

FIGURE 3. A VUV profile measurement by NO signal vs. slit position in the horizontal (left-right) and vertical (up-down) direction.

of a Gaussian beam, the profile in the trajectory of Up-Down, Left-Right can be fitted to an integral of a Gaussian. Its fitted Gaussian profile is shown as solid lines in Fig.3. The beam profile was found to have a size of $\sim$10.5 (vertical) $\times$ $\sim$1.5 mm^2 (horizontal). Each slit was moved independently from the fully opened position to the fully closed position while measuring the NO signal. With the center of the slit corresponding to 20.0 mm, the VUV was determined to be located about 1.6 mm downstream and 6.5 mm higher in the case of Fig.3.

DETECTION LIMIT

If the UHV (Ultra High Vacuum) chamber is carefully treated, the main components of the residual gas are well known to be H_2 molecules. The W foil heated up was used as a catalyzer to dissociate the H_2 molecules into H atoms through the interaction between the molecules and the hot W surface. Therefore, the H atoms resonantly ionized correspond to neutral hydrogen atoms dissociated from H_2 molecules near the W surface, and then evaporated into the vacuum space where the Lyman-α and 355 nm laser beams pass. It was found that the H atoms dissociated from the residual H_2 molecules in the UHV chamber (3×10^{-10} mbar at 1200 K) are sufficient to give an adequately large signal without introducing any H_2 gas. The number of hydrogen ions extracted and detected by the MCP is then the ratio of the amplitude $\times$ time-width of the MCP raw signal obtained in the condition between on-resonance and off-resonance wavelength, since no more than one hydrogen ion per laser pulse is expected to reach the MCP in the case of the off-resonance wavelength.

The detection limit of the present Lyman-α resonant ionization method can be evaluated using the following procedure.

Assuming the probability of H_2 dissociation at the W surface at 2000 K to be 10% in 9.5×10^{-8} mbar of H_2 gas, about 3×10^7 H atoms exist in a volume of about $0.6\ cm^3$, which is the effective volume of extraction along the laser passage. Since we obtained an MCP signal of 4.5V with 80 ns at 2000 K, in contrast with that of 2 mV with 20 ns, we can evaluate an amount of 9000 H atoms per laser pulse are extracted at 2000 K. Therefore, the extraction and detection efficiency is calculated to be 3×10^{-4}. Since the residual gas in the UHV is assumed to be consisted of hydrogen molecules entirely, the detection limit of the present Lyman-α resonant ionization method is estimated to be much better than 3×10^3 H/cm^3. In the case of the laser induced fluorescence technique using the Lyman-α [7] or the Balmer-α [8] photons, the lowest detection limit is reported to be as low as the order of 10^7 -$10^9\ cm^{-3}$, or $10^7\ cm^{-3}$, respectively. In the case of the two-photons induced Balmer-α emission by Bokor *et al.* [9] the detection limit is reported to be of the same order as that by the three-photon ionization by Bjorklund *et al.* [10], $3 \times 10^9\ cm^{-3}$. In the case of the laser absorption spectroscopy by Stutzin [11], using the Lyman-β laser, a line density ($\int_{path}$ n(l)dl, where n is number density) of $2 \times 10^{12}\ cm^{-2}$ is reported.

Thus, the present result is 3-5 orders of magnitude better than the other methods and is the highest that we know. Also, the present result is improved by more than 2.5 times compared with that of our previous report [12]. The method of applying Lyman-α resonant ionization spectroscopy can therefore be recognized as an ultra sensitive and efficient mass-selection technique.

ACKNOWLEDGEMENTS

The authors would like to acknowledge Drs. K. Kato, J.P. Marangos, and Messrs. T. Shiraishi, M. Fujino and Okada for their contributions to helpful discussions as well as the installation of the laser system.

REFERENCES

1. Y. Miyake, K. Shimomura, A. P. Mills, and K. Nagamine, Hyperfine Interactions 106 (1997) 237.
2. Y. Miyake, J. P. Marangos, K. Shimomura, P. Birrer and K. Nagamine, Nucl. Instr. and Meth. B **95**, 265 (1995).
3. K. F. Canter and P. H. Lippel, W. S. Crane and A. P. Mills, Jr. Studies of Positrons in Solids Surfaces and Atoms (World Scientific Singapore 1986).
4. J.P. Marangos, N. Shen, H. Ma, M.H.R. Hutchinson and J.P. Connerade; J. Opt. Soc. Am. B **7**, 1254 (1990).
5. Y. Miyake, K. Shimomura, Y. Matsuda, R.J. Scheuermann, P. Bakule, S. Makimura, P. Strasser, S.N. Nakamura, K. Ishida, T. Matuszaki, I. Watanabe, and K. Nagamine, Physica B 289-290 (2000) 666-669

6. H. Matsuda and H. Wollnik, Nucl. Inst. and Methods 103 (1972) 117.

7. T. Kajiwara, M. Inoue, T. Okada, K. Muraoka and M. Akazaki, Rev. Sci. Instrum. **56**, 12 (1985)

8. P. Gohil and D.D. Burgess, Plasma Phys. **25**, 1149 (1983)

9. J. Bokor, R.R. Freeman, J.C. White and R.H. Stortz, Phys. Rev. A **24**, 612 (1981)

10. G.C. Bjorklund, C.P. Ausschnitt, R.R. Freeman and R.H. Storz, Appl. Phys. Lett. **33**, 54 (1978)

11. G.C. Stutzin, A.T. Young, A.S. Schlachter, J.W. Stearns, K.N. Leung, W.B. Kunkel, G.T. Worth and R.R. Stevens, Rev. Sci. Instrum. **59**, 8 1363 (1988) and **59**, 8 1479 (1988)

12. Y. Miyake, K. Shimomura, A.P. Mills, Jr, , J.P. Marangos and K. Nagamine, Resonance Ionization Spectroscopy 47-53 (1998)

SESSION II

The FTICRMS Laser Microprobe : Speciation of Surfaces in Resonant Laser Ablation / Ionization (RLA)

Jean-François Muller*, Gabriel Krier, Frédéric Aubriet,
Lionel Vernex-Loset and Benoît Courrier

Laboratoire de Spectrométrie de Masse et de Chimie Laser, Université de Metz, IPEM,
Technopôle 2000 – 1 boulevard Arago, 57078 Metz cedex 3, France

Abstract. The coupling of a tunable laser with a laser microprobe FT-ICR-MS prototype developed at the University of Metz made it possible to carry out the first RLA-FTICRMS experiment, which combines the selectivity and the sensitivity of two photon resonant laser ablation/ionisation and the high accuracy mass measurement. This technique is a tool for the direct analysis of organic (or inorganic) traces on surfaces. The selectivity of the element's detection depends on the choice of resonant line, and on the screen effect of gazeous plume emitted by the laser pulse. In fact, the resonant two photons processes are most effective when the laser power density is lower than 5.10^8 W/cm^2 because the gaseous cloud is relatively transparent in this condition.

INTRODUCTION

The Resonant Laser Ablation/Ionization ensures selective ionization of chemical elements or organic molecules generally by biphotonic processes when using only one laser pulse on a solid sample. This technique was introduced and developed since 1984 in our laboratory with the LAMMA 500 instrument [1,2]. At this time, we demonstrated the ability of this technique for the selective detection of cadmium, copper, iron and lead in various polymeric thin films. On the other hand, we showed that the minimum of energy per laser pulse for the PAH's desorption from environmental particles decreases significantly at low wavelength ($\lambda < 240$ nm) which corresponds for all PAH to a two photon ionization process after laser desorption.

The process is complex and allows three steps: i) at the selected wavelength, the solid material must have a relatively high absorption coefficient. Under this condition, the threshold of laser ablation is weaker ($< 10^7$ W/cm^2); ii) when the rate of ablation reaches a certain photon density (typically $5.10^7 - 1.10^8$ W/cm^2), the gaseous plume located at the top of the irradiated solid's surface absorbs a part of the last photons coming from the laser pulse (screen effect) which limits the efficiency of laser ablation process; iii) simultaneously, the selected atoms (for example, chromium Cr$^\circ$) present in the plume are ionized according to the wavelength for two photons process [357.89 nm for Cr$_6^\circ$; 2.hv (3.464 eV) = 6.928 eV > 6.76 eV (E$_I$ (Cr+)].

However, the increases of selectivity obtained by RLA-MS are lower than those obtained by the RIMS technique. But, in standard mass spectrometry, it is impossible to eliminate the isobaric interference [3].

---

* corresponding author

CP584, *Resonance Ionization Spectroscopy 2000: 10th Int'l. Symp.*, edited by J. E. Parks and J. P. Young
© 2001 American Institute of Physics 0-7354-0024-5/01/$18.00

For this reason, the coupling of a tunable laser with the FTICR mass spectrometer limits the interference by the high accuracy mass measurement (R = m/Δm > 20.000 to 200.000). We will describe in this paper firstly the experimental device and performances, secondly the influence of the power density on the process involved.

EXPERIMENTAL

The analyzed material was studied using a laser microprobe FTICR mass spectrometer that has been described in detail elsewhere [4,5]. This instrument is a modified dual cell Nicolet Instrument FTMS 2000 (3.18 Teslas) coupled to a reflection laser interface: the laser beam is normal to the sample's surface. A new sample probe fitted with motorized micromanipulators in the three spatial directions allows a spatial accurracy of less than 5 μm.

The laser beam diameter on the sample, located inside the dual cell, can be adjusted for a specific wavelength when using the displacement of internal focalization lenses and by external adjustable telescope. The range of the power density on the surface's sample varies from 5.10^5 W/cm^2 to 10^{10} W/cm^2. The dye laser (Quantel TDL 90) is pumped by a Nd-Yag laser (Quantel Brillant B) delivering a nominal energy of 400 mJ/pulse at 532 nm. Different dyes such as Rhodamine 590, 610, 640 ; LDS 698, 740 Exciton Company Inc. were used. After doubling the frequency (KDP crystal), it is possible to obtain wavelengths ranging between 275 nm and 405 nm. By additional mixer crystal (UV radiation and residual IR) the complementary range between 218.5 nm to 278 nm is reached.

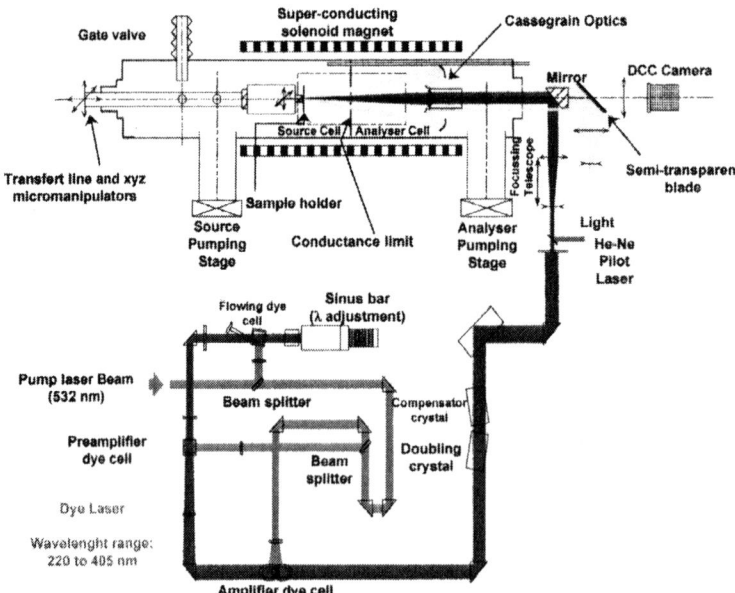

FIGURE 1. Coupling of the tunable dye laser TDL 90 with a FTICRMS laser microprobe

RESULTS AND DISCUSSION

1- Evolution of ion intensity versus the wavelength

The sample analyzed by FTICRSMS consists of polystyrene film doped with 500 ppm of copper and chromium. These two elements are introduced in the form of acetylacetonate complexes. The wavelength changes from 355 nm to 360 nm and the relative intensities of Cr^+ and Cu^+ ions are measured. It is clear from the results displayed in figure 2 that at the wavelength 357.87, the $^{52}Cr^+/^{63}Cu^+$ ratio is maximum for 357.87 nm. This wavelength corresponds to a biphotonic process. The first photon (3.464 eV) allows the transition from the fundamental state $^7S_3$ toward the excited state $^7P_4$. The second photon (3.464 eV) induces the chromium atom ionization (the total energy is higher than the chromium ionization energy (6.76 eV). Under moderate photon density ($I \approx 10^8$ W/cm^2), a moderate exaltation of Cr^+/Cu^+ ratio is observed, due to the fact that the ablation laser plume contains a large amount of neutral styrene's molecules. These induce a screen effect and it will decrease also the total ion amount by collision processes in the plume.

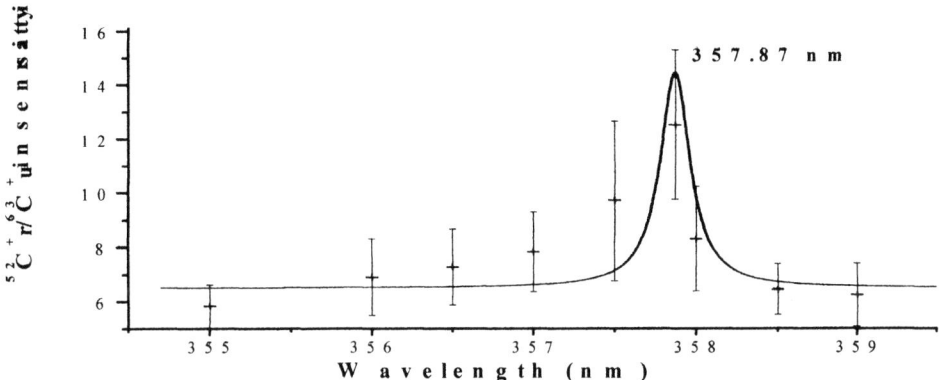

FIGURE 2. Evolution of the $^{52}Cr^+/^{63}Cu^+$ ion signal ration vs. wavelength in the neighborhood of 357.87 nm. Polystyrene doped with 500 ppm of copper and chromium

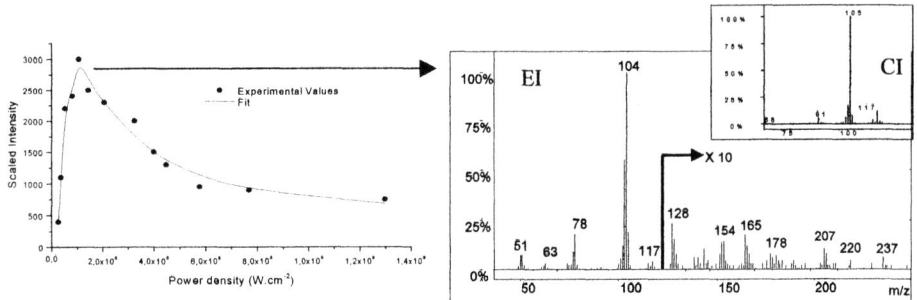

FIGURE 3. Ablation curve of polystyrene at 355 nm versus photodensity. The ion current is relative to $M^{\circ+}$ of styrene (m/z = 104)

FIGURE 4. Mass spectra of styrene ejected by laser ablation of polystyrene

In a different experiment (laser ablation cell directly coupled with ion trap) we demonstrated that the polystyrene laser ablation at 355 nm (figure 3) induces mainly the styrene (photodepolymerization process) with a power density equal to 10^8 W/cm^2. Styrene is clearly detected by electron impact ionization mode [M$^{\circ+}$: m/z = 104] or by chemical ionization mode (reactive gas : CH$_3$-CN) [(M + H)$^+$: m/z = 105] (figure 4).

2- Influence of the power density

To precise the role of the laser power density, we chose a doped glass sample containing chromium (2 %) tin, lead and several earth rare elements. The power laser density varies from $4.5.10^{10}$ to $1.1.10^8$ W/cm^2 on the surface. The experiments in resonant and non resonant laser ablation/ionization were respectively carried out under laser wavelengths of 357,9 and 355nm (figure 5a and 5b respectively). It appears clearly that the selectivity of the resonant mode of ionization increases when the power density used decreases. Exaltation of the peaks of chromium ions becomes very significant when the power density reaches the range values between 10^8 and 5.10^8 W/cm^2. On the contrary, at $4,5.10^{10}$ W/cm^2, the spectra at 357,9 nm and 355 nm (resonant wavelength) are similar where earth rare ions dominate. The relative intensity of each ion depends clearly of the microplasma temperature (Saha – Eggert equation).

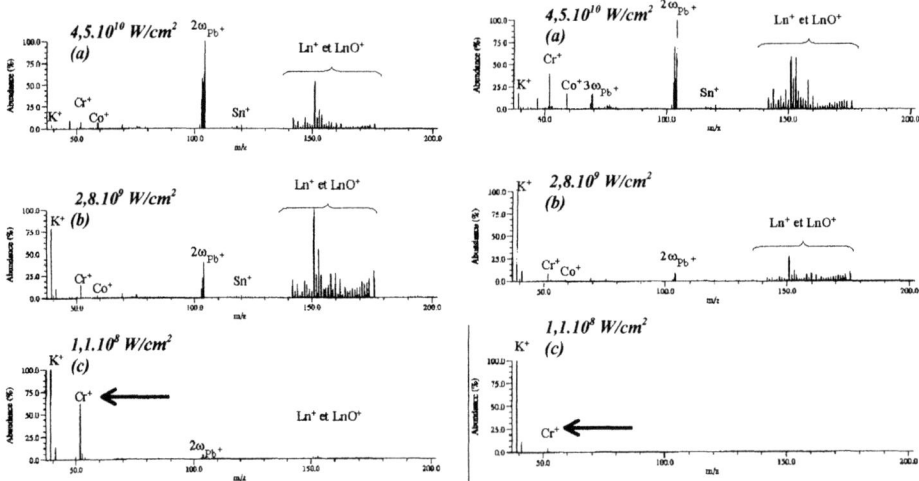

FIGURE 5a. FTICRMS analysis of a doped glass containing 2% of chromium in weight by laser ablation/ionization using the line of resonance of chromium at the wavelength of 357,9 nm Ln indicates the various lanthanides (neodymium, samarium, europium and gadolinium)

FIGURE 5b. FTICRMS analysis of a doped glass containing 2% of chromium in weight by laser ablation/ionization using the wavelength of 355 nm Ln indicates the various lanthanides (neodymium, samarium, europium and gadolinium)

If the power density is lower, the plume is relatively cold and transparent, which is in favor of the biphotonic absorption process. On the contrary, under high power density, the gazeous phase is a microplasma with photonic emission (plasma ignition). The high temperature reached induced a screen (or bremstahlung inverse) effect by an increase of the electronic and ionic density [6]. The yields of the resonant ablation/ionization selectivity decrease significantly when increasing the laser power density (figure 5b).

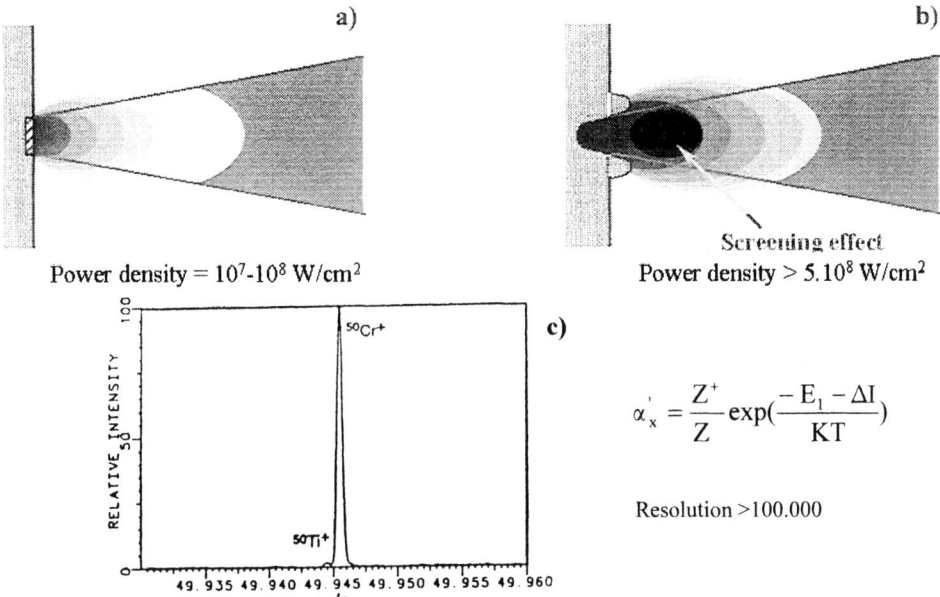

a)

Power density = 10^7-10^8 W/cm^2

b)

Screening effect
Power density > 5.10^8 W/cm^2

c)

$$\alpha'_x = \frac{Z^+}{Z} \exp(\frac{-E_1 - \Delta I}{KT})$$

Resolution >100.000

FIGURE 6. Laser ablation process a)- at relatively low power density (10^7-10^8 W/cm^2) ; the plume is transparent ; b)- at high power density , the screening effect is effective ; c)- high resolution spectrum with the separation of ^{50}Ti$^+$ isotope and ^{50}Cr$^+$ isotope at 7.10^7 W/cm^2

For example, at relatively low power density (7.10^7 W/cm^2), the chromium detection limit in ciment matrix is quite lower than 10 ppm.

CONCLUSIONS

The combination of laser ablation/ionization of solid material in resonant and/or non resonant processes with TOFMS or FTICRMS is a flexible and versatile technique. It allows easy elimination of the spectral interferences. At low laser power density, the ablation rate depends clearly on the relative transparency of the plume. It is the best condition for resonant ablation/ionization conditions. Under moderate laser power density (typically 5.10^7 – 5.10^8 W/cm^2), the selectivity is very dependent on the matrix effect. On the contrary, if the laser power density is high, plasma temperature controls the ionization process and the resonant selectivity disappears. On the other hand, at higher power density, all elements existing with a concentration higher than the 100-500 ppm range are detected. The cluster ions distribution (both positive and negative ions) allows the speciation of major compounds in the solid material.

ACKNOWLEDGMENTS

The authors would like to thank the Lorraine Region for financial support.

REFERENCES

1. Muller, J.F., Verdun, F., Krier, G., Lamboulé, M., Gondouin, S; , Tourmann, J.L.,Muller, D., Lorek, S., C.R. Acad. Sci., série II, **16**, 299, (1984)
2 Verdun, F., Krier, G., Muller, J.F., Anal. Chem., **59**, 1383, (1987)
3 Arlinghaus, H.F., Thonnard, N., Spaar, M.T., Sachleben, R.A., Larimer, F.W., Foote, R.S., Woichik, R.P., Brown, G.M., Slopp, F.V., Jacobson, K.B., Anal. Chem., **63**, 402, (1991)
4 Muller, J.F., Pelletier, M. , Krier, G., Weil, D., Campana, J., Microbeam Analysis, Russel P.E. Ed., San Franisco, Press Inc, 311, (1989)
5 Pelletier, M., Krier, G., Muller, J.F., Weil, D., Johnston, M., Rapid Comm. Mass Speectrom., **2**, 146, (1988)
6 Catherinot, A., Anglerand, B., Aubreton, J., Champeaux, C., Germain, C., Girault, C., Laser Processing Treatment and Desorption, Ed. Conde et al., Academic Publishers, Lisbonne, (1995)

Electronic Processes during Ion-Beam Sputtering of Metals studied by Resonance Laser Ionization Mass Spectrometry

Vicky Philipsen, Jeroen Bastiaansen, Erno Vandeweert,
Peter Lievens, and Roger E. Silverans

*Laboratorium voor Vaste-Stoffysica en Magnetisme, K.U.Leuven,
Celestijnenlaan 200 D, B-3001 Leuven, Belgium.*

Abstract. The electronic processes governing keV ion-beam sputtering of metals are studied by double resonance laser ionization mass spectrometry of the ejected metal atoms. We present the population partitions over the ground and metastable states of sputtered Cu and Sr atoms and their state-selective flight-time distributions. These results support the model indicating resonant electron transfer from the metal to the sputtered particle as a key mechanism to populate the electronic states.

INTRODUCTION

The interaction of energetic ions with metal surfaces results in the emission of atoms in the atomic ground state, but also in excited atomic states. Information on the electronic processes responsible for the final atomic state of the released atoms is embedded in the emitted atoms in the form of their probability of ejection into a specific atomic state and their kinetic energy (1-7). These observables can be accessed by state-selective resonant laser ionization mass spectrometry (RIMS).

In this contribution we demonstrate the use of RIMS for the investigation of ground and metastable state Cu and Sr atoms after ion-beam sputtering of clean polycrystalline targets. We completed the measurements of the population partitions and of the state-selective flight-time distributions of ion-beam sputtered ground and metastable state atoms. The combined data provide evidence for resonant electron transfer between the bulk and the escaping particle as an important mechanism during the emission process.

EXPERIMENTAL PROCEDURE

The measurements are performed in a vacuum chamber with a base pressure of 5×10^{-10} hPa. An ion gun produces 15 keV Ar^+ ions directed onto the centrally located target foil at 45° incidence. The ion-beam can be operated in continuous mode or pulsed mode (typical duration 300 ns) with an average current density of 0.5 $\mu A/mm^2$ on the target. Before each experiment the sample is sputter cleaned across an area of 12 mm^2 by continuous ion bombardment.

CP584, *Resonance Ionization Spectroscopy 2000: 10th Int'l. Symp.*, edited by J. E. Parks and J. P. Young
© 2001 American Institute of Physics 0-7354-0024-5/01/$18.00

Linearly polarized light is provided by two independent laser systems: an optical parametric oscillator (tuning range 225-1600 nm) pumped by a Nd:YAG laser and a dye laser pumped by another pulsed Nd:YAG laser. Both laser systems are equipped with frequency doubling units. The two lasers operate at a repetition rate of 10 Hz and are synchronized. The laser pulses have a duration of about 6 ns and a bandwidth of about 10 GHz. Maximum pulse energies range from a few mJ in the UV up to 50 mJ in the visible wavelength region. The laser beams intersect the atomic plume perpendicularly to the axis of ion extraction and 4 mm in front of the sample surface.

The sputtered atoms are state-selectively ionized via double resonance two-color two-step laser ionization into an autoionizing state. The laser-ionized particles are extracted into a time-of-flight mass spectrometer and detected. Further experimental details can be found in ref. (8).

The autoionizing states needed in efficient ionization schemes are for most elements not available in literature. Therefore, a spectroscopic exploration of the continuum structure of the atom is indispensable to locate the autoionizing states (9).

RESULTS

Population partitions

The quantitative determination of the population partition of atoms sputtered in ground and metastable states, can be done based on the procedure of double resonance two-color two-step ionization: the same intermediate state can be used to probe the ground and different metastable states (as far as the optical selection rules permit)(5). This requires only the saturation of the excitation step, which is already attained at moderate laser pulse energies. The ratios of the measured photoion signals then immediately reflect the relative population on the atomic states. This procedure could not be followed for Sr I because the ground state differs in parity from the first metastable multiplet. In this case the use of double-resonant ionization schemes was even more essential, since both the excitation and ionization step needed to be in saturation to be able to deduce the population partition. The measured population distributions over the ground and metastable states of Cu and Sr atoms sputtered from polycrystalline pure targets

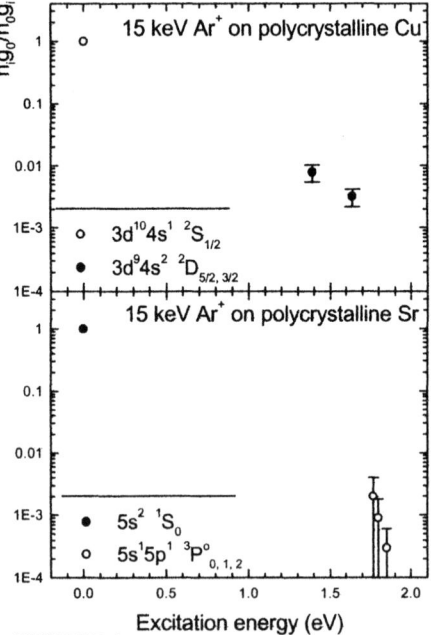

FIGURE 1.
Population distribution of Cu and Sr atoms sputtered from polycrystalline pure targets produced by keV Ar^+ ions. The populations (n_i) are given relative to the ground state population (n_0) and corrected for the statistical weight of each state (g_i).

are presented in Fig. 1. The populations are given relative to the ground state population as a function of the excitation energy and corrected for the statistical degeneracy. Distinction is made in the symbols representing the atomic states: filled circles denote electronic configurations with a filled outer-electron shell ([Ar] $3d^9 4s^2$ for Cu and [Kr] $5s^2$ for Sr), while open circles represent electronic configurations with a partially filled outer-electron shell ([Ar] $3d^{10} 4s^1$ for Cu and [Kr] $5s^1 5p^1$ for Sr).

From these experiments we observe a substantial population on all metastable states. Comparison with earlier evaporation studies already demonstrated that this must be due to non-thermal mechanisms during the sputtering process (7).

State-selective flight-time distributions

The state-selective flight-time distributions of the atoms are obtained by measuring the photoion intensities as function of the delay time between the ion pulse and the laser pulses. Due to the high sensitivity achieved by employing double resonance two-color two-step ionization schemes we were able to determine the flight-time distributions of all states present in the population partitions.

Fig. 2 exemplifies for each element a flight-time distribution for atoms sputtered in states with a closed and open outer-shell electronic configuration. For a better comparison, all distributions are normalized to their maximum. For the sputtered Cu atoms a clear difference is apparent between the flight-time distribution from the state with an open outer-shell electronic configuration and this from the states with a closed outer-shell electronic configuration: for the 'open' state the maximum is shifted towards a shorter flight time and the total distribution rises and falls off steeper than for the 'closed' states. In previous

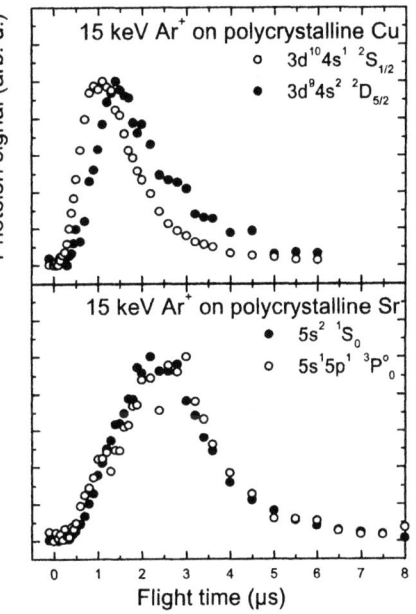

FIGURE 2. State-selective flight-time distributions of Cu and Sr atoms sputtered from pure polycrystalline foil by keV Ar^+ ions.

investigations we found the same behavior with respect to the electronic configuration for state-selective flight-time distributions of sputtered Co and Ni atoms (5-7). The flight-time distributions of Sr atoms sputtered in the ground and metastable states are identical, regardless of the electronic configuration of the state.

DISCUSSION

The attempts to identify the electronic mechanisms involved in the emission of metastable atoms following ion-beam sputtering of clean metal surfaces have led to a range of models (1-7) including collision-induced excitation, non-radiative de-excitation and local heating of the electronic subsystem in the collision cascade area.

Recently we introduced, based on the combined results on ion-beam sputtering of Co and Ni, multichannel resonant electron transfer (RET) as being an important electronic process during sputtering (5-7). We restrict ourselves here to the essential aspects of the RET model and interpret the combined data on sputtering of Cu and Sr in relation to these concepts. When a particle is sputtered from the metal surface, it will escape as a positive ion with a probability to be neutralized by a conduction electron tunneling from the bulk to the ion. This probability is determined by (*i*) the energy difference between the atomic energy level and the Fermi level, (*ii*) the coupling strength between the atomic state and the metal conduction electrons, and (*iii*) the particle's escape velocity perpendicular to the metal surface.

Concerning ion-beam sputtering of Cu atoms, the first consideration implies that all atomic states with excitation energies below 3.1 eV can be populated through resonant electron transfer from the bulk conduction band to these atomic states (see Fig. 1). The observed relatively low population on the metastable states reflects the difference in coupling strength to the metal conduction band states for the atomic ground state and the metastable states. The correspondence between the Cu-bulk electron configuration (calculated to be $3d^{9.5}4s^{0.7}4p^{0.7}$ (10)) and the electron configuration of the atomic state favors a transfer of the neutralizing electron into the $3d^{10}4s^1$ ground state above a transfer into one of the metastable states characterized by a $3d^9 4s^2$ configuration. The influence of the particle's escape velocity on the atomic state can be examined in the state-specific flight-time distributions of the Cu atoms (Fig. 2). The faster the sputtered particle escapes from the surface, the shorter the interaction time with the surface and the more probable that a tunneling electron will arrive in the atomic ground state, which is strongly coupled to the bulk. All these experimental data on sputtered Cu atoms fit nicely into the RET model.

The experimental results on ion-beam sputtering of Sr atoms are also weighted against the main ingredients of RET. The ground state and the first excited multiplet of the Sr atom lie well within the 3.1 eV wide energy window defined by the energy difference between the excitation energy of the atomic ground state and the Fermi energy of the Sr bulk. The population partition of ion-beam sputtered Sr atoms (Fig. 1) shows a relative population of only 10^{-3} on the metastable states. Since the band structure of pure Sr metal is known to be highly *s*-like (11), the neutralizing electron will thus preferentially end up in the $5s^2$ atomic ground state. The observed population partition can thus be interpreted in the RET model. The neutralization probability however seems to be independent of the particle's velocity as can be seen from the flight-time distributions in Fig. 2. This can be an indication that additional factors, e.g., the density of states, have to be considered in the electron transfer probability. Moreover, due to the low surface binding energy of this metal, local distortions of the lattice and a low average velocity of the ensemble of sputtered particles could be reflected in the overall neutralization probability.

In conclusion, we exploited RIMS as an unmatched technique for assessing the important observables involved in the emission process during ion-beam sputtering of metals. The combined population partitions and state-selective flight-time distributions of Cu and Sr atoms sputtered from polycrystalline pure targets were interpreted within the framework of the RET model.

ACKNOWLEDGEMENTS

This work is financially supported by the Fund for Scientific Research – Flanders (Belgium) (F.W.O.), the Flemish Concerted Action Research Program (G.O.A.) and the Interuniversity Poles of Attraction Program (I.U.A.P.) - Belgian State, Prime Minister's Office - Federal Office for Scientific, Technical and Cultural Affairs. E.V. and P.L. are Postdoctoral Fellows of the F.W.O.

REFERENCES

1. B.J. Garrison, N. Winograd, R. Chatterjee, Z. Postawa, A. Wucher, E. Vandeweert, P. Lievens, V. Philipsen, and R.E. Silverans, *Rap. Comm. in Mass Spectrom.* **12** (1998) 1266-1272.
2. A. Wucher and Z. Sroubek, *Phys. Rev. B* **55** (1997) 780-786.
3. A. Cortona, W. Husinsky, and G. Betz, *Phys. Rev. B* **59** (1999) 15495-15505.
4. Z. Sroubek and J. Lorincik, *Surf. Rev. Lett.* **6** (1999) 257-264.
5. E. Vandeweert, V. Philipsen, W. Bouwen, P. Thoen, H. Weidele, R.E. Silverans, and P. Lievens, Phys. Rev. Lett. **78** (1997) 138-141.
6. R.E. Silverans and P. Lievens, in *Resonance Ionization Spectroscopy*-1998, edited by J.C. Vickerman et al., AIP Conference Proceedings 454, New York, 1998, 197-205.
7. V. Philipsen, J. Bastiaansen, P. Lievens, E. Vandeweert, and R.E. Silverans, *Vacuum* **56** (2000) 273-278; E. Vandeweert, J. Bastiaansen, V. Philipsen, P. Lievens, and R.E. Silverans, *Nucl. Instr. and Meth. in Phys. Res. B* **164-165** (2000) 795-802.
8. E. Vandeweert, J. Bastiaansen, V. Philipsen, P. Lievens, and R.E. Silverans, submitted to *Phys. Rev. B*.
9. V. Philipsen, J. Bastiaansen, G. Verschoren, P. Lievens, E. Vandeweert, R.E. Silverans, and H.H. Telle, *Spectrochim. Acta Part B* **55** (2000) 1539-1550.
10. O. Jepsen, D. Glötzel, A.R. Mackintosh, *Phys. Rev. B* **23** (1980) 2684-2696.
11. B. Vasvari, A.O.E. Animalu, and V. Heine, *Phys. Rev.* **154** (1967) 535-539.

Evaluation of a System for Trace and Particle Analysis Based on Resonance Ionization of Sputtered Atoms

N. Erdmann[a*1], A. Benninghoven[b], M. Betti[a], T. Gouder[a], C. Grüning[c], F. Kollmer[b], P. Lievens[d], F. Miserque[a], V. Philipsen[d], R.E. Silverans[d], E. Vandeweert[d]

[a]European Commission Joint Research Center, Institute for Transuranium Elements, Postfach 2340, D-76125 Karlsruhe, Germany
[b]Physikalisches Institut, Westfälische Wilhelms-Universität Münster, Wilhelm.Klemm-Str. 10, D-48149 Münster, Germany
[c]Institut für Kernchemie, Universität Mainz, D-55099 Mainz, Germany
[d]Laboratorium voor Vaste-Stoffysica en Magnetisme, K.U. Leuven, Celestijnenlaan 200D, B-3001 Leuven, Belgium

Abstract. For the analysis of actinide containing micron-size particles, a new experimental setup has been evaluated which uses sputtered neutral atoms instead of secondary ions. The atoms are post-ionized with lasers, either non-resonantly or resonantly. First experiments on non-resonant ionization of micron-size uranium oxide particles have shown that the sputter yields for secondary neutral atoms are much higher than for secondary ions, and uranium particles < 0.5 μm can easily be detected and measured. The use of two-color, two-step resonant ionization of the sputtered neutral uranium atoms from thin films was investigated. Several excitation schemes were tested and a significant population on several low-lying metastable states after ion sputtering was observed.

INTRODUCTION

Nuclear facilities are controlled worldwide by IAEA and EURATOM safeguards to prevent undeclared nuclear activities. One of the main concerns is the characterization of so-called "hot particles", i.e. particles containing actinides as well as fission products. They have dimensions varying between nm (nanometers) up to tenths of μm (microns). Such particles are also of high interest for nuclear forensic analysis as well as for a risk assessment of contaminated areas.

Instrumental analytical techniques like secondary ion mass spectrometry (SIMS) have been applied for their localization, chemical and isotopic characterization [1]. However, the problem of the separation of isobars like $^{238}Pu/^{238}U$ or $^{241}Am/^{241}Pu$ in such kind of micro-particles is still unsolved.

---

[1] Present address: Environment Institute, European Commission Joint Research Center, T.P. 290, I-21020 Ispra (VA), Italy

CP584, *Resonance Ionization Spectroscopy 2000: 10th Int'l. Symp.*, edited by J. E. Parks and J. P. Young
© 2001 American Institute of Physics 0-7354-0024-5/01/$18.00

Resonance ionization spectroscopy (RIS) is known to be a good solution to overcome the problems of isobaric interferences due to its extremely high element selectivity, especially if more than one resonant transition is involved. For the actinides, this technique has mainly been applied to the analysis of bulk samples [2].

In this work, a new experimental setup for particle analysis has been evaluated which uses sputtered neutral particles instead of the secondary ions. These are post-ionized with lasers, non-resonantly (sputtered neutrals mass spectrometry = SNMS) as well as resonantly (sputter initiated RIS = SIRIS).

Proposed setup

A combined SNMS/SIRIS system for the analysis of sub-micron particles should consist of a pulsed ion gun with a narrow (≤ a few hundred nm) spot size, two pulsed laser systems, one for non-resonant ionization plus a tunable system for resonant ionization, and a time-of-flight (TOF) mass spectrometer. Several advantages of such a system over the SIMS method are expected:

- Typically, higher yields are obtained for secondary neutral particles than for secondary ions, which would lead to an increase in sensitivity and lower the detection limit towards smaller particles (≈ 0.2 μm).
- The sputter yields for neutral particles are known to be more comparable for different elements, also less "matrix effect" is observed in SNMS than in SIMS, which would make a quantification of the elemental composition in a sample possible.
- The SIRIS option would offer the advantages of RIMS: high element selectivity and sensitivity plus isotopic composition determination. This can be applied for the investigation of trace elements in microparticles, where no chemical separation can be performed. One example would be the determination of traces of plutonium in uranium particles without isobaric interferences ($^{238}U/^{238}Pu$ or $^{241}Am/^{241}Pu$).

Since no such instrument including all options was commercially available, the different components were tested separately to evaluate a possible setup from commercial components. For this purpose, experiments on non-resonant ionization of micron-size particles were performed at the University of Münster, Germany. Investigations on resonant ionization were performed at the K.U. Leuven, Belgium.

EXPERIMENTAL

The experiments at the University of Münster were performed with an SNMS setup [3] consisting of a pulsed gallium liquid metal ion gun with a spot size of a few hundred nm and a gridless reflector time-of-flight (RTOF). For ionization of the neutral species, an ArF excimer laser (Lambda Physik, Göttingen, Germany) with a pulse repetition rate of 20 Hz was used. The laser wavelength of 193 nm, which corresponds to a photon energy of ≈ 6.4 eV, is short enough to ionize uranium (ionization energy ≈ 6 eV) in a one-step process. The instrument can also be used in SIMS mode.

All experiments at the K.U. Leuven were performed with their RIS setup, as described in [4], consisting of a pulsed Ar^+ ion gun and a RTOF mass spectrometer. For resonant ionization, two tunable pulsed lasers, a dye laser with a frequency doubling unit and an optical parametric oscillator system were used.

RESULTS

Results obtained for non-resonant ionization of ion-beam sputtered uranium atoms

Monodisperse uranium oxide particles [5] of different ^{235}U enrichment (0.5% depleted, 1% and 10% ^{235}U enriched) and size (1 μm and 0.5 μm), deposited on graphite foils, were investigated. Mass spectra in the mass range for uranium and its oxides were acquired. The strongest signals were observed on the atomic mass, while UO^+ and UO_2^+ peaks were also present. For the same dose of primary ions, SNMS and SIMS ion counts were compared. It could be observed that the signal of the sputtered atoms was 200 times larger than that of secondary ions (on the atomic mass). In the case of UO^+, the SNMS signal was still 10 times higher than the SIMS signal.

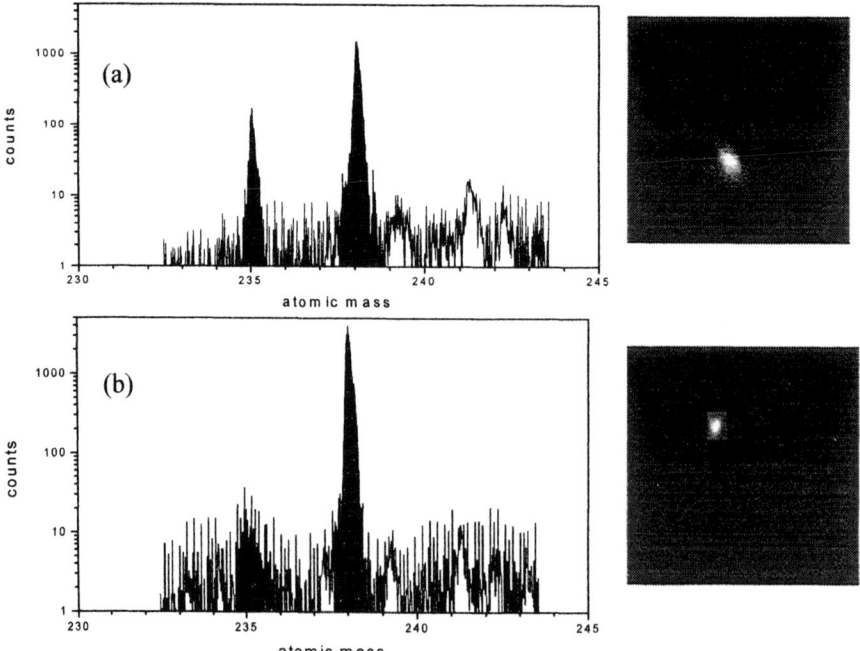

FIGURE 1. TOF mass spectra and SNMS images (15 × 15 μm²) of monodisperse uranium oxide particles: (a) 1 μm size, 10% ^{235}U and (b) 0.5 μm size, depleted uranium.

The $^{235}U/^{238}U$ isotopic ratios obtained from the acquired mass spectra were in agreement with the certified values. For example, for the 0.5 µm depleted uranium particles, a $^{235}U/^{238}U$ ratio of 0.0048 ± 4.6% was determined, which is in excellent agreement with the certified value of 0.00492. Simultaneous to the acquisition of the mass spectra, images were taken for different masses by rastering the ion beam over a small area on the sample. The SNMS images for a 1 µm enriched uranium particle (10% ^{235}U) and a 0.5 µm depleted uranium particle, acquired on atomic mass 238, are shown in figure 1, together with the mass spectra. The signal obtained from the 0.5 µm particles (the integrated number of counts shown in figure 1(b) resulted from only a fraction of the particle) indicates that the SNMS system would be capable of measuring much smaller uranium particles.

Results for resonant ionization of uranium atoms after ion sputtering

Because the Ar^+ ion gun spot size of the RIMS system did not allow for the analysis of single particles, samples for resonant ionization investigations were prepared by sputter-deposition of 0.5 µm thick layers of depleted metal uranium on Ta foils. They were transported under vacuum conditions. Slight oxidation of the uranium metal occurred.

Resonant transitions to more than 10 different intermediate levels [7], starting from the atomic ground state, were tested. A great variation of transition strengths could be observed, which in the future will make a more detailed analysis and comparison of different excitation schemes necessary to find an optimum scheme. The excitation laser was scanned over a small wavelength range in the vicinity of two strong transitions from the ground state. This scan was repeated twice, first by using only one laser for excitation and ionization (a so called one-color, two-step excitation) and then with the second tunable laser ionizing non-resonantly into the continuum above the ionization potential (using a two-color, two-step ionization scheme). In the latter case, the excitation laser power was attenuated so the resonant transitions were just saturated. These scans are shown in figure 2. Several additional resonances, which could be assigned to transitions from metastable levels above the ground state, were observed. The corresponding energies of the 1st and 2nd excited level are 620.2 cm^{-1} and 3800.8 cm^{-1} above the ground state.

From this scan it can be estimated that the 1st and the 2nd excited level are considerably populated upon ion-beam sputtering, which is clearly different from the population after thermal evaporation, where a Boltzmann energy distribution can be expected. This will lead to a decrease in sensitivity for resonant ionization compared to non-resonant ionization, which is not state-selective. The most efficient RIS excitation schemes will start from the atomic ground state.

Furthermore, using the two-color scheme, it can be observed that the non-resonant background, which is present in the one-color scan due to the high UV laser power (necessary to achieve sufficient non-resonant ionization), can strongly be reduced. A high elemental selectivity will thus be achieved by using a two-color scheme.

Keeping the excitation laser fixed to a transition from the ground state ($5f^3 6d7s^2$, J = 6) to an intermediate state at E = 30795.4 cm^{-1} (J = 5), the ionization laser was scanned above the first ionization potential. Several autoionizing states for two-step resonant excitation were observed, which are described in [4]. Using ionization into an autoionizing state will strongly enhance the selectivity of the RIS process.

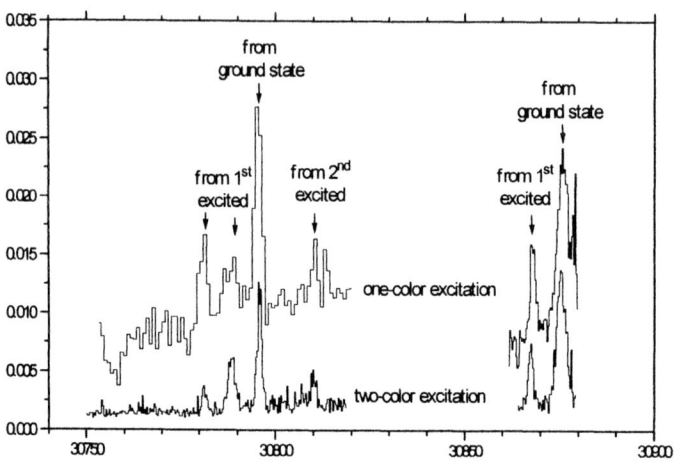

FIGURE 2. Scan of the excitation laser for two-step resonance ionization of uranium atoms after ion sputtering. Several transitions starting from the lowest-lying states of uranium can be observed.

CONCLUSION

It could be shown that the use of SNMS instead of SIMS could strongly enhance the sensitivity for the detection of uranium oxide micro-particles, which will make the analysis of particles much smaller than 0.5 μm possible. First experiments on Th/Ta mixtures have indicated the capability of SNMS for quantifying elemental compositions. Investigations on resonant ionization have shown that the use of a two-color, two-step resonant ionization will strongly enhance elemental selectivity. This will occur, however, at the expense of sensitivity, since all low-lying metastable atomic states are populated after ion sputtering. At present, the selective detection of plutonium traces in micron-size uranium particles with SIRIS seems possible. Since a rather large fraction of sputtered neutral particles is present in the form of uranium oxide, the contribution of dissociation of molecules to the SNMS atomic signal and an estimation of the efficiency of the RIS process will have to be more carefully studied in the future.

REFERENCES

1. Tamborini, G. and Betti, M., "Characterization of radioactive particles by SIMS", *Microchim. Acta* **593**, 1-7 (1999).
2. Erdmann, N., Grüning, C., Trautmann, N., Waldek, A., Huber, G., Kunz, P., Nunnemann, M., Passler, G., "Trace Analysis of Plutonium by Resonance Ionization Mass Spectroscopy" in: *Resonance Ionization Spectroscopy 1998*, edited by J.C. Vickerman, I. Lyon, N.P. Lockyer, and J.E. Parks, AIP Conference Proceedings 454, New York 1998, pp. 279-282
3. Kollmer, F., Kamischke, R., Schnieders, A., Ostendorf, R., and Benninghoven, A., "TOF-SIMS and Laser-SNMS analysis of sub-µm particles", in preparation
4. Vandeweert, E., Philipsen, V., Lievens, P., Silverans, R.E., Grüning, C., Miserque, F., Betti, M., Gouder, T., and Erdmann, N., "Two-color two-step laser ionization spectroscopy of uranium sputtered from thin metallic films", contribution to these conference proceedings.
5. Erdmann, N., Betti, M., Stetzer, O., Tamborini, G., Kratz, J.V., Trautmann, N., and van Geel, J., "Production of monodisperse uranium oxide particles and their characterization by scanning electron microscopy and secondary ion mass spectroscopy", *Spectrochim. Acta B*, **55**, 1565-1575 (2000).
6. Gouder, T., "Electronic structure of uranium overlayers on magnesium and aluminium", *Surface Science* **382**, 26-34 (1997).
7. Blaise, J. and Wyart, J.F., "Energy Levels and Atomic Spectra of Actinides", *International Tables of Selected Constants* **20** (Tables Internationes de Constantes), Paris, 1992.

Trace Element Analysis of Precious Metals in Minerals by Time-of-Flight Resonance Ionization Mass Spectrometry (TOF-RIMS)

Stamen S. Dimov*, Stephen L. Chryssoulis* and Robert H. Lipson†

*AMTEL, 100 Collip Circle, Suite 205, UWO Research Park, London, Ontario, N6G 4X8, Canada
†Department of Chemistry, The University of Western Ontario, London, Ontario, N6A 5B7, Canada

Abstract. The TOF-RIMS Mass Spectrometer, developed at AMTEL, has been successfully applied for quantitative trace element analysis of Au, Pd and Rh in sulphides, iron oxides and silicates. A series of experiments were conducted in order to test the applicability and determine the detection sensitivity of several one and two step resonant ionization schemes related to Au, Pd and Rh. The experimental data are compared with the corresponding theoretical estimates. Attained minimum detection limits are in the 8-17 ppb range with precision of ± 15%. A comparative study of minerals by Dynamic SIMS and TOF-RIMS in the overlapping range of detection sensitivities for Au, Pd and Rh shows good correlation of the quantified data.

INTRODUCTION

Detection and quantification of precious metals (Au, Pd, Rh and Pt) in rock forming minerals is important for establishing the attainable recovery of PGE (platinum group elements) by sulphide flotation. To achieve this goal, sub-part per billion (ppb) detection limits are required plus the ability to quantify results. Existing commercially available quantitative microbeam techniques (EPMA, PIXE, SIMS, TOF-LIMS) have limits of detection in the 0.1-200 ppm range. An alternative to ultra-trace element analysis is the technique of time-of-flight resonance ionization mass spectrometry (TOF-RIMS). The analytical procedure consists of three steps: (i) the solid sample is vaporized by laser ablation to form a plume containing neutral atoms, (ii) the neutral atoms of specific atomic elements in the plume are resonantly excited and ionized by another laser and finally (iii) ions created in the plume are mass analyzed in a time-of-flight mass spectrometer. It has been demonstrated that the decoupling of the processes of laser ablation and ionization of the neutral species [1] minimizes matrix effects and provides better quantitative results.

This paper describes the development of a commercial TOF-RIMS instrument for trace element analysis of precious metals in minerals. The system was optimized for detection of Au, Pd and Rh using one- and two-colour resonant ionization schemes. While some of these schemes have been previously studied by other groups, [2] (for Au detection) others (for Rh and Pd detection) are utilized here for the first time.

CP584, *Resonance Ionization Spectroscopy 2000: 10ᵗʰ Int'l. Symp.*, edited by J. E. Parks and J. P. Young

EXPERIMENTAL

TOF-RIMS Mass Spectrometer

The TOF-RIMS mass spectrometer used in this work incorporates two powerful pulsed Nd: YAG laser sources that are used for separate laser ablation and ionization [3]. One of the lasers is operated at wavelength $\lambda = 266$ nm for laser ablation, while the second laser can provide outputs at $\lambda = 532$ nm, 355 nm, or 266 nm after appropriate optical harmonic generation. These outputs can be used either for non-resonant multiphoton ionization (NRMPI) or to pump a tunable dye laser that provides resonant excitation and ionization. A narrow-band dye laser (Lumonics, Model HD-300B) and a harmonic doubler (INRAD, Model AT-III) delivers tunability from 950 nm down to 198 nm in the ultraviolet (UV). This wavelength range practically ensures the applicability of the TOF-RIMS technique to every element of the periodic table except He and Ne.

The design of the mass spectrometer system allows for fast switching of the measurement mode: TOF-LIMS with NRMPI or TOF-RIMS.

Optimal Choice of the Resonant Excitation Scheme

A variety of one-, two- and three-colour resonant ionization schemes (RIS) for Au, Pd and Rh have been theoretically evaluated using the approach outlined by Saloman [2]. In this paper we report the experimental studies only for one- and two-colour schemes that provide both sufficient ionization efficiency and implementation simplicity using commercial instrumentation.

Gold

Fig. 1 shows the energy level diagram of Au and two of the optical schemes used for resonant excitation and ionization:

$\omega_1 + \omega_1$: $5d^{10}6s\ ^2S_{1/2} \rightarrow 5d^{10}6p\ ^2P_{1/2}^{\circ}$ ($\lambda_1 = 267.595$ nm) $\rightarrow$ Au$^+$ ($\lambda_1 = 267.595$ nm)

$\omega_1 + \omega_2$: $5d^{10}6s\ ^2S_{1/2} \rightarrow 5d^{10}6p\ ^2P_{1/2}^{\circ}$ ($\lambda_1 = 267.595$ nm) $\rightarrow$ Au$^+$ ($\lambda_2 = 266$ nm)

The first resonant ionization scheme consists of one resonant step followed by photoionization using two photons of the same energy. In this case, the two-step excitation and ionization process is provided by one tunable laser which wavelength, in exact resonance with the atomic gold $^2S_{1/2} \rightarrow\ ^2P_{1/2}^{\circ}$ transition at $\lambda = 267.6$ nm (fundamental frequency, ω_1).

The second resonant ionization scheme involves using an additional laser source at λ=266 nm (fourth harmonic of the Nd:YAG laser; fundamental frequency, ω_2) to improve the ionization efficiency in the second step. The laser fluences necessary to saturate the excitation and photoionization steps defined by these two schemes are shown in Table 1.

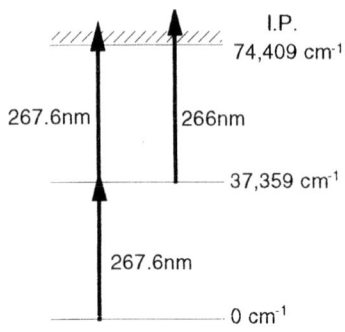

FIGURE 1. RIS schemes for resonant excitation and ionization in gold.

TABLE 1. Data for RIS Schemes for Au

Scheme		$\Delta\omega_D$ [rads^{-1}]	A_{21} [s^{-1}]	σ_{21} [cm^2]	σ_I [cm^2]	E_{sat} [mJ/mm^2]	P_{sat} [W/cm^2]
$\omega_1+\omega_1$	excitation	2×10^{10}	1.1×10^8	6.27×10^{-13}		3.26×10^{-3}	65.16
	ionization				24×10^{-18}	0.31	6.2×10^6
$\omega_1+\omega_2$	excitation	2×10^{10}	1.1×10^8	6.27×10^{-13}		3.26×10^{-3}	65.16
	ionization				10.3×10^{-18}	0.72	1.44×10^7

$\Delta\omega_D$ is the Doppler width of the transition, A_{21} is the Einstein transition probability, σ_{21} is the absorption cross-section of the transition, E_{sat} is the energy required to saturate the transition, P_{sat} is the power required to saturate the transition

Palladium

Palladium has an ionization potential of 67,236 cm^{-1} [4] and offers several suitable intermediate levels for resonant excitation and subsequent ionization using either one- or two-colour type schemes (Fig. 2).

As shown in Fig. 2 ionization can be realized by either a photon at λ=276.3 nm or by the more powerful 266 nm source. The lifetime of the 5p $^3P^o_1$ intermediate resonant atomic state is 5 ns [5]. The estimated saturation parameters are shown in Table 2. Reference data about the energy level diagram of Pd, the ionization potential, radiation lifetimes and oscillator strengths were taken from [4,5].

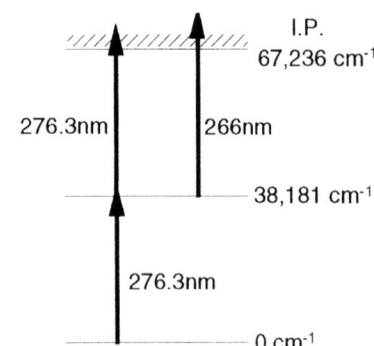

FIGURE 2. RIS schemes for resonant excitation and ionization in palladium.

$\omega_1 + \omega_1$: $4d^{10}\,{}^1S_0 \rightarrow 4d^9 5p\,{}^3P_1^{\,o}$ ($\lambda_1 = 276.39$ nm) $\rightarrow$ Pd$^+$ ($\lambda = \lambda_1 = 276.39$ nm)

$\omega_1 + \omega_2$: $4d^{10}\,{}^1S_0 \rightarrow 4d^9 5p\,{}^3P_1^{\,o}$ ($\lambda_1 = 276.39$ nm) $\rightarrow$ Pd$^+$ ($\lambda_2 = 266$ nm)

TABLE 2. Data for RIS schemes for Pd

Scheme		$\Delta\omega_D$ [rads^{-1}]	A_{21} [s^{-1}]	σ_{21} [cm^2]	σ_I [cm^2]	E_{sat} [mJ/mm^2]	P_{sat} [W/cm^2]
$\omega_1+\omega_1$	excitation	2.78×10^{10}	2.07×10^7	9.05×10^{-14}		4.1×10^{-3}	82.25
	ionization				9.48×10^{-18}	0.76	1.52×10^7
$\omega_1+\omega_2$	excitation	2.78×10^{10}	2.07×10^7	9.05×10^{-14}		4.1×10^{-3}	82.25
	ionization				8.48×10^{-18}	0.88	$176. \times 10^7$

Rhodium

The ionization potential of Rh is 60197 cm^{-1} [4]. A resonant ionization scheme of the type $\omega_1 + \omega_1$ is possible using transitions from either the 5s $^4F_{9/2}$ ground state or from the metastable 5s $^4F_{7/2}$ term, 1529.27 cm^{-1} above the ground state.

I. $\omega_1 + \omega_1$: $4d^8 5s\ a^4F_{9/2} \rightarrow 4d^8 5p\ z^2G_{9/2}^{\,o}$ ($\lambda_1 = 316.32$ nm) $\rightarrow$ Rh$^+$ ($\lambda = \lambda_1 = 316.32$ nm)

II. $\omega_1 + \omega_1$: 5s $a^4F_{7/2} \rightarrow 5p\ z^2F_{7/2}^{\,o}$ ($\lambda_1 = 328.148$ nm) $\rightarrow$ Rh$^+$ ($\lambda = \lambda_1 = 328.148$ nm)

Although this metastable state is expected to be populated after laser ablation, the overall efficiency of this approach will nevertheless be comparatively low. The preferred scheme for trace element analysis of rhodium is therefore a stepwise excitation and ionization of the $\omega_1 + \omega_2$ type where again the powerful 266 nm laser output is used in the second photoionization step.

An energy level diagram and the two-colour schemes for Rh resonant excitation and ionization are shown in Fig. 3. They include resonant transitions starting from the ground state 5s $a^4F_{9/2}$ and two different intermediate resonance levels:

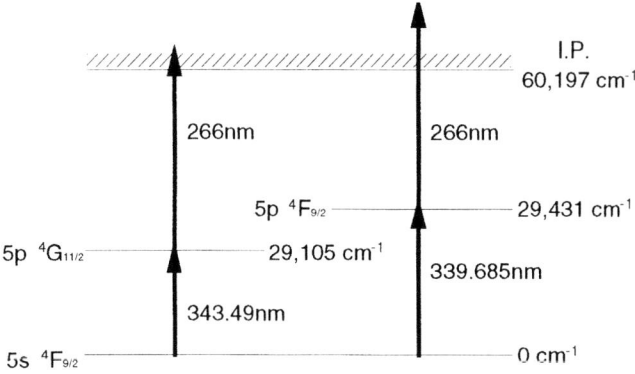

FIGURE 3. RIS schemes for resonant excitation and ionization in rhodium.

5p $z^4G_{11/2}^{\,o}$ and 5p $z^4F_{9/2}^{\,o}$ at excitation wavelengths of $\lambda = 343.49$ nm and $\lambda = 339.6$ nm, respectively.

I. $\omega_1 + \omega_2$: $4d^8 5s\ a^4F_{9/2} \rightarrow 4d^8 5p\ z^4G_{11/2}^{\,o}\ (\lambda_1 = 343.489\ \text{nm}) \rightarrow \text{Rh}^+\ (\lambda_2 = 266\ \text{nm})$

II. $\omega_1 + \omega_2$: $4d^8 5s\ a^4F_{9/2} \rightarrow 4d^8 5p\ z^4F_{9/2}^{\,o}\ (\lambda_1 = 339.685\ \text{nm}) \rightarrow \text{Rh}^+\ (\lambda_2 = 266\ \text{nm})$

The lifetimes of the 5p $^4G_{11/2}^{\,o}$ and 5p $^4F_{9/2}^{\,o}$ intermediate terms are 10.1 ns and 9.2 ns [6], respectively. The estimated saturating laser fluences for the schemes above mentioned are given in Table 3.

TABLE 3. Data for RIS schemes for Rh

Scheme		$\Delta\omega_D$ $[\text{rads}^{-1}]$	A_{21} $[\text{s}^{-1}]$	σ_{21} $[\text{cm}^2]$	σ_I $[\text{cm}^2]$	E_{sat} $[\text{mJ/mm}^2]$	P_{sat} $[\text{W/cm}^2]$
I. $\omega_1 + \omega_1$	excitation	267×10^{10}	1.22×10^8	7.13×10^{-13}		2.6×10^{-3}	52.49
	ionization				11.55×10^{-18}	0.54	1.09×10^7
II. $\omega_1 + \omega_1$	excitation	2.58×10^{10}	2.36×10^7	1.57×10^{-13}		2.28×10^{-3}	45.53
	ionization				12.43×10^{-18}	0.49	9.75×10^7
I. $\omega_1 + \omega_2$	excitation	2.46×10^{10}	1.31×10^8	1.0×10^{-12}		1.9×10^{-3}	38.02
	ionization				8.48×10^{-18}	0.68	1.36×10^7
II. $\omega_1 + \omega_2$	excitation	2.49×10^{10}	6.41×10^7	4.73×10^{-13}		1.98×10^{-3}	39.65
	ionization				8.26×10^{-18}	0.71	1.42×10^7

RESULTS AND DISCUSSIONS

A series of experiments were conducted in order to test the applicability of the proposed resonant ionization schemes and to determine their detection sensitivities. Plots of the normalized Pd peak heights versus ionization laser energy for the $\omega_1 + \omega_1$ and $\omega_1 + \omega_2$ types schemes are shown in Fig. 4. Similar dependences of the ionization efficiencies were obtained for the schemes applied for Au and Rh detection. Clearly the $\omega_1 + \omega_2$ type scheme provides saturation of the TOF-RIMS signal thus becoming the preferred option for the quantification of low trace concentrations of precious metals in minerals.

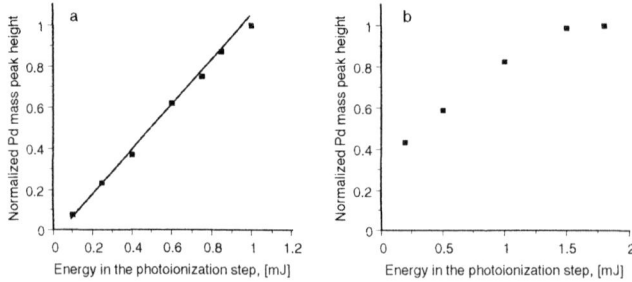

FIGURE 4. Dependence of the TOF-RIMS ions signal for Pd on: a) the energy of the laser source used in the $\omega_1 + \omega_1$ resonant ionization scheme and b) the energy of the photoionization source used in the $\omega_1 + \omega_2$ resonant ionization scheme.

To determine the minimum detection limits (MDL) of the TOF-RIMS mass spectrometer when using the $\omega_1 + \omega_2$ type scheme, calibration curves were established using homogeneous CANMET and NIST reference standards covering up to three orders of magnitude in concentrations in different matrices (Au, Pt, Ag and Cu). Attained minimum detection limits (MDL = 2σ of the background) are 8 ppb for Au, 15 ppb for Pd and 17 ppb for Rh.

The reproducibility of the TOF-RIMS measurements was estimated using a series of 10 consecutive measurements, where for each, 50 laser shots were averaged using a gold reference standard. The relative standard deviation (RSD) of the means is ~12%.

TRACE ELEMENT ANALYSIS OF PRECIOUS METALS IN MINERALS

The TOF-RIMS instrument has been applied for trace element analysis of Au, Pd and Rh in sulphide minerals (pyrrhotite, pentlandite, pyrite, aresenopyrite and chalcopyrite), iron oxide (hematite and magnetite) and silicates with concentrations in the ppm to low ppb range. The same mineral particles were analyzed with Dynamic SIMS using CAMECA 3f SIMS mass spectrometer. The TOF-RIMS data are in a good agreement (relative mean deviation: ± 16%) with the data obtained for Au, Pd and Rh by Dynamic SIMS within the overlapping sensitivity range of both instruments.

CONCLUSIONS

This study demonstrates the ability of the TOF-RIMS technique to provide routine quantitative ultra trace element analysis of precious metals in minerals. The absence of large matrix effects when using separate ablation and postionization steps was documented by analyzing standard reference samples with different matrices and by the comparable results obtained in minerals. The reported minimum detection limits for precious metals is in the 8-17 ppb range.

ACKNOWLEDGEMENTS

This research is supported by the National Research Council (NRC) of Canada.

REFERENCES

1. Dimov, S.S., and Chryssoulis, S.L., *Spectrochimica Acta*, **53B**:399-406, (1988).
2. Saloman, E.B., *Spectrochimica Acta*, **43B**:37-64, (1990).
3. Dimov, S. S., Chryssoulis, S. L., "Microbeam Techniques in the Mineral Industry" in *Analytical Technology in the Mineral Industries*, edited by L. J. Cabri et al, TMS Conference Proceedings, San Diego, pp: 209-219, (1999).
4. Callender, C.L., Hackett, P.A. and Rayner, D.M., *J. Opt. Soc. Am. B*, **5**, 614-618, (1988).
5. Banmann, M., and Liening, H., *Physics Letters*, **36A**:239-330, (1971).
6. Duquette, D.W. and Lawler, J.E., *J. Opt. Soc. Am. B*, **2**, 1,948-1,952, (1985).

Counting Radioactive Noble Gas Atoms : Lasers, Accelerators or Decay Counters ?

Bernhard E. Lehmann

Physics Institute, University of Bern, Sidlerstr. 5, CH-3012 Bern, Switzerland

Abstract. Radon-220 and Radon-222 in environmental air samples can be measured on-line with high temporal resolution by state-of-the-art alpha counting. Argon-37, Krypton-85 and Argon-39 are measured routinely by low level decay counting using high-pressure gas proportional counters in an underground laboratory. For Krypton-81 decay counting is not possible in environmental samples. Therefore, various attempts have been made to use atom counting techniques for this isotope. Laser Resonance Ionization Spectroscopy was used to count Kr-81 atoms from old groundwater samples. In a more recent study groundwater dating in the Great Artesian Basin in Australia was accomplished using Cyclotron Accelerator Mass Spectroscopy. Collinear Beam Spectroscopy using Photon Burst Detection and Atom Trap Trace Analysis were successfully used to detect both Kr-81 and Kr-85 in natural Kr samples. Most of these new techniques, however, need to be further improved to become routine tools in environmental studies.

INTRODUCTION

Naturally occurring or man-made radioisotopes have been used in recent years in growing numbers in environmental studies. They serve as tracers which participate in minute concentrations in the transport processes in the atmosphere, in the water cycle or in the exchange processes between these reservoirs and living species. Concentrations of radionuclides and their variations in time and space not only map pathways in our environment but in many applications it is also possible to determine the relevant time constants using the half-life of a particular radionuclide. The most advanced analytical techniques have to be used because the concentrations of these tracers are usually extremely small down to a few atoms in a water or air sample.

NOBLE GAS RADIOISOTOPES

Because they are chemically inert, noble gas radioisotopes are particularly useful for dating purposes. Once an air sample is isolated from the source reservoir the activity decreases according to the well-known decay laws; no chemical or biological processes have to be considered as possible competing loss channels. In Table 1 the half-lives and possible fields of applications are summarized for the six nuclides discussed in this contribution.

CP584, *Resonance Ionization Spectroscopy 2000: 10th Int'l. Symp.*, edited by J. E. Parks and J. P. Young
© 2001 American Institute of Physics 0-7354-0024-5/01/$18.00

TABLE 1. Noble Gas Radionuclides Used in Environmental Studies.

	Half-Life	Fields of Applications
Radon-220	55.6 s	Calibration of Near-Surface Turbulent Gas Transport
Radon-222	3.82 d	Trace Gas Transport in Soil, Atmospheric Mixing
Argon-37	35.0 d	Tropospheric Mixing, Water-Rock-Interactions
Krypton-85	10.76 y	Atmospheric Circulation, Dating Young Groundwater
Argon-39	269 y	Groundwater Dating, Deep Ocean Circulation
Krypton-81	229 000 y	Dating of Very Old Groundwater and of Polar Ice

TABLE 2. Production Processes and Typical Atmospheric Activities

	Production in the Environment	Activity in Air (Bq/m^3)
Radon-220	Decay of ^{224}Ra (natural ^{232}Th-series)	100
Radon-222	Decay of ^{226}Ra (natural ^{238}U-series)	10
Argon-37	Cosmic Rays : $^{40}Ar(n,4n)^{37}Ar$, $^{36}Ar(n,\gamma)^{37}Ar$ Neutrons in Rocks : $^{40}Ca(n,\alpha)^{37}Ar$	0.003
Krypton-85	Fission of ^{238}U (Nuclear Fuel Processing)	1.4
Argon-39	Cosmic Rays : $^{40}Ar(n,2n)^{39}Ar$, $^{40}Ar(n,np)^{39}Cl \rightarrow {}^{39}Ar$ Neutrons in Rocks : $^{39}K(n,p)^{39}Ar$	0.017
Krypton-81	Cosmic Rays : $^{80}Kr(n,\gamma)^{81}Kr$, spallation of $^{82,83,84,86}Kr$	0.0000017

The activities listed in Table 2 are typical average numbers. They can of course vary depending on the specific situation. For all nuclides the range can go from zero (when all the atoms in an old sample have decayed) to values above the ones presented in the list. For example :

- Rn-220 activities of 100 Bq/m^3 or higher in air only exist very close to the ground surface (lowest approx. 20 cm) in calm nights without wind. Atoms cannot reach high altitudes in the troposphere before they decay.
- Rn-222 activities of 10 Bq/m^3 are typical for outdoor air ; indoors activities can be 2-3 orders of magnitude higher and in soil gas values are typically 20 000 Bq/m^3 or higher.
- The range of Ar-37 activities found in air extracted from groundwaters is rather large. Up to 10 Bq/m^3 have been measured in rock formations with high U- and/or Ca-concentrations where subsurface production is enhanced.
- Higher Kr-85 activities are regularly found in plumes downstream of nuclear reprocessing plants.
- The Ar-39 activity in atmospheric air is constant. However, in subsurface environments with high U- and/or K-concentration subsurface production can increase the activity by up to a factor of 5.
- The Kr-81 activity of atmospheric air is constant, subsurface sources appear to be negligible.

DECAY COUNTING VS ATOM COUNTING

Decay counting techniques measure the concentrations of the radioactive atoms of interest in an environmental sample by counting the number of α-, β- or γ-signals in a defined time interval. In contrast, atom counting techniques do not wait for the atoms to decay. In Figure 1 the typical average activities (Bq/m³) in air from Table 1 are compared to the respective concentrations (atoms/m³) in air. Marked areas indicate where the different analytical approaches are adequate : A) Alpha-Decay-Counting; B) Low Level Gas Proportional Counting; C) Atom Counting.

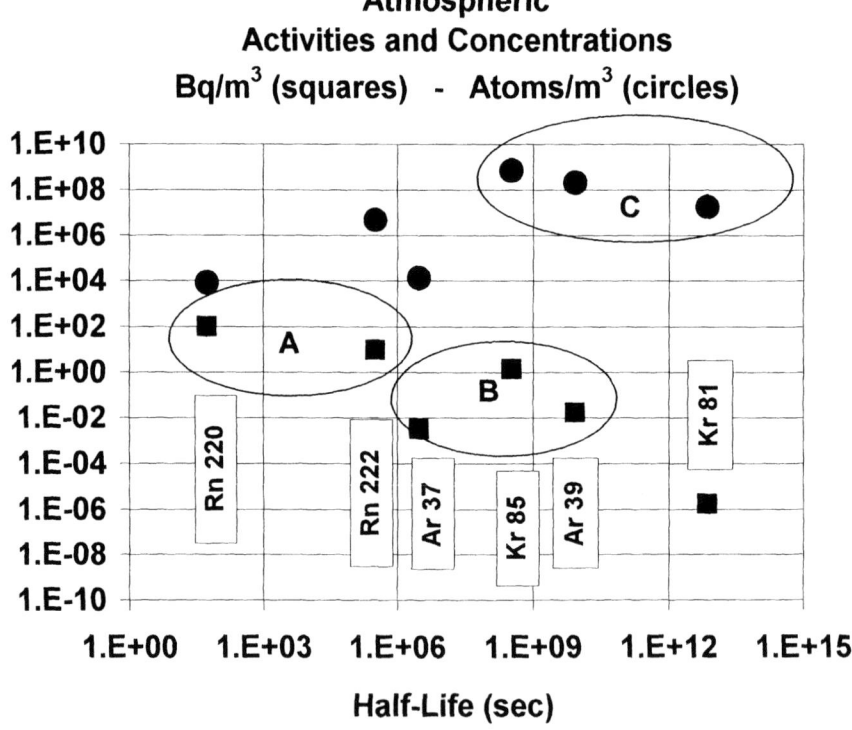

Figure 1 : Atmospheric activities in Bq/m³ (squares) from Table 1 and corresponding atmospheric concentrations atoms/m³ (circles) vs the half-life (s).

LOW LEVEL DECAY COUNTING

State-of-the-art alpha-counters using digital signal processing for pulse shape analysis in a flow-through ionization chamber can be used for on-line monitoring of [220]Rn and [222]Rn in tropospheric air or soil gas. Detection limits are about 3 Bq/m^3 in a 10 min interval [1].

With low level gas proportional counters activities of [37]Ar, [85]Kr and [39]Ar can be measured in air or gas samples extracted form ground- or oceanwater. In order to achieve the necessary very low background count rates, detectors in our underground laboratory in Bern, are shielded by 30m of sandstone from cosmic rays, the walls of the laboratory are built with special low activity concrete, old lead shieldings surround counters that are made from high-purity copper and are operated in anticoincidence arrangements inside guard counters [2],[3].

Typical parameters for groundwater dating are as follows [4] :

- Krypton-85 : From ~300 liters of water 0.02 cm^3 of Kr are separated (yield 85%) which correspond to an activity of < 0.025 Bq or count rates of < 0.018 cps (counting efficiency 70%). Counters of 16 cm^3 volume are operated at 2-5 bars using P-10 gas (90 % Ar, 10% CH$_4$, + Kr sample). Standard counting time is several days (< 1500 counts per day).
- Argon-39 : From a minimum of 2000 liters of water ~ 0.7 liters of Ar are separated to yield an activity of < 0.0012 Bq (count rate < 0.0009 cps with 70% counting efficiency). Counters of 100 cm^3 volume are operated at ~ 8 bars using again P-10 gas. Standard counting time : up to 6 weeks (< 70 counts per day).

For Krypton-81 the specific activity in natural Kr is too low for these techniques (< 1 decay per week per m^3 of air). This nuclide is to be measured by atom counting techniques. For Kr-85 and Ar-39 such techniques may (or may not) become competitive in terms of sensitivity (sample size), precision, time required for sample preparation and analysis, and cost.

ACCELERATOR MASS SPECTROMETRY

As shown in Figure 1 atomic concentrations of the three longer lived noble gas radionuclides are not very different (on the order of 10^7 to 6x10^8 atoms per m^3 of air). However, the isotopic interference from the stable isotopes in the same samples differs considerably. Table 3 lists these isotope ratios (number of radioactive atoms / number of stable atoms) for the typical activities in Table 1. In addition the calculated number of radioactive atoms per liter of water in equilibrium with air (at T = 15 °C) is given.

TABLE 3. Isotope Ratios and Concentrations in Water Samples

	Isotope Ratio	Concentration (atoms per liter of water)
Krypton-85	$< 2.2 \times 10^{-11}$	50 000
Argon-39	$< 8.0 \times 10^{-16}$	7 500
Krypton-81	$< 5.6 \times 10^{-13}$	1 200

Such low isotope ratios cannot be measured by conventional mass spectrometers. Accelerator Mass Spectrometry using Tandem Accelerators is also not feasible because noble gases do not easily form negative ions. It is however possible, to use cyclotrons or linear accelerators that operate with positive ions [5]. In a recent study [6] ^{81}Kr concentrations were measured in air extracted from 4 very large groundwater samples (16 000 liters each) from the Great Artesian Basin in Australia. With measured activities down to 30% of the atmospheric equilibrium value and corresponding groundwater ages up to 400 000 years this study for the first time demonstrated that the production rate for this isotope in rocks indeed must be extremely small as was estimated in theoretical studies [7].

RESONANCE IONIZATION SPECTROSCOPY

Optical techniques make use of laser light which can be tuned to selected electronic transitions in the respective noble gas atoms. In Figure 2 various schemes are outlined.

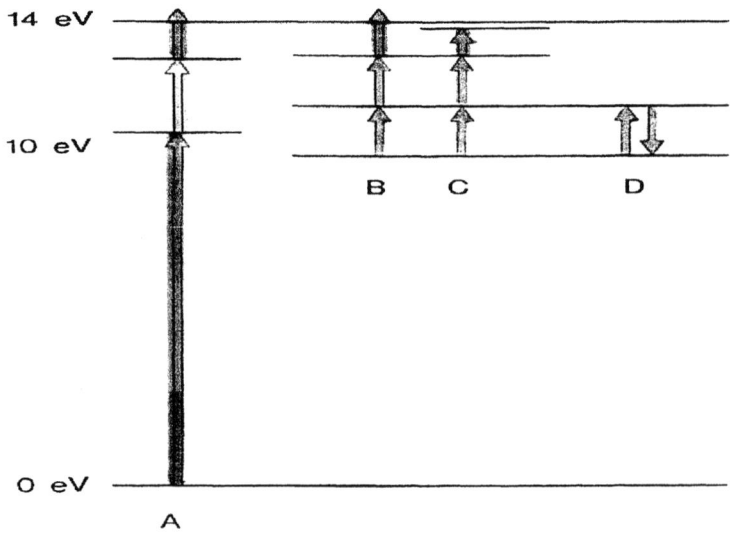

Figure 2 : Optical excitation schemes for the selective detection of Kr atoms.

In Resonance Ionization Spectroscopy (RIS) Kr atoms can be selectively excited and ionized from the groundstate (Fig.2, scheme A) with laser pulses at 116 nm, 558 nm and 1064 nm, the latter being e.g. the fundamental wavelength of a Nd:YAG laser which is used to pump various dye lasers and other non-linear optical processes (doubling, mixing, 4-wave mixing) to produce the required wavelengths [8]. This method has a very good elemental selectivity but is generally not isotope selective because the bandwidths of the pulsed lasers are larger than isotope effects in the noble gas spectra. Isotope selectivity is achieved in combination with mass-spectrometers (quadrupole, time-of-flight TOF), however, the very small isotope ratios (Table 3) require rather complicated isotope pre-enrichment procedures prior to the final measurement in e.g. a RIS-TOF system. In spite of these difficulties both ^{81}Kr and ^{85}Kr have been analyzed in Kr gas extracted from natural groundwater and polar ice samples [9], [10]. Such a system is currently operated at the Institute of Rare Isotope Measurements (IRIM) at the University of Knoxville.

COLLINEAR BEAM SPECTROSCOPY

Isotope effects are much larger in collinear beam spectroscopy where atoms are accelerated to a constant energy and therefore different isotopes move at different velocities giving rise to Doppler-shifted absorption spectra. In addition, by using quasi-resonant charge exchange of Kr ions in an alkali vapour cell (Cs, Rb, K) a metastable state at about 10 eV can effectively be populated (Fig.2, schemes B, C and D) from which a selective excitation can start using cw visible or near-infrared laser light. Such a scheme was proposed by Aseyev and collaborators in 1991 [11]. Ionization can again be achieved by lasers (Fig.2, scheme B) or else by field ionization from a high Rydberg state (Fig.2, scheme C). The later approach is currently pursued by a group at Texas A&M [12].

Photons rather than ions are detected in scheme D of Figure 2. In this approach an optical two-level system is selected where an atom can be cycled while flying though the laser beam. Several photons are generated per atom giving rise to a "photon burst" whenever an atom of a selected isotope passes through the detector. In a collaboration between Colorado State University and the Los Alamos National Laboratory ^{85}Kr was recently measured at ambient levels with such a method [13]. In this study it was also pointed out that a replacement of standard photomultipliers which have only a 2% quantum yield in the 811nm infrared region by large area avalanche photodiodes (LAAD) is very advantageous. Such devices can be operated in a single-photon counting mode with counting quantum efficiencies up to 25% when cooled to liquid nitrogen temperature.

At our institute in Bern we work on a similar approach [14]. In our current detector a series of 8 short detectors (4 cm length) each equipped with a LAAD should generate one count each from a passing ^{81}Kr atom. Optical excitation is achieved with narrow-band cw diode laser light transported by an optical fiber to the fast beam (10 keV) of neutral metastable Kr atoms. This system has been demonstrated to be sensitive enough to allow work with small Kr samples (3×10^{-3} cm^3 STP) representing approx. 50 liters of groundwater by recirculating Kr gas with a small turbomolecular pump between ion source and beam-line in a closed vacuum system.

ATOM TRAP TRACE ANALYSIS

Scientists at Argonne National Laboratory have recently published their results on using a magneto-optical trap for detecting ^{81}Kr and ^{85}Kr atoms in natural Kr gas[15]. The optical scheme D of Figure 2 was also used in this approach. In contrast to the collinear beam experiments where the interaction time between a metastable atom and the laser beam in the detector is only on the order of a few μs, metastable atoms in atom trap can be looked at for up to one second. Accordingly, the number of photons re-emitted per atom is about six orders of magnitude larger (typically 10^7 photons vs 10 photons for a characteristic optical cycling time of about 60 ns). Extremely high isotope selectivity (essentially background free counting) was demonstrated. However, at present the efficiency of transferring atoms from a Kr gas sample into a metastable state and into an atom trap is rather small (1 in 10^7). Improvements are necessary to make the technique practical for reasonable size groundwater samples.

REFERENCES

1. Lehmann B.E., Lehmann M., Neftel A., Gut A., Tarakanov S.V., Geophysical Research Letters, Vol. 26, No.5, 606-610 (1999)
2. Oeschger H. and Wahlen M., Annual Review of Nuclear Science, Vol.25, 423-463 (1975)
3. Loosli H.H., Earth and Planetary Science Letters, 63, 51-62 (1983)
4. Loosli H.H., Lehmann B.E., Smethie W.M. Jr., „Noble Gas Radioisotopes" in *Environmental Tracers in Subsurface Hydrology*, edited by P.Cook and L.Herczeg, Kluwer Academic Publishers (2000)
5. Kutschera W., Paul M., Ahmad I., Antaya T.A., Billquist P.J., Glagola B.G., Harkewitcz R., Hellstrom M., Morrissey D.J., Pardo R.C., Rehm K.E., Sherill B.M., Steiner M, Nucl. Instr. Methods B29, 241-248 (1994)
6. Collon P., Kutschera W., Loosli H.H., Lehmann B.E., Purtschert R., Love A., Sampson L., Anthony D., Cole D., Davids B., Morrissey D.J., Sherrill B.M., Steiner M., Pardo R.C., Paul M., Earth and Planetary Science Letters 182, 103-113 (2000)
7. Lehmann B.E., Davis S.N., Fabryka-Martin J., Water Resources Research 29, 2027-2040 (1993)
8. Hurst G.S., Payne M.G., Kramer S.D., Chen C.H., Phillips R.C., Allman S.L., Alton G.D., Dabbs J.W., Willis R.D., Lehmann B.E.., Reports on Progress in Physics 48, 1333 (1985)
9. Lehmann B.E., Oeschger H., Loosli H.H., Hurst G.S., Allman S.L., Chen C.H., Kramer S.D., Payne M.G., Phillips R.C., J.Geophys. Res. 90, B13, 11549 (1985)
10. Thonnard N., Willis R.D., Wright M.C., Davis W.A., Lehmann B.E., Nuclear Instruments and Methods B29, 398 (1987)
11. Aseyev S.A., Kudryavtsev Y.A., Lethokov V.S., Petrunin V.V., J.Phys.B : At.Mol.Opt.Phys. 24, 2755-2763 (1991)
12. Li X., Lassen J., Schuessler H.A., RIS-2000 proceedings, this volume (2000)
13. Fairbank W.M. Jr, Hansen C.S., LaBelle R.D., Pan X.J., Zhang Y., Chamberlin E.P., Nogar N.S., Miller C.M., Feary B.L., Oona H., Proc.Soc.Photo-Opt.Instr.Eng. Vol. 3270 (1998)
14. Ludin A., Lehmann B.E., Appl.Phys. B 61, 461-465 (1995)
15. Chen C.Y., Li Y.M., Bailey K., O'Connor T.P., Young L., Lu Z.-T., Science, Vol. 286, 1139-1141 (1999)

Towards trace detection of ^{85}Kr by two-step excitation and field ionization

Xinghua Li, Jens Lassen and Hans A. Schuessler

Dept. of Physics, Texas A&M University
College Station, TX 77843-4242

Abstract. We have implemented a new laser-based detection scheme for the rare krypton isotopes. The scheme utilizes two-step excitation of metastable krypton atoms in a nearly perfect three-level system followed by field ionization. The detection scheme was realized experimentally with a mass-separated atom beam. A total detection efficiency of about 10% is obtained.

INTRODUCTION

Ultra sensitive trace isotope detection of ^{85}Kr and ^{81}Kr has many important applications such as radioactive dating and nuclear non-proliferation monitoring. Due to their extremely low isotopic abundance in the range 10^{-11} to 10^{-13}, ultra-sensitive detection techniques must be capable of identifying these rare isotopes in a natural sample. Presently a number of different techniques are being developed for this purpose. While the Low-Level Counting (LLC) [1] technique can be used to detect ^{85}Kr in Kr samples from several hundred liters of water, such a technique is no longer applicable for ^{81}Kr due to its low concentration. Accelerator Mass Spectrometry (AMS) [2] was used to measure the ^{81}Kr concentration in atmospheric krypton. However, as with all mass spectrometry methods, the interference from isobaric ^{81}Br background was not addressed. Resonance Ionization Spectrometry (RIS) has been used for detecting noble gas atoms with isotopic selectivity [3]. Using RIS, Hurst and coworkers were able to count ^{81}Kr atoms in an enriched sample [4]. Atom trap trace analysis [5] has recently been applied to the detection of ^{81}Kr and ^{85}Kr with a detection efficiency of 10^{-7}.

In this paper we present a new approach based on two-step laser excitation of krypton metastable atoms in a mass-separated atom beam. Charge-exchange of a noble gas ion beam in an alkali vapor cell is known to produce metastable noble gas atoms with very high efficiency. By means of two-step excitation in a nearly perfect three-level system, the population can be efficiently transferred to several high-lying Rydberg levels. There upon the atoms are field ionized and the ions can be counted individually. An energy filter, in combination with the energy offset by field ionization, provides efficient background ion elimination.

CP584, *Resonance Ionization Spectroscopy 2000: 10th Int'l. Symp.,* edited by J. E. Parks and J. P. Young

THE EXCITATION SCHEME

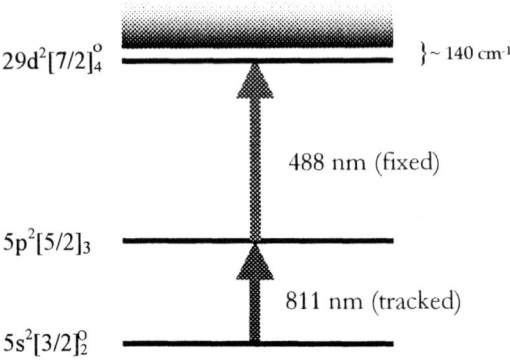

$29d^2[7/2]_4^o$ $\Big\}\sim 140\ cm^{-1}$

488 nm (fixed)

$5p^2[5/2]_3$

811 nm (tracked)

$5s^2[3/2]_2^o$

FIGURE 1. The excitation scheme.

The excitation scheme is shown in Fig. 1. The metastable krypton atoms are produced by charge exchange in Rb-vapor from a fast Kr ion beam. Subsequently they are excited by a Ti:Sapphire laser at 811 nm to the intermediate level in the first step. Because the only decay channel for atoms in the $5p$ level is back to the metastable level, the transition is a recyclable (closed) transition. By means of Doppler-tuning of the atom beam energy the atoms can be further excited to various high-lying Rydberg levels with a single-mode Ar^+ laser at 488 nm. For efficient transfer of population, it is necessary that the transition to the Rydberg level is also recyclable. Among various Rydberg levels, the $29d^2[7/2]_4$ level closely matches the requirement. The strongest transition to this level is from the $5p^2[5/2]_3$ level. Transition rates to other levels are at least four times weaker in intensity due to the λ^{-2} rule. More importantly, this transition has the largest oscillator strength among all possible transitions from the intermediate level to the $29d^2$ and $31s^2$ manifolds.

There exists a detailed treatment of a three-level atom interacting with two near-resonant monochromatic fields by Whitley and Stroud [6] using quantum electro-dynamic theory (QED). According to this theory the maximum fraction of population in the Rydberg level is

$$\rho_{33} = (2r^2 + 1) / 2(r + 1)^2,$$

where $r=(A_{5p-5s}/2A_{29d-5p})^{1/2}$, and A_{5p-5s} and A_{29d-5p} are the Einstein A coefficients of the two transitions. Based on the known values of lifetimes for the intermediate and Rydberg levels (30 ns and $\sim$7 μs, respectively), we estimate that as much as 84% of the population can be effectively transferred to the Rydberg level under optimum conditions.

Using the QED theory, it is possible to calculate the double-resonance line shape. In Fig. 2 we present the results for the peak shape at the double-resonance scheme for Kr under steady-state excitation conditions. It is found that maximum population transfer occurs when both lasers are in exact resonance with the transitions.

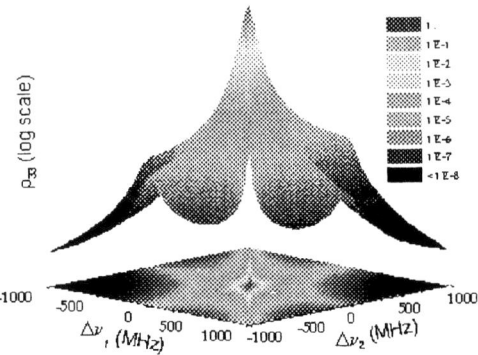

FIGURE 2. Peak shape of the double-resonance scheme for krypton. The laser powers are 15 mW and 1.0 W for the first and second steps respectively. The lifetime of the Rydberg level is assumed to be 7 μs.

EXPERIMENTAL SETUP

A hollow cathode discharge ion source (Danfysik 911A) generates a beam of Kr ions, which were accelerated to about 10 keV and mass-separated with a low-resolution bending magnet ($m/\Delta m \sim 200$). The mass-selected ion beam is deflected by a few degrees and co-propagates with laser light from a Ti:Sapphire ring laser and a single-mode Ar^+ laser. The beam-line is evacuated by turbo-molecular pumps to 10^{-6} torr pressure. The ion beam enters an alkali vapor charge exchange cell for neutralization. A large fraction of the ions is thereby transferred into a metastable level. The remaining ions are deflected and the fast neutral atoms enter the interaction region with the lasers of about 0.8 meters. Thereafter the atoms pass through a collisional ionization [7] cell and a field ionizer [8]. The Rydberg atoms can be either collisionally ionized or field ionized. The product ions, after passing through a 135° energy filter, are detected with a Faraday cup or a channeltron. Field ionization occurs at a well-defined field strength and potential and also adds an additional offset to the ion energy. This effect is used to the advantage where the background ions are significantly reduced through the use of energy discrimination.

Since field ionization can only be applied to high-lying Rydberg atoms, collisional ionization is used for checking the excitation of atoms from the metastable level into the intermediate level during system alignment. Excitation of the atoms in the intermediate level to the Rydberg level is achieved by Doppler-tuning of the atom beam energy and simultaneous tracking of the first-step laser wavelength.

RESULTS

The experimental procedure is as follows: first the *5s-5p* transition is located using the collisional ionization technique. Since it is a closed transition, an enhancement in ion current/counts is observed whenever the metastable atoms are excited to the intermediate level. In the next step the simultaneous tracking of the acceleration voltage with respect to the laser wavelength of the Ti:Sapphire laser and the energy filter is established. By scanning the atom beam energy, transitions of the atoms in the intermediate level to the Rydberg levels are detected. In Fig. 3 we present the spectrum of a ^{84}Kr beam in the 5-12 keV energy region detected with field ionization. As expected the transition to the $29d^2[7/2]_4$ is the strongest among all possible transitions. This transition is used in the ultra-sensitive detection of krypton isotopes. As a test we present the result of such a measurement for stable krypton isotopes. The spectrum is shown in Fig. 4.

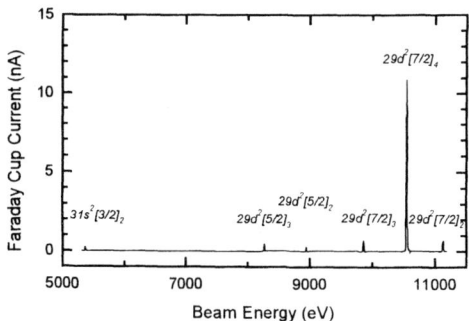

FIGURE 3. Transitions of ^{84}Kr that can be accessed through 488 nm laser pumping in 5-12 keV energy region. All transitions start from the same intermediate $5p^2[5/2]_3$ level. The upper levels of the transitions are shown in the figure.

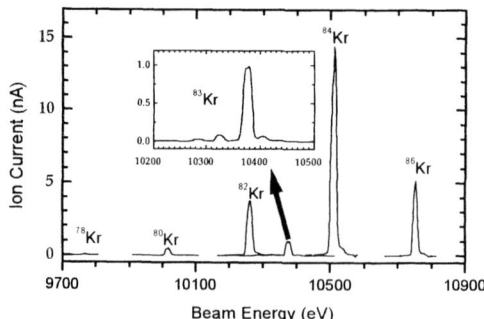

FIGURE 4. Detection of stable isotopes of krypton in a natural gas sample using the proposed excitation scheme. The inset shows the spectrum of ^{83}Kr with hyperfine structures.

From Fig. 4 it is possible to deduce the isotopic concentration of the stable isotopes. The obtained relative abundances are (from ^{78}Kr to ^{86}Kr): .003, 0.035, 0.26, 0.29, 1.0, 0.35. The result compares reasonably well with the natural isotopic abundances: .006, .039, 0.20, 0.20, 1.0 0.30. Note that this result was obtained from one run only and no effort was made to accurately measure the isotope ratio. Also the system is optimized for ^{84}Kr detection only.

We have also investigated the overall detection efficiency versus the different charge exchange media. It is found that rubidium gives slightly higher detection efficiency than cesium. For a small charge exchange ratio, the Rydberg current increases linearly with the neutralization ratio, reaching a maximum at around 80%. With an ion beam current of 500 nA and 80% charge exchange ratio, an overall detection efficiency of about 10% is achieved.

OUTLOOK

We have proposed and demonstrated a new sensitive detection scheme for rare krypton isotopes. The technique utilizes two-step laser excitation of metastable krypton atoms in a fast atom beam. By using closed transitions in the first and second steps and a powerful single-mode Ar^+ laser to saturate the weak transition to the Rydberg level, population is efficiently transferred to the high-lying Rydberg level. Through field-ionization and energy filtering we suppressed the background ions by orders of magnitude. Initial results using rubidium as a charge exchange medium gave a total efficiency of 10%. The major problem of our detection system is that the first transition alone cannot be detected for Kr-85, only the two-step excitation is expected to have sufficient S/N ratio. Therefore currently we are trying to locate the [85]Kr transition to the Rydberg level using a slightly enriched [85]Kr sample (10^{-8}). Once the transition has been found, we can proceed to do [85]Kr detection in natural krypton samples.

ACKNOWLEDGMENTS

This research is supported by the Texas Advanced Research Program under grant 010366-082.

REFERENCES

1. H. H. Loosli and H. Oeschger, *Earth Planet. Sci. Lett.* **7**, 67-73 (1969).
2. P. Collon, T. Antaya, B. Davids, M. Fauerbach, R. Harkewicz, M. Hellstrom, W. Kutschera, D. Morrissey, R. Pardo, M. Paul, B. Sherrill and M. Steiner, *Nuclear Instrum. Methods* **B123**, 122-127(1997)
3. G. S. Hurst, M. G. Payne, S. D. Kramer, C. H. Chen, R. C. Philips, S. L. Allman, G. D. Alton, J. W. Dabbs, R. D. Willis, B. E. Lehmann, *Rep. Prog. Phys.* **48**, 1333 (1985).
   B. E. Lehmann, H. Oeschger, H. H. Loosli, G. S. Hurst, S. L. Allman, C. H. Chen, S. D. Kramer, M. G. Payne, R. C. Philips, *J. Geophys. Res.* **90**, B13, 11547 (1985).
4. S. D. Kramer, G. S. Hurst, C. H. Chen, M. G. Payne, S. L. Allman, R. C. Philips, B. E. Lehmann, H. Oeschger, H. H. Loosli, R. D. Willis and N. Thonnard, *Nuc. Instrum. Methods* **B17**, 395-401 (1986).
5. C. Y. Chen, Y. M. Li, K. Bailey, T. P. O'Connor, L. Young and Z. T. Lu, *Science* **286**, 1139-1141 (1999).
   K. Bailey, C. Y. Chen, X. Du, Y. M. Li, Z. –T. Lu, T. P. O'Connor and L. Young, *Hyperfine Interactions* **127**, 515-518 (2000).
6. R. M. Whitley and C. R. Stroud, Jr., *Phys. Rev.* **14**, 1498-1513 (1976).
7. R. Neugart, W. Klempt, and K. Wendt, *Nuc. Instrum. Methods* **B17**, 354-359 (1986).
8. K. Stratmann, R. Hohmann, H.J. Kluge, S. Kunze, J. Lantzsch, L. Monz, E. W. Otten, G. Passler, J. Stenner, K. Wendt, K. Zimmer, *Rev. Sci. Instrum.* **65**, 1847 - 1852 (1994)

SESSION III

From Single Ion to Single Pion

Ken Ledingham

*Department of Physics and Astronomy, University of Glasgow, Glasgow G12 8QQ, UK
and AWE plc, Aldermaston, Reading Berks RG7 4PR*

Abstract. The first RIS conference held in Gatlinburg, Tennessee in 1981 demonstrated conclusively that tuned lasers could be used to detect, particularly atoms, in very small quantities even down to single atom levels. Since then the field has matured as the lasers have developed. Recently a number of laboratories have shown that molecules can be detected with high intensity, short pulse lasers with minimum fragmentation. A natural extension of this work is to bathe atoms in short pulse beams of terawatt and petawatt lasers (greater than the total electrical output of the USA). At these intensities the ponderomotive forces are so great that the atomic electrons are accelerated to multi MeV energies and we enter the regime of laser induced nuclear physics.

INTRODUCTION

This paper will describe what happens to atoms, molecules and nuclei as they are irradiated in laser light varying from very low intensities of about a few Wcm^{-2} up to ultra-high intensities of about $10^{21} Wcm^{-2}$. Thus we shall be discussing laser regimes which change in intensity over 21 orders of magnitude.

In the late 1970s the field of resonant ionization spectroscopy was established using tuned pulsed lasers typically with ns pulse widths [1] in the low intensity regime for transitions between atomic and molecular levels. For transitions to the continuum or off resonant transitions, the laser intensities were increased by between 7-10 orders of magnitude to yield measurable ionization rates. This is the regime where multiphoton processes are all important i.e. a perturbative photon-by photon based picture of the interaction of individual electrons [2].

The last few years have witnessed a dramatic upsurge in short pulse laser technology with its concomitant high achievable peak powers. This has been brought about by the development of chirped-pulse amplification (CPA)[3,4]. The pulse widths are typically between 50-150fs and with intensities between $10^{13-16} Wcm^{-2}$. At these intensities the electric field of the applied laser radiation is comparable with the Coulomb field strength between atomic electrons. The picture of multiphoton ionization is no longer applicable and must be replaced by a tunneling ionization picture in which the electrons tunnel with ease through the atomic and molecular potentials that have been extensively modified by the laser field.

When the laser intensity is further increased to between $10^{17-21} Wcm^{-2}$ the free electrons can quiver in the laser focus totally relativistically with ponderomotive energies of MeV. Of current interest are a number of mechanisms proposed which can cause the electrons in plasmas to be accelerated to many times the ponderomotive potential such as wakefield acceleration [5].

CP584, *Resonance Ionization Spectroscopy 2000: 10th Int'l. Symp.*, edited by J. E. Parks and J. P. Young

This paper will deal with each of these laser regimes in turn giving one or two experimental examples in each case. The survey is by no means exhaustive and for ease of presentation, all the examples cited were carried out in the author's laboratory.

Experiments in the laser regime 10^{0-10} Wcm^{-2} (Multiphoton picture)

Fig 1a shows the most simple two photon resonant ionization scheme with a single laser colour. Much of the original work in the early eighties were carried out using such a scheme with a single ns pulse dye laser. Although most people quoted their results with respect to pulse energies in J, the pulse intensities were for the resonant step typically about 10Wcm^{-2} and for the off resonance ionization step to the continuum about 10^{7}Wcm^{-2}. Of course if the laser had sufficient power then no resonant steps were required to reach ionization. Fig 1b shows a multiphoton step-by-step off resonance absorption to the continuum via virtual states. For an atom or molecule with an ionization potential of a few eV, laser intensities of about 10^{10} Wcm^{-2} are necessary requiring tight focussing.

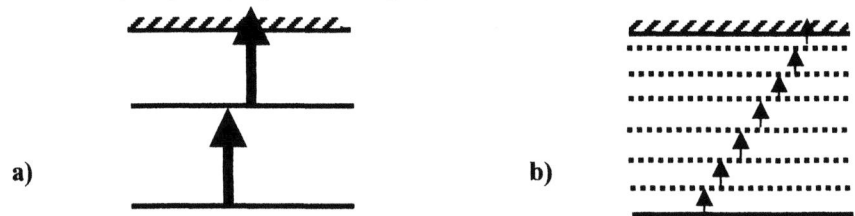

FIGURE 1. The resonant step in a) requires a laser power of about 10Wcm^{-2} while the ionization step is about 10^{7}Wcm^{-2}. The off resonant multiphoton scheme shown in b) with an atom of a few eV and a wavelength of 1μ requires a laser intensity of about 10^{10}Wcm^{-2}.

The number of experiments using resonant ionization spectroscopy is legion but a very interesting one carried out in an number of laboratories world wide passed a laser beam through cesium vapour in an ionization chamber. When the laser wavelength was scanned in the UV a number of s-np transitions were excited and subsequently ionised resulting in the ionization yield shown in Fig 2

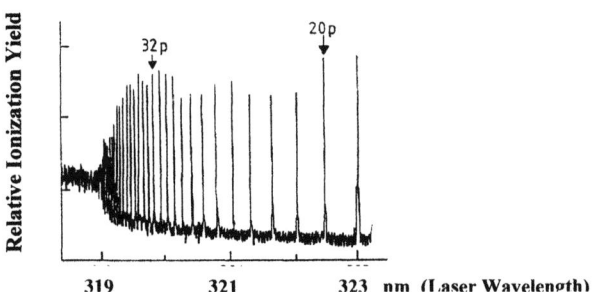

FIGURE 2. The relative ionization yield via s-np transitions after passing a tunable laser beam through cesium vapour in an ionization counter

Experiments in the laser regime 10^{13-16} Wcm^{-2} (Tunneling Picture)

Some years ago we and others noticed that a number of polyatomic molecules when irradiated with short pulse lasers <90fs at 750-790 nm and intensities up to about 10^{15} Wcm^{-2}, not only did not fragment to any great degree but provide multiply charged parent ions. This surprising observation is found in both linear and ring structured molecules and is very similar to the behaviour of inert atoms like Xenon under the same irradiation conditions. This is shown in Fig. 3. for deuterated benzene

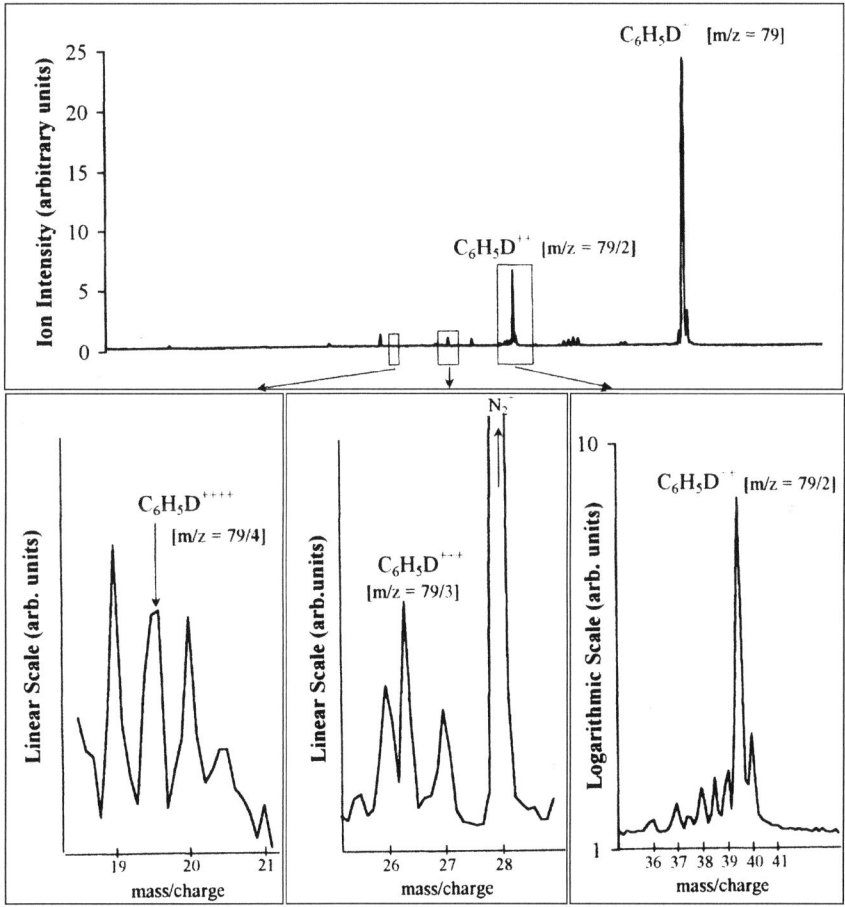

FIGURE 3. Deuterated benzene vapour irradiated at 2×10^{15} Wcm^{-2} showing peaks at parent mass 79, doubly charged ,triply charged and quadrupled charge.

It was also noticed that when the wavelength was reduced to 375 nm, the fragmentation generally increased although at our pulse widths the parent ion was still the dominant peak. What is happening here is that dissociative lifetimes normally in the ps range were being bypassed by the high intensity light source. This is very different behaviour from irradiating molecules with ns pulses at $10^9 Wcm^{-2}$ where low mass fragments dominate the spectrum.

This led to femtosecond laser time-of-flight mass spectrometry (FLMS) for uniform quantitative analysis of molecules being investigated. Various samples of molecular gases and vapours have been studied, using ultra-fast ($\sim$50 fs) laser pulses with very high intensity (up to 1.6×10^{16} Wcm^{-2}) resulting in tunnel ionisation. Some of these molecules have high ionisation potentials, requiring up to ten photons for non-resonant ionisation. The relative sensitivity factors (RSF)[6] have been determined as a function of the laser intensity and it has been demonstrated that for molecules with very different masses and ionisation potentials, uniform ionisation has been achieved at the highest laser intensities. Quantitative laser mass spectrometry of molecules is therefore a distinct possibility using laser intensities in the range 10^{13-16} Wcm^{-2}.

If the laser intensity is increased to above 10^{16}Wcm^{-2} then molecules become multiply charged and coulomb explosion takes place with the fragments preferentially flying off along the direction of the E field of the laser. This has been carried for a number of low and medium mass molecules [7].

Experiments in the laser regime 10^{17-21}Wcm^{-2} (Relativistic Picture)

The interaction of ultra-intense focused laser beams with solid targets is a new field of research resulting in the production of exotic plasma conditions similar to the conditions which exist in the interior of some stellar objects. The lasers generate very high energy electrons and ions which can subsequently produce γ-rays, positrons, neutrons and pions. The results obtained from these studies have major implications to fundamental plasma physics and high energy accelerator physics as well as important technological potential for the production of compact sources of neutrons, positrons and isotopes.

The experiment used the ultra-intense beam-line of the VULCAN Nd: glass laser incident on a tantalum target, 1.75mm thick, at 45° with p-polarised light within an evacuated target chamber [8]. Briefly the Vulcan laser delivered pulses on target up to 50 J with pulse lengths about 1 ps. A 95% reflecting turning mirror located after the recompression gratings was used to steer the beam on target focused by a f/1.7, 22cm focal length on-axis parabolic mirror. This turning mirror permitted 5% transmission of the laser energy for measurements of the laser spectrum, the pulse duration by a single shot autocorrelator and the focal spot quality by an equivalent plane monitor. Equivalent plane images showed that the diameter of the focal spot was 9μm, containing 35% of the laser energy when the target holder was placed in the beam. The focused intensity on target for each shot was monitored by a penumbral imaging camera.

A number of (γ,n) reactions from different targets placed behind the Ta target were carried out. The characteristic positron activity from neutron deficient nuclei was measured in the following way. Positrons slow down in materials and annihilate at rest with electrons producing two back to back photons of 511 keV energy. These were detected with high efficiency and signal to noise ratio using either two 3"x3" or 2"x2" NaI scintillators operated in coincidence. The absolute efficiency of these systems were determined using a calibrated ^{22}Na source. The activity of the sources were corrected to time zero and knowing the (γ,n) cross sections, the number of high energy photons produced by each laser shot could be determined from which could be estimated the number of initial high energy electrons. Typically $10^{9\text{-}10}$ γ-rays were generated from the highest activity sources ~5kBq. Fig 4 shows the activity measured as a function of time for a number of isotopes. Because of the differing Q values for the isotope production, the activities can be used to establish the shape of the bremsstrahlung spectrum and hence the temperature of the plasma. Thus nuclear activation can be used as an additional and powerful plasma diagnostic technique. At present for laser intensities of about 10^{19}Wcm^{-2}, isotopes with half-lives with up to about a day can be readily detected for medium A isotope. For large A isotopes, longer lived activities can be measured since the (γ,n) cross sections increase with A..

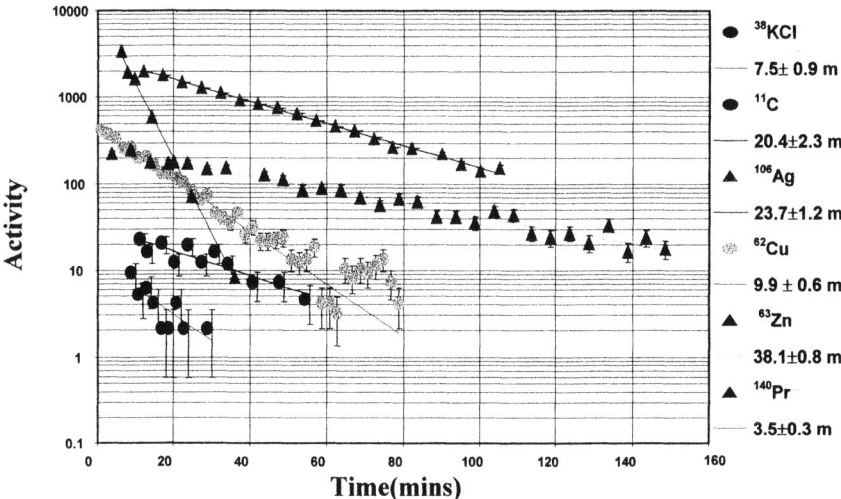

Figure 4. The activities of the isotopes listed below as a function of time. The measured half-lives agree well with the accepted values.

In summary with respect to nuclear physics, it has been shown that following the interaction of an intense laser beam of about 10^{19} Wcm^{-2} with a high Z target a sufficiently intense high energy γ-ray beam is generated such that photo-nuclear reactions can be carried out. ^{11}C, ^{38}K, ^{62,64}Cu, ^{63}Zn, ^{106}Ag, ^{140}Pr, and ^{180}Ta have been produced in measurable quantities up to activities of about 5kBq at the highest laser intensities. In addition photo-fission of ^{238}U has been demonstrated from the most abundant fission fragments. Some ten years ago Boyer, Luc and Rhodes [9] estimated theoretically that for 248 nm radiation at an intensity of 10^{21}Wcm^{-2}, some 10^{6} fissions could be generated per pulse in uranium 1cm thick. This is in reasonable agreement with fission yields from U in the present experiment at 10^{19}Wcm^{-2} intensity of 1.053μ radiation.

What could the applications for such laser induced nuclear technology be? This field of work has many potential scientific and technological applications that will become increasingly accessible as the intensity of the laser light available at RAL increases. Some of these possibilities are:

- Laser production of short lived radioactive isotopes from a table-top laser especially as the technology is progressing to miniaturise these intense lasers. This can be especially useful for producing short lived positron emitters for Positron Emission Tomography (PET), an important diagnostic procedure being held back because of the need for nuclear accelerators.
- Laser Production of Protons for Proton Radiography – an exciting way forward for this important therapeutic technique.
- Laser induced fission (our original submission) and our experiments to date have already demonstrated that this is no longer a theoretical dream.
- If we generate fission then we have an intense point source of neutrons and fission fragments with many possible applications.
- Laser induced particle physics. We have already shown that a laser can produce high-energy electrons. When VULCAN is upgraded in the next year or two so that the laser intensities can be scaled by a factor 0f 100, we shall be able to generate photons of energies such that particle production will be possible (the pions of the title of this paper). This is the realm of the particle physicists.
- Laser induced fast electron ignitor fusion that is of immense importance.
- A radiography technique for the study of dense materials.

ACKNOWLEDGEMENTS

I would like to acknowledge all the hard work and dedication of the staff and students in my group at Glasgow over the years since its inception in 1983. More recently in carrying out the nuclear work at the Rutherford Appleton Laboratory, I have collaborated with dedicated teams from the plasma physics groups at Imperial College London, RAL and AWE, without whom none of this would have been possible. They carried out all the important work and I am only the messenger.

REFERENCES

1. Hurst, G.S., Payne, M.G., Kramer, S.D. and Young, J.P. *Rev.Mod.Phys.* **51,** 767-819, 1979.
2. Protopapas, M., Keitel, C.H., and Knight, P.L. *Rep.Prog.Phys.* **60,** 389-486, 1997
3. Strickland, D. and Morou, G. *Opt.Commun.* **56,** 219-221,1985
4. Perry, M.A. and Morou, G. *Science* **264,** 917-924, 1994. 54.
5. Tajima, T.and Dawson, J.M. *Phys.Rev.Lett.* **43,** 267-270, 197
6. Fang, X., Ledingham,K.W.D., Graham, P., Smith,D.J., McCanny, T., Singhal, R.P., Langley, A.J., Taday, P.F. *Rapid Comm.Mass.Spectrom.* **13,** 1390-1397, 1999
7. Graham, P., Ledingham, K.W.D., Singhal, R.P., McCanny,T., Hankin,S.M., Fang,X., Tzallas,P., Kosmidis, C. Taday, P.F. 7 and Langley, A.J. *J.Phys.B.At.Mol.Opt.Phys.,* **33,** 3779- 3794, 2000
8. Ledingham, K.W.D., Spencer, I., McCanny, T., Singhal, R.P., Santala, M.J.K., Clark, E., Watts,T., Beg,F.N., Zepf, M., Krushelnick, K., Tatarakis, M., Dangor, B., Norreys, P.A., Allott, R., Neely, D., Clarke, R.J., Machacek, A.C., Wark, J.S., Cresswell, A.J., Sanderson, D.C.W., and Magill *Phys.Rev.Lett.* **84,** 899-902, 2000
9. Boyer, K., Luk, T.S., and Rhodes, C.K., *Phys.Rev.Lett.* **60,** 577-580, 1988

High Intensity Laser Generation of Proton Beams for the Production of β⁺ Sources used in Positron Emission Tomography

I Spencer[1], K W D Ledingham[1,3], R P Singhal[1], T McCanny[1],
E L Clark[2,3], K Krushelnick[2], M Zepf[2], F N Beg[2], M Tatarakis[2],
C Escoda[2], M Norrefeldt[2], A E Dangor[2], P A Norreys[4], R J Clarke[4],
R M Allott[4]

1 Department of Physics and Astronomy, University of Glasgow, Glasgow G12 8QQ, UK
2 Blackett Laboratory, Imperial College of Science, Technology and Medicine, London SW7 2BZ, UK
3 Radiation Physics Department, AWE plc, Aldermaston, Reading, RG7 4PR, UK
4 Central Laser Facility, Rutherford Appleton Laboratory, Chilton, Oxon, OX11 0QX, UK

Abstract. Protons of energies up to 37 MeV have been generated using ultra-intense laser-solid interactions. These protons can be used to induce nuclear reactions in materials to produce β⁺ emitting nuclei of relevance to the nuclear medicine community for Positron Emission Tomography, namely ¹¹C and ¹³N via (p,n) and (p,α) reactions. Activities of the order of 100 kBq have been measured from a single laser pulse. The possibility of using ultra-intense lasers as a substitute to cyclotrons for isotope production is discussed.

INTRODUCTION

In the last three years radioactive nuclei (decaying by electron capture/positron emission) have been generated by high intensity laser induced energetic gamma rays (1-5). These isotopes have been created by (γ,mn) reactions. The development of this technique for "table-top" production of radio-nuclides for medical applications is the natural extension of this work. However the nuclear medical community are more interested in short-lived β⁺ emitters for Positron Emission Tomography (PET) produced via proton induced reactions, since the radioactive isotopes can be separated from their stable carriers using fast chemistry. The technique of using lasers to generate proton beams for the production of PET sources has previously been demonstrated theoretically (6).

When a high-intensity laser pulse interacts with matter, the electric field component of the pulse causes electrons to be accelerated in the direction of the E-field. Space charge separation effects then cause ions to be accelerated. Recent experiments have used both gas targets (7) and solid targets (8). These experiments have reported proton energies of up to 18 MeV at laser intensities of 5×10^{19} Wcm⁻². This paper describes how these laser-produced energetic protons can be used to carry out nuclear reactions on various nuclei, in order to produce radioisotopes of relevance to the nuclear medicine community.

CP584, *Resonance Ionization Spectroscopy 2000: 10ᵗʰ Int'l. Symp.*, edited by J. E. Parks and J. P. Young
© 2001 American Institute of Physics 0-7354-0024-5/01/$18.00

EXPERIMENTAL

The experiment used the Chirped Pulse Amplification (CPA) (9-10) beam line of the VULCAN Nd:Glass laser (11) incident on a thin aluminium or CH foil target, at 45° incidence with p-polarised light within an evacuated target chamber. This system delivered pulses of energies up to 120 J, duration 0.9 – 1.2 ps with a wavelength of 1.053 μm. The beam was p-polarised, and when focused to a spot size of radius 6μm on the target using a f/4 off-axis parabolic mirror at 45°, yielded intensities of up to 10^{20} Wcm^{-2}. This intensity was determined by measuring the pulse energy using a joulemeter, the pulse duration from a single shot autocorrelator, and the focal spot quality by an equivalent plane monitor. The contrast ratio between the peak and ASE energies was measured by a third order autocorrelator and found to be 1: 10^{-6}. This is sufficient to generate a pre-plasma on the target surface, a few picoseconds before the arrival of the main pulse. The main pulse then interacts with this pre-plasma, causing electrons, and subsequently protons, from hydrocarbon impurities on the target surface, to be accelerated.

To characterise the proton beam, and to produce radioisotopes, different activation samples were placed in the target chamber, behind the target, in the "straight through" direction at a distance 10 mm from the target, and in front of the target, directed along the target normal, at a distance 50 mm from the target, in the "blow-off" direction. The time taken to extract the activation samples after irradiation was a few minutes, as the chamber must be returned to atmospheric pressure before the samples could be removed. These samples were then transferred to a nuclear laboratory for analysis. Hence, the lifetimes of the activated samples that could be measured were limited to a few minutes or more. These practicalities are a temporary problem, and will be resolved in future experiments. A schematic of the experimental arrangement is shown in Figure 1.

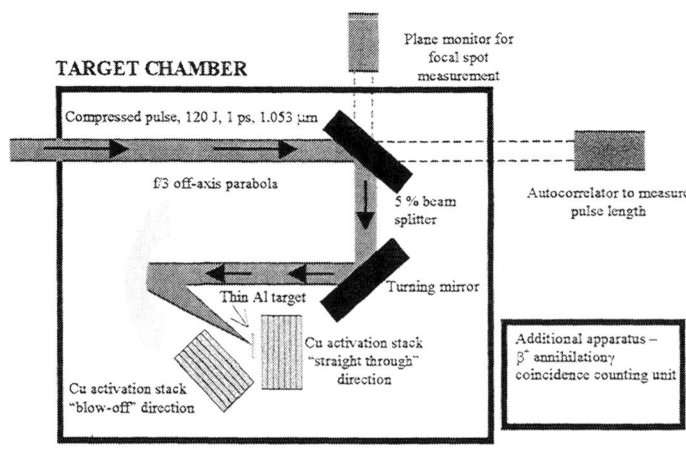

Figure 1 Schematic of experimental arrangement

RESULTS AND DISCUSSION

The energy spectrum, and spatial distribution of the accelerated protons were determined by placing activation stacks in both the "blow-off" and "straight through" directions. These consisted of 9 pieces of copper foil, 100 μm thick, plus a 1mm piece.

Copper undergoes the nuclear reaction $^{63}Cu(p,n)^{63}Zn$ when bombarded with protons of energies greater than 4.15 MeV. The ^{63}Zn isotope produced is radioactive, and decays via β^+ emission with a half life of 38.1 mins. β^+ slow down in solids, and annihilate at rest with free electrons, producing two back-to-back photons of energy 511 keV. These were detected using two 3"x3" NaI scintillation detectors, operated in coincidence. The efficiency of the coincidence system was measured using a calibrated ^{22}Na source, and hence the absolute activity, i.e. the number of (p,n) reactions in the copper pieces in the stack could be determined. Knowing the cross section (12), the number of protons incident on each piece of copper in the stack can be found, and hence a proton energy spectrum can be built up. Each piece of copper in the stack represents an energy "bin", because protons are slowed down in copper. Figure 2 shows typical energy spectra of the emitted protons from the front and rear of the target. It is clear that more energetic protons are observed in the "straight through" direction, having energies up to a diagnostic limit of 37 MeV. Assuming the proton energy distribution is proportional to exp(-E/T) where E is the proton energy and T is the proton temperature, the temperatures for the front and back proton beams are 1.4 and 2.9 MeV respectively. The first copper foils in each stack are shown as insets in Figure 2. The white circles observed are the result of aluminium debris from the target, and correspond to the dimensions of the proton beam, and hence provide spatial information about the proton beam.

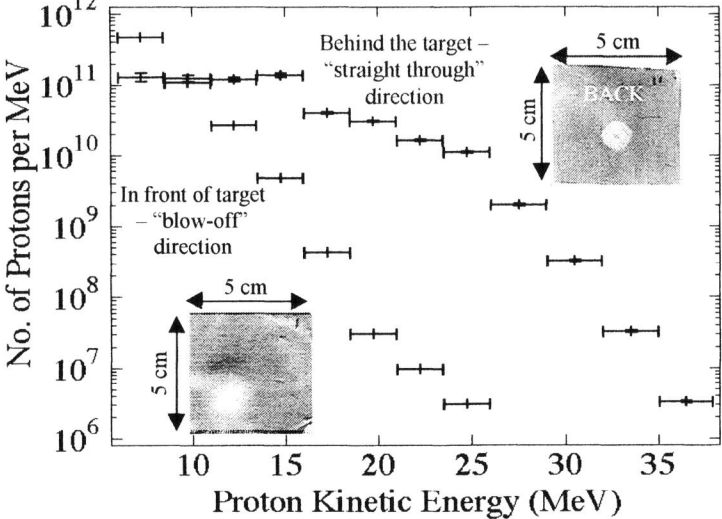

Figure 2 Proton energy spectra obtained from copper activation stacks in front of and behind the target. Insets show the first copper pieces in each stack, containing spatial information

The hazy image in the blow-off direction indicates that the protons are less well focussed. The sharp edges in the straight through direction are caused by the formation of multi-megagauss magnetic fields (8) during the laser-solid interaction. The proton beam in the straight through direction was emitted in a cone of 26.5° (1/2 angle). The intensity on target for this shot was 9.7×10^{19} Wcm^{-2}.

Different samples were placed in the target chamber and were bombarded with energetic protons from the intense laser-solid interaction. Samples were chosen to produce isotopes of relevance to the nuclear medicine community.

First, a boron sample was placed in front of the target, in the "blow-off" direction, prior to the laser irradiation. The activity of the boron sample was then measured in the coincidence unit following the laser interaction. From the decay curve obtained (Figure 3) it is clear from the measured half-life of 20.31 $\pm$0.4 mins that the isotope produced was ^{11}C (accepted value 20.34 mins) which is used in medical imaging (Positron Emission Tomography, or PET). The boron nuclei undergo (p,n) reactions to form ^{11}C, and the cross section for this reaction is shown in Figure 3 (inset). By extrapolating the curve, it can be seen that the activity at the time of laser irradiation was 200 kBq.

A sample of silicon nitride (Si_3N_4) was placed in front of the target. The activity as a function of time was again measured following laser irradiation, and the decay curve is shown in Figure 3. Again, the isotope ^{11}C was produced, but this time via the reaction ^{14}N(p,α)^{11}C. The activity at time zero (laser irradiation) was two orders of magnitude lower than the boron activation, at 2 kBq, but this may not be surprising, since the cross-section for this reaction is much lower (Figure 3 inset).

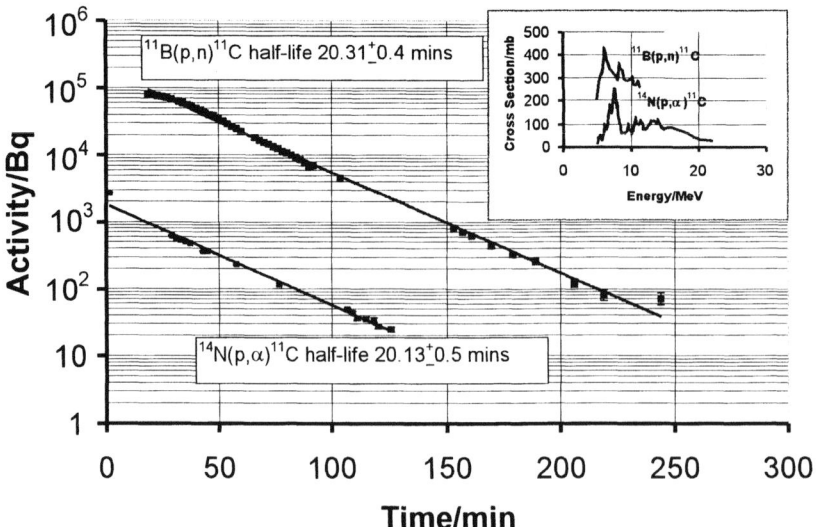

Figure 3 Decay curves for the activated boron and silicon nitride samples, showing that the isotope ^{11}C was produced, via ^{11}B(p,n)^{11}C and ^{14}N(p,α)^{11}C. (Inset) the experimentally measured cross-sections (12) for these reactions.

Finally, a sample of silicon oxide (SiO_2) was placed in front of the target to undergo nuclear activation, and its activity measured in the usual way. From the decay curve (Figure 4) it is clear that two isotopes are present, and from the half-life measurements, they are ^{13}N and ^{11}C, with accepted half lives of 9.97 mins and 20.34 mins respectively. The ^{13}N component is produced via the reaction $^{16}O(p,\alpha)^{13}N$ and the cross section for this reaction is shown in Figure 4 (inset). It is believed that the ^{11}C component is produced from activated carbon nuclei, which are present due to hydrocarbon impurities on the sample surface. From the cross sections of the nuclear reactions that are likely to generate this isotope, it can be seen that the reaction $^{12}C(p,p+n)^{11}C$ is most likely (see Figure 4 (inset)). High energy γ-rays and neutrons are also produced in the laser-solid interaction. The presence of two isotopes presents no problem for applied purposes, because the nitrogen and carbon components can be separated via fast chemistry.

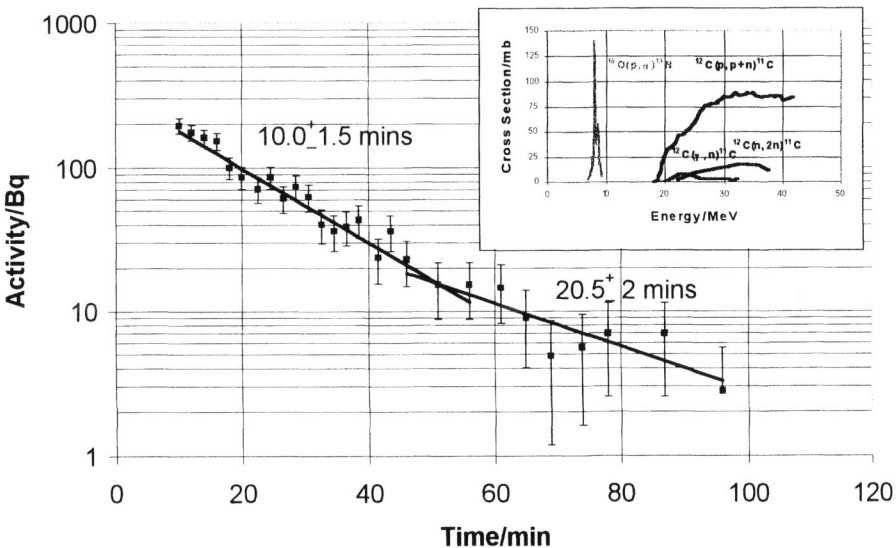

Figure 4 Decay curve for the activated silicon oxide sample. Two isotopes are present, ^{13}N, from $^{16}O(p,\alpha)^{13}N$ reactions, and ^{11}C, most likely from $^{12}C(p,p+n)^{11}C$ reactions. The carbon nuclei are present due to hydrocarbon impurities on the target surface. (Inset) the cross sections for these reactions.

CONCLUSIONS

It has been shown that protons of energies up to 37 MeV can be generated when an ultra-intense laser interacts with solid targets. These energetic protons can then be used to induce (p,n) and (p,α) reactions in various materials, to produce radioisotopes that are used in nuclear medicine. Activities of 2×10^5 Bq of the isotope ^{11}C have been measured.

Although ^{11}C is a useful isotope for PET, the favoured isotope at the moment is ^{18}F. The reaction usually employed to produce this isotope is ^{18}O(p,n)^{18}F. This reaction has approximately the same integrated cross section as the reaction ^{11}B(p,n)^{11}C which was utilised in our experiments. Hence it is feasible that the VULCAN laser could produce 2×10^5 Bq of ^{18}F per shot, operating at 10^{20} Wcm^{-2}.

A typical patient ^{18}F dose for PET is 5 mCi (1.85×10^8 Bq) although up to 20 mCi (7.4×10^8 Bq) sources are necessary so that fast chemistry can be performed to separate the tracer from the inactive carrier. Assuming that a VULCAN type laser could deliver 10 Hz, then the integrated activity after 500 seconds is about 1×10^9 Bq, or 27 mCi. If this source were generated by a 1 J pulse typical of table-top lasers then a kHz repetition rate would be required. Further improvement in the production rate will be achieved by optimising the laser pre-pulse, the target material and surface treatment. Also, it has been shown that the proton spectrum from the back of the target is more energetic, which could also lead to higher activities. Yamagiwa and Koga (6) have estimated that ^{18}F yields as high as 10^{14} Bq (2.7×10^3 Ci), two orders of magnitude higher than standard cyclotrons, can be generated if the laser intensity were increased to 10^{21} Wcm^{-2}.

Finally it has been shown that beams of cyclotron produced energetic protons can be used to treat tumours, with greater efficiency than conventional X-radiotherapy, because protons have more suitable dE/dx curves than gamma-rays. It has been shown that the proton beams produced in our experiments are heavily collimated, as seen in the angular distribution data presented, and hence the potential for laser induced proton therapy is currently being investigated in collaboration with proton oncologists.

In summary, it has been shown that high-intensity lasers can be used to produce short-lived positron-emitting nuclei, and the commercial applications of this new science are well within reach in the next few years.

ACKNOWLEDGEMENTS

The authors would like to acknowledge the support of the engineering and target area staff of the Central Laser Facility of the Rutherford Appleton Laboratory. This work was supported by the Engineering and Physical Sciences Research Council.

REFERENCES

1. Norreys, P A *et al, Physics of Plasmas,* **6,** 2150 (1999)
2. Phillips, T W *et al, Review of Scientific Instrumentation,* **70,** 1213 (1999)
3. Ledingham, K W D, and Norreys, P A, *Contemporary Physics,* **40,** 367 (1999)
4. Ledingham, K W D *et al, Physical Review Letters,* **40,** 899 (2000)
5. Cowan, T E *et al, Physical Review Letters,* **84,** 903 (2000)
6. Yamagiwa, M, and Koga, J, *Journal of Physics D,* **32,** 2526 (1999)
7. Krushelnick, K *et al, Physical Review Letters,* **83,** 737 (1999)
8. Clark, E L *et al, Physical Review Letters,* **84,** 670 (2000)
9. Strickland, D, and Morou, G, *Optics Communications,* **56,** 219 (1985)
10. Perry, M A, and Morou, G, *Science,* **264,** 917 (1994)
11. Danson, C N *et al, Jornal of Modern Optics,* **45,** 1653 (1998)
12. EXFOR www database – http://www-nds.iaea.or.at/exfor

The Role of Electronic Excitation in Chemistry of Laser Ablated Actinide Ions

John K. Gibson and Richard G. Haire

Chemical and Analytical Sciences Division, Oak Ridge National Laboratory,
P.O. Box 2008, Oak Ridge, TN 37831-6375, USA

Abstract. Insights into the electronic structures and energetics of actinide atoms and ions are key to understanding and predicting the properties of this series of elements, which are radioactive. Many members of this series are man-made. We have studied several chemical reactions of free actinide ions (An^+) to better understand the electronic nature of this unique group of elements and enhance the general understanding of the periodic table. This understanding enables better predictions of the character of heavy actinide and transactinide elements, where relativistic effects are expected to play an increasingly significant role. The basis for these studies along with a summary and interpretation of the key results is described, together with the first results for the shortest-lived nuclide studied to date by this technique, Bk-249.

BACKGROUND

The actinide (An) elements present electronic structures which uniquely comprise quasi-valence 5f electrons. In contrast to the essentially localized 4f electrons found in the homologous lanthanide (Ln) elements, the actinide 5f-electrons dramatically change in character when proceeding across the series. For early actinide metals, such as Np, bonding of the itinerant 5f electrons is clearly manifested by low symmetry structures and anomalously low melting points. Proceeding across the series beyond Pu, the 5f electrons become energetically and spatially more constrained, and their direct participation in chemical bonding in condensed phase systems decreases. However, the 5f electrons remain important in determining chemical and physical properties throughout the series. The most important "indirect" 5f effect is the energy required to excite a 5f electron to a valence orbital, where it can become engaged in bonding; this promotion is typically from a 5f to a 6d orbital [1].

Most chemical studies of the actinides have been with the condensed phase, where electronic structures are perturbed and obfuscated by secondary interactions. Gas-phase metal ion - molecule reactions represent uncomplicated systems for probing effects of ion electronic structures and energetics on their chemistry. Due to the absence of alternate interactions and reaction pathways, types of bonding which may not appear in the condensed phase often emerge in gas phase complexes. Gas-phase ion chemistry is particularly well-suited to scarce and radioactive transuranic actinides, because extensive chemistry can be performed with sub-milligram amounts of material.

CP584, *Resonance Ionization Spectroscopy 2000: 10th Int'l. Symp.*, edited by J. E. Parks and J. P. Young

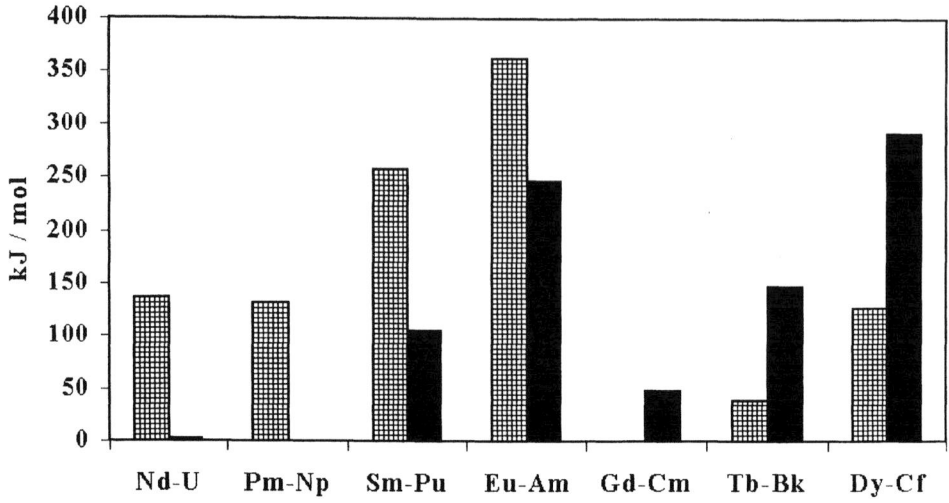

FIGURE 1. Promotion energies, $\Delta_p E$, to excite monopositive ions, M^+, from the ground state to the divalent state: $[Xe]4f^{n-2}5d^16s^1$ for the Ln^+ (hatched bars) or $[Rn]5f^{n-2}6d^17s^1$ for the An^+ (solid bars).

We use the Laser Ablation with Prompt Reaction and Detection (LAPRD) technique to study gas-phase actinide ion - molecule reactions. One focus for elucidating chemical behavior has been on An^+-induced dehydrogenation of alkenes, presumed to proceed by C-H bond activation via metal ion insertion and subsequent H_2-loss. This is shown in Equations (1) and (2) for 2-butene.

$$CH_3-CH=CH-CH_3 + An^+ \rightarrow H-An^+-CH_2-CH=CH-CH_3 \tag{1}$$
$$H-An^+-CH_2-CH=CH-CH_3 \rightarrow An^+-\{CH_2=CH-CH=CH_2\} + H_2 \tag{2}$$

It is the product complex product ion in Eqn. (2) which is identified experimentally but the initial rate-controlling insertion step in Eqn. (1) determines the net reaction cross section and product yield. For Eqn. (1) to proceed, two chemically active valence electrons must be present at the metal center to form the H-An^+-C activated complex. In the case of the lanthanides, the 4f electrons are ineffective at C-H activation and a $Ln^+\{[Xe]4f^{n-2}5d^16s^1\}$ configuration is required for dehydrogenation [2]. Accordingly, relative dehydrogenation efficiencies of Ln^+ exhibit an inverse correlation with the promotion energies for exciting the ion from its ground state to the $5d^16s^1$ "divalent" configuration, $\Delta_p E[Ln^+]$. These energies vary from zero for reactive Gd^+ to 360 kJ mol^{-1} for inert Eu^+ [3-5].

Three central questions will be addressed here regarding actinide electronic structures and energetics. (1) Is the insertion mechanism model applicable to C-H activation by Ln^+ also applicable to An^+? (2) Are the quasi-valence 5f electrons chemically inert in dehydrogenation reactions? (3) Do the spectroscopically determined 5f-to-6d promotion energies, $\Delta_p E[An^+]$, effectively predict reaction efficiencies of the actinide ions and are estimated $\Delta_p E[An^+]$ accurate?

In Fig. 1 are shown $\Delta_p E[M^+]$ for selected Ln^+ and homologous An^+. The values for the Ln^+ are from Martin et al. [3], except for the Brewer estimate for Pm^+ [4]; those for the An^+ are from Blaise and Wyart [5]. As seen in Fig. 1, the $\Delta_p E[M^+]$ for Np^+ and Gd^+ are zero (i.e., $d^1 s^1$ ground states), and only 3 kJ mol^{-1} for U^+. Fig. 1 illustrates important features that motivate investigating the gas-phase chemistry of An^+. It is clear that the trend in $\Delta_p E[An^+]$ is very different from that for $\Delta_p E[Ln^+]$. The $\Delta_p E[An^+]$ vary from (essentially) zero for Th^+, Pa^+, U^+ and Np^+ to a predicted value of 520 kJ mol^{-1} for No^+ [4]. As the first four An^+ have two non-5f valence electrons in low-lying states, it is necessary to examine the chemistry of heavier An^+ to directly assess if 5f electrons participate in C-H activation. If 5f-to-6d promotion is a requisite for C-H activation, large variations in reactivity would should occur between the transuranium An^+; Np^+ should be highly reactive whereas Cf^+ should have a low reactivity like Sm^+. We have previously reported on the gas-phase chemistry of Th^+ and U^+ [6], Np^+ and Pu^+ [7], Am^+ [8], Cm^+ [9] and Cf^+ [10]. We summarize here these results and present initial results for Bk^+ in the context of the above considerations. The $Bk^+\{5f^8 7s^2\}$ state, 84 kJ mol^{-1} above ground, is significantly more stable than the $Bk^+\{5f^8 6d^1 7s^1\}$ configuration, which is 148 kJ mol^{-1} above the ground state. Accordingly, studies with Bk^+ provide an opportunity to assess whether the $7s^2$ closed valence shell can activate C-H bonds.

The LAPRD experimental technique has been described in detail elsewhere [6-11] and only a very brief synopsis is included here. Metal ions are laser ablated from a solid target typically containing a few atomic percent of one or more actinides and/or lanthanides (as oxides) dispersed in copper. The ablated ions traverse a 3 cm reaction zone containing an indeterminate pressure (< 1 mbar) of reactant gas. After a delay of ~35 μs, positive ions are injected into a reflectron time-of-flight mass spectrometer to determine amounts and compositions of product ions. The efficacy of this approach in assessing ground state metal ion chemistry was established by demonstrating good agreement between results from Fourier transform ion cyclotron mass spectrometry [2] and LAPRD [11] experiments with Ln^+. The initial Bk^+ results reported here were obtained using ~1 mg of Bk-249, which beta-decays to Cf-249 with a half life of ~320 days.

RESULTS AND DISCUSSION

A time-of-flight mass spectrum following ablation of a Bk-249 target into 1-butene is shown in Fig. 2. The target contained Bk-249, Pu-242 and Tb-159, which allowed for direct comparison of reaction cross sections and abundance distributions for these different ions under identical experimental conditions. Only monopositive ions were detected ($z = 1$), so that the m/z scale on the top of each portion of the spectrum corresponds to ion mass. The M^+ and MO^+ were directly ablated and appeared in approximately the same abundances in the absence of the 1-butene. The vertical mV scales allow for direct comparison of peak intensities; the bottom (1 mV) spectrum in the top two panels of Fig. 2 is a 5X vertical magnification of the top (5 mV) spectrum. The

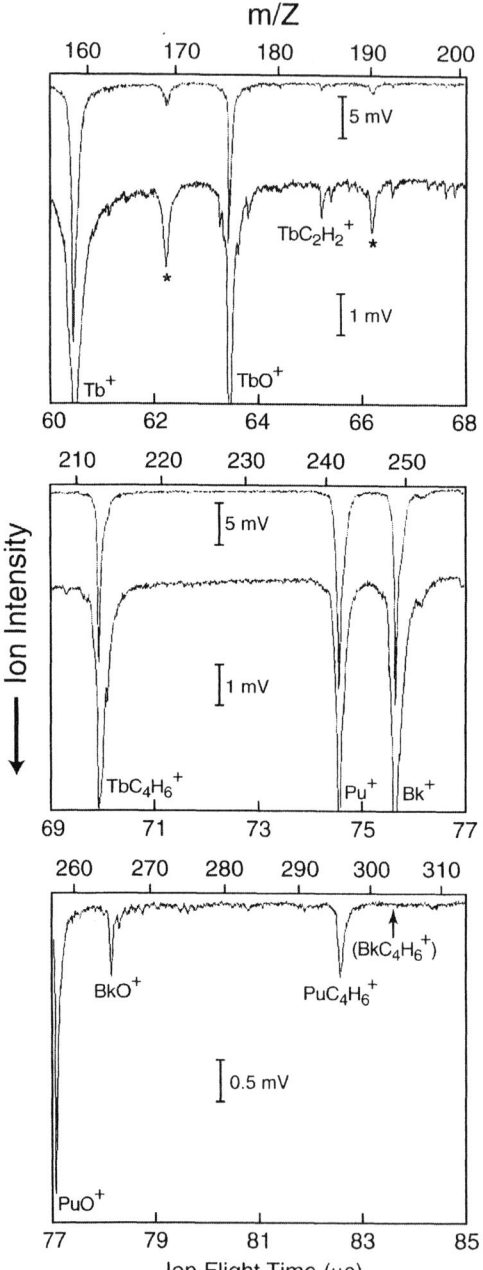

ORNL 99-04353/rra

asterisked peaks at m/z 169 and 191 were not assigned but their presence does not affect the main interpretations. The arrow in the top right portion of the bottom panel indicates where the hypothetical $BkC_4H_6^+$ peak would have appeared. The spectrum reveals that Tb^+ is significantly more effective than Pu^+ at dehydrogenation (C-H activation) to produce the $MC_4H_6^+$ product complexes, and that Bk^+ is essentially inert in this regard. The $\Delta_pE[M^+]$ values are 39, 104 and 148 kJ mol^{-1} for Tb^+, Pu^+ and Bk^+, respectively [3,5]. These results indicate that promotion to the "divalent" $5f^86d^17s^1$ configuration of Bk^+ is requisite for insertion into a C-H bond, and that $\Delta_pE[Bk^+]$ is significantly larger than $\Delta_pE[Pu^+]$, as predicted from the spectroscopically measured energy levels [5]. The results demonstrate that the 5f electrons of Bk^+ are localized and chemically inert towards organometallic sigma bond formation; and that the $7s^2$ electrons of Bk^+ must be spin-unpaired prior to insertion into a C-H bond. The data in Fig. 2 illustrate that substantial chemical differences exist between homologous lanthanide and actinide elements, which may not be apparent from condensed phase experiments. Thus, Bk and Tb behave similarly in the condensed phase but their gas-phase chemistries are entirely disparate.

These results for the reactivity of Bk^+ with 1-butene illustrate the LAPRD technique as applied to heavy actinides. The value of gas-phase metal ion reactions for illuminating actinide electronic structures and energetics, particularly as they affect chemistry, is clear. Extensive LAPRD studies of

FIGURE 2. Time-of-flight mass spectrum for ablation of berkelium, plutonium and terbium oxides into 1-butene (C_4H_8). The mass regions are shown sequentially in the three frames.

$$\text{Pu}^+ + \text{c-C}_6\text{H}_{10} \longrightarrow \text{PuC}_6\text{H}_6^+ + 2\text{H}_2$$

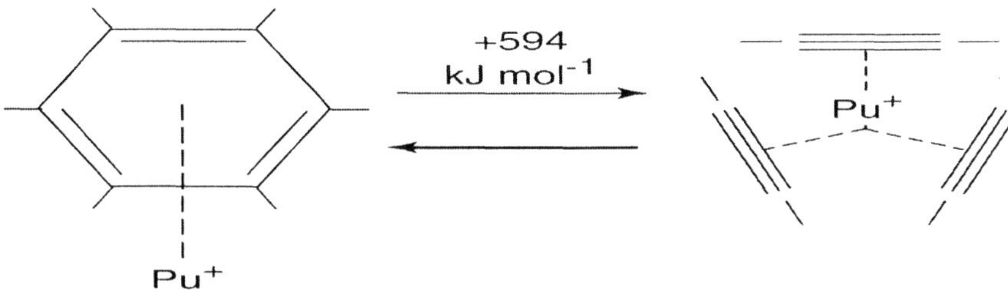

FIGURE 3. Potential structures of the PuC$_6$H$_6^+$ product of dehydrogenation of cyclohexene by Pu$^+$.

several An$^+$ have been carried out in our laboratory and the results have been reported [6-10,12]. The emphasis has been on reactions with simple and functionalized hydrocarbons to evaluate electronic effects. Dehydrogenation by C-H activation is generally a primary reaction channel but C-C activation can also result in cracking, as illustrated by the appearance of the TbC$_2$H$_2^+$ peak in Fig. 2. Consolidating the An$^+$-induced dehydrogenation results obtained to date provides the ordering of dehydrogenation activities in Equation (3).

$$\text{Th}^+ \approx \text{U}^+ \approx \text{Np}^+ > \text{Cm}^+ \gtrsim \text{Pu}^+ > \text{Bk}^+ > \text{Am}^+ \approx \text{Cf}^+ \tag{3}$$

When $\Delta_p E[\text{M}^+] \geq 200$ kJ mol^{-1}, it was found with Ln$^+$ that the minuscule product yields preclude differentiating relative reactivities by LAPRD [11]. Similarly, with the actinide elements, both Am$^+$ and Cf$^+$ were found to exhibit very low reactivities and their electronic energetics could not be differentiated.

The efficiencies of C-H activation by the actinide ions included in Eqn. (3) are in accord with the $\Delta_p E[\text{An}^+]$ in Fig. 1. The results of this series of experiments have established that C-H activation by An$^+$ proceeds by insertion to generate a transient C-An$^+$-H intermediate. Two non-5f valence electrons must be available at the metal center to form the sigma-type organometallic bonds, at least for the transneptunium An$^+$. Because the ground state or very low-lying configurations of Th$^+$, U$^+$ and Np$^+$ have two non-5f valence electrons, their high reactivities can not illuminate the role of the 5f electrons in that region of the series. However, the anomalously high dehydrogenation activity of UO$^+$ does suggest a direct role of the uranium 5f electrons [6].

PROSPECTS

It is anticipated that it will be feasible to carry out metal ion - molecule reactions of both Pa$^+$ and Es$^+$ by LAPRD, thereby providing a systematic study of elementary gas-phase ion chemistry for the first ten transactinium elements. Although the information derived from these LAPRD studies is of great value in establishing the role of electronic structures and energetics in actinide chemistry, several issues can not be addressed by

LAPRD. One is illustrated in Figure 3: only the compositions of product complexes such as $PuC_6H_6^+$ are determined, but not the structures or bonding. Ion trap mass spectrometry offers the potential for a more expansive inquiry into actinide ion chemistry. Product structures and bonding can be probed by collision induced dissociation; reaction kinetics and mechanistics can be examined by varying both reaction times and reactant pressures; and secondary reactions can be studied by sequential mass spectrometry (MS^n). We intend to employ Quadrupole Ion Trap and/or Ion Cyclotron Resonance Mass Spectrometry to permit enhanced capabilities for probing actinide chemistry. Gas-phase actinide ion chemistry has been demonstrated as a non-spectroscopic approach to obtain information on electronic structures and energetics. In this regard, Cornehl et al. [2] have derived new information on the electronic energetics of Ho^+ based on its reactivity. Ion trap studies should allow one to better define relative actinide excitation energies and provide a basis to determine unknown energy levels. Greater sensitivity afforded by an ion trap could even allow studies of Fm^+, for which no spectroscopic information is available. Studying the chemistry by these techniques of even heavier ions of accelerator-produced actinides and/or transactinides may enable experimental investigation of relativistic effects on the chemistry (electronic energetics) of these elements.

ACKNOWLEDGMENT

This research was sponsored by the Division of Chemical Sciences, Geosciences, and Biosciences, Office of Basic Energy Sciences, U.S. Deptartment of Energy, under contract DE-AC05OR22725 with Oak Ridge National Laboratory, managed and operated by UT-Battelle, LLC.

REFERENCES

1. Katz, J.J., Morss, L.R., and Seaborg, G.T. "Summary and Comparative Aspects of the Actinide Elements," in *The Chemistry of The Actinide Elements*, edited by J.J. Katz, G.T. Seaborg, and L.R. Morss, Chapman and Hall, London, 1986, pp. 1121-1195.
2. Cornehl, H.H., Heinemann, C., Schroder, D., Schwarz, H., *Organometallics*, **14**, 992-999 (1995).
3. Martin, W.C., Zalubas, R., Hagan, L., *Atomic Energy Levels - The Rare Earth Elements*, U.S. Department of Commerce, Washington, D.C., 1978.
4. Brewer, L., *J. Optical Soc. Am.*, **61**, 1666-1682 (1971)
5. Blaise, J., Wyart, J.-F., *Energy Levels and Atomic Spectra of Actinides*, Tables Internationales de Constantes, Paris, 1992.
6. Gibson, J.K., *Organometallics*, **16**, 4214-4222 (1997).
7. Gibson, J.K., *J. Am. Chem. Soc.*, **120**, 2633-2640 (1998).
8. Gibson, J.K., *Organometallics*, **17**, 2583-2589 (1998).
9. Gibson, J. K., Haire, R.G., *J. Phys. Chem. A*, **102**, 10746-10753 (1998).
10. Gibson, J.K., Haire, R.G., *Int. J. Mass Spec.* (in press).
11. Gibson, J.K., *J. Phys. Chem.*, **100**, 15688-15694 (1996).
12. Gibson, J.K., *Inorg. Chem.*, **38**, 165-173 (1999).

SESSION IV

Conservation of Information:
Quantum Mechanics and RIS

G. Samuel Hurst

2314 W. Gallaher Ferry Road, Knoxville, TN 37932, USA

Abstract. Basic concepts in probability and information theory have been used to examine some longstanding problem in quantum physics. In particular, a new concept called the *conservation of information* appears to offer insight into laser interferometer experiments done at the University of Rochester and at the University of California at Berkeley. These interferometer experiments demonstrate that information controls quantum interference and presumably the wave function. New "gedanken" experiments proposed here deal explicitly with the collapse of the quantum mechanical wave function by information alone. Use would be made of laser fluorescence and resonance ionization spectroscopy to explore more fully the control of the wave function with information; as well as the temporal relationship of information generated locally with information in a distant system. If one assumes the conservation of information at all time and in any reference frame, teleportation of atoms emerges as a natural consequence.

INTRODUCTION AND SUMMARY

Consideration of certain games involving probability leads to the idea that Shannon's statistical information [1] can be divided into two parts. One of these is "missing information" (MI), just as Boltzmann used the term to denote the information that is missing "out there" apart from the mind. Missing information is an abstract concept and is closely related to entropy. The second part, relative information (RI), is the information "in here" in the mind. Relative information is also based on probability, but the algorithm that transforms probability to information is relative to knowledge already stored in the mind. It is natural to suggest that the sum of MI and RI is a constant in these games, prompting the more general question of the *conservation of information* in the physical world where observations are made. Some recent laser interferometer experiments reinforce the information conservation concept and invite speculations on its generality.

New "gedanken" experiments, centered on the collapse of the quantum mechanical wave function are described to extend and further illustrate the conservation of information concept. One of these involves the use of cold atoms with long de Broglie wavelengths in situations not unlike those considered in the Einstein, Podolsky, and Rosen EPR experiments, but where single atoms are detected using RIS.

CP584, *Resonance Ionization Spectroscopy 2000: 10th Int'l. Symp.*, edited by J. E. Parks and J. P. Young

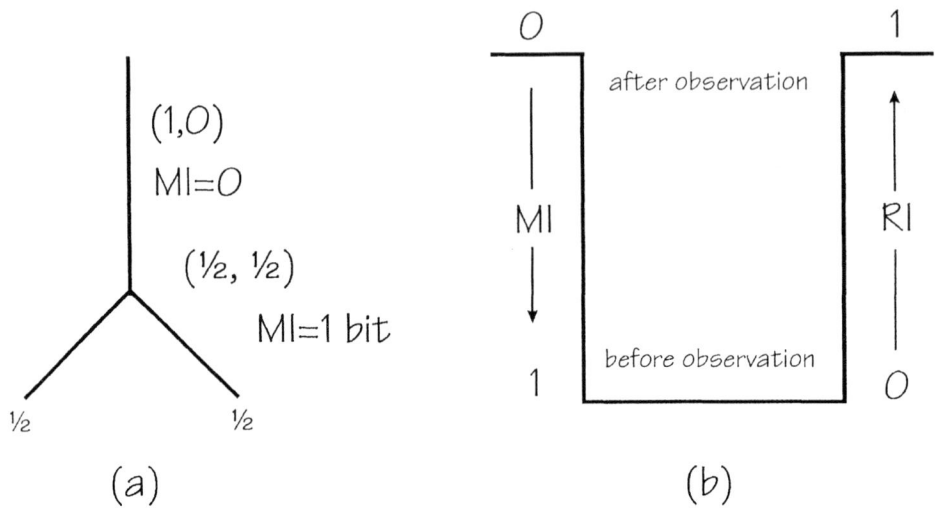

FIGURE 1. Shannon's Information illustrated for the binary case (part a) and the Information Well, using Boltzmann's "missing information" (part b).

SYMMETRIC AND ASYMMETRIC INFORMATION

"Information" is one of the most elusive words in the English language, but here the word has a much more precise and technical meaning. Shannon [1] based the technical definition of information on a probability distribution, which uniquely determines the quantity H in the expression:

$$H = -\sum_i P_i \ell n P_i \qquad (1)$$

where the ℓn is understood to be the base 2 logarithm. Thus, when the probability distribution is (1/2,1/2) Boltzmann's missing information, MI, is 1 bit, and when the probability distribution is collapsed at (1,0) or (0,1), then MI equals 0 (see Fig 1a). It will be convenient to think of missing information as a well depth, as shown in Fig 1b. Before an observation is made, in the binary example above, MI equals 1 bit. After the observation, MI equals 0 bits. Meanwhile, the observer's relative information, RI, has increased from 0 bits to 1 bit.

The following example illustrates the meaning of relative information: Monty Hall hosted the game show *Let's Make a Deal*, for many years on North American television. The game involved three doors; behind two were goats and behind another an expensive car. Monty Hall used an important rule: after the host had chosen a door he *must* show a goat in one of the two other doors, thus creating two distinct probability sets because of asymmetric information. For an observer not knowing the set genealogy, the probabilities

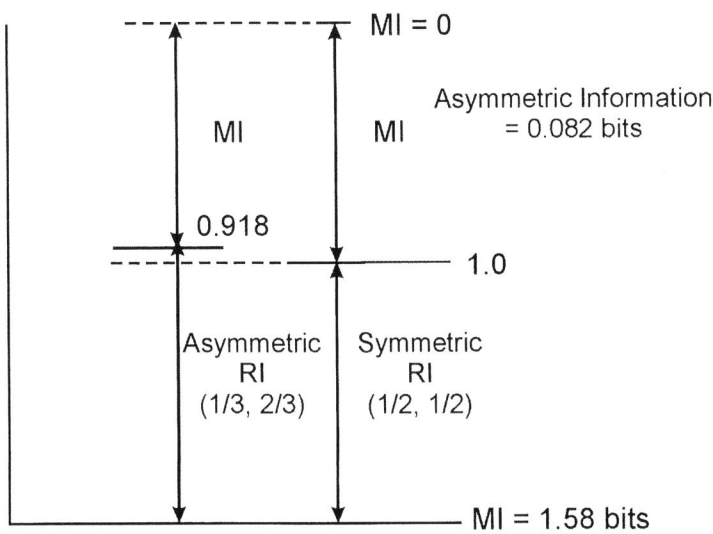

FIGURE 2 . Missing Information (MI) and Relative Information (RI) for the stages of the Monty Hall game MI and RI are related through conservation of information.

of the prize in either of the two closed doors are (1/2,1/2), the symmetric case. An observer using all the knowledge available knows that the probabilities of the car is 1/3 for the closed door in the first set and is 2/3 for the remaining closed door in the second set; this is the asymmetric case. Fig. 2 is a plot of the information quantities involved. At the initial stage with 3 closed doors, MI equals 1.58 bits for both observers. At the interesting stage where the observers must make a decision on which door is most likely to have the prize, MI for the symmetric case is exactly 1 bit, but for the observer with asymmetric information MI is 0.918 bits. Thus RI for the symmetric case is 0.580 bits and RI is 0.662 bits for the asymmetric case.

CONSERVATION OF INFORMATION

The two observers have differing relative information because we have tacitly assumed the conservation of information where the sum of MI and RI is a constant. This seems quite natural, as a consequence of resolving Shannon's information into missing information and relative information. But there is also a philosophical significance; conservation of information connects the observer, with his relative information "in here," to the abstract world "out there," This has interesting consequences in the classical and in the quantum world.

It is instructive to argue for the conservation of information by analogy with the conservation of energy as illustrated in Fig. 3. When a particle in a gravitational field falls freely, the sum of its potential energy, PE, and its kinetic energy, KE, is a constant, as pictured on the left of the figure. When rearranged in the well, shown at the center, there

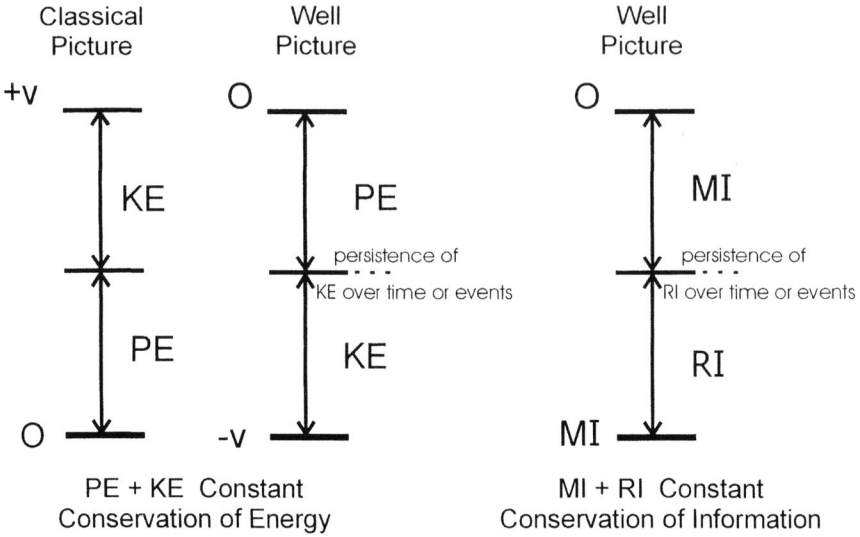

FIGURE 3. Illustration of the close parallel of "conservation of energy" (left) and the "conservation of information" (right).

is a close parallel with the conservation of information, which is shown on the right. Both MI and PE are abstractions that are assumed to exist independently from an observation. But RI and KE are fruits of observation; we observe a particle with KE, and we make an observation to define RI. There is another analogy: after a particle has experienced a change in its PE, its KE may remain constant for repeated observations until dissipative forces, such as friction, causes a change in KE. Many observers can have the same value of RI without changing the MI. This "persistence" of both KE and RI over time or events is illustrated as constants with the dashed lines in Fig. 3.

INFORMATION AND THE HEISENBERG UNCERTAINTY PRINCIPLE

Feynman has discussed wave interference by a single slit.[2]. Assume that a wave is traveling in the x-direction and then attempt to define a position in the orthogonal direction, y, by using a slit of width 2w. If the wavelength of the oncoming photon is substantially greater than 2w, there is an interference pattern on a screen placed after (or beyond) the slit. And if we try to increase path information by squeezing the slit down to width w, the interference pattern spreads even further in the y-direction. In making the change from 2w to w an observer has gained 1 bit per photon in position information. Feynman is careful to point out that the spreading on the screen in the y-direction is really a loss in momentum information; further, the gain in position information comes at the expense of momentum information in accord with the Heisenberg uncertainty principle. In fact, it is quite reasonable that the gain in position information is exactly compensated by the loss in momentum information. However, this exact compensation is not crucial to our argument for the conservation of information due to the following argument: When a photon is actually detected at some y-position on the screen, the wave function that represented the probability distribution prior to the measurement collapses to a detector point. The observer has gained RI equal to the total amount of MI. This gain in position information, and the simultaneous loss in momentum information, can be regarded as a virtual transition that does not alter the final measurement of RI. This argument holds, as do interference experiments, even if only 1 photon at a time is traveling through the slit.

SOME LASER INTERFEROMETRY EXPERIMENTS

Support for conservation of information comes from some laser interferometry experiments conducted during the past several years and making the popular press [3, 4]. Mandel, at the University of Rochester, directed a single laser beam into a beam splitter. Each of the split beams went through down-converters so that one "signal" beam from each converter could be directed to a detector to record an interference pattern that is produced when slight changes are made in one of the path lengths. The two idler beams from the down-converters were combined in a single detector, and in this case, the usual interference pattern was observable. However, if either of the idler beams were blocked, the interference pattern disappeared. This is astonishing, for these idler beams were not directly involved; only the signal beams went through the interferometer. *It appears that information alone is controlling the interference pattern and is collapsing the wave function!*

With an idler beam blocked it is possible to tell, by reverse ray-tracing, from which arm of the apparatus the signal beam came. Thus, 1 bit of information on the path taken by each photon is gained, and since RI increases by 1 bit, MI must decrease by 1 bit to satisfy information conservation, collapsing the wave function. There can be no interference pattern when MI equals 0. We now see by example how the observer and the observed are coupled through the conservation of information.

Professor Raymond Chiao [5] at the University of California at Berkeley works on superluminal processes and has approached laser interferometry in a different way from Mandel's. Chiao uses a polarizing filter in one arm of the interferometer and that makes the interference pattern disappear, but when a second filter is added, the interference pattern reappears. Thus, it seems that relative information can be erased. This invites another comparison with the conservation of energy: Imagine a particle in a smooth bowl starting with PE at the top, with KE at the bottom and then climbing back to the top of the bowl with the original PE and no KE. Similarly, in the Berkeley experiment there is MI only when the photons and their paths are identical, then a single filter reveals the RI, which goes back to MI when two filters are used. John Horgan refers to this as the erasure of information [3].

The conservation of information concept provides no clue about any underlying mechanisms that may be involved in the collapse of the wave function. This would be asking too much, for even the conservation of energy does not reveal its secrets. However, one supposes that some type of interaction is involved as information is acquired.

91

POSSIBLE EXPERIMENTS ON THE COLLAPSE OF WAVE FUNCTIONS

At the RIS-96 meeting at Penn State I speculated (at a workshop) that information control could be used to separate materials and even isotopes if lasers with isotopic selectivity were used to acquire RI on the path of atoms in an atom interferometer apparatus. This concept subsequently appeared in a publication of the Korean Physical Society [6]. From a more fundamental view, collapse of the wave function through the use of information is put to test. Imagine an atom that is cold enough that its de Broglie wavelength is much longer than the width of a laser beam through which it propagates. If an interaction occurs between a photon and the cold atom then it is possible to get information on the path of the atom and this will collapse the wave function. Note that it is not necessary to actually measure the scattered photon, as long as it is *possible* to make a measurement. By this, I mean that it must be possible to chase down the wandering photon before it has degraded to heat when RI fades into entropy. The interactions could be brought into vivid evidence by using enough photons so that the average number of interactions is much larger than 1 and thus collapses nearly all of the atoms with long de Broglie waves into the laser beam itself. Large increases in the population of atoms confined to the laser beam would be a clear indication of collapse of the wave function, and studies of the average number of atoms in the laser beam as a function of photon flux would give the cross section for the interaction.

We can now propose possible experiments on the collapse of the wave function using cold atoms, laser fluorescence, and resonance ionization spectroscopy. From this base, it is possible to speculate on the possibility of teleportation of atoms. Thus far, teleportation has been demonstrated only with photons [7]. Fig. 4 shows the crux of this idea. Cold atoms make their way toward a double slit with equal probability of travel along each of two classical paths. But they are not classical atoms because their de Broglie wavelengths are long compared to the spacing between the slits. One may speak only of the probability that they are on path 1 or path 2; they travel both paths at once. A pulsed laser is used to stimulate fluorescence from the atom if it is on path 1, thus detection, or even the possibility of a detection, collapses the wave function to path 1. Imagine two observers, Peter and Paul, who are *distant* observers. They may be located at an arbitrary distance from the atom source and the *local* observers located at the left of the slits. A pulsed RIS laser, shared by Peter and by Paul, produces resonance ionization of the cold atom that may or may not have come their way.

Paul's knowledge of the collapse of the atomic wave functions first came from the transmission of the photons stimulated from an atom of path 1 when they reached his photon detector. Paul's photon signal is crucial; it is, in fact, the way that RI is transmitted to Paul. This RI, and the MI measured by Paul, should be consistent with the equation:

$$RI + MI = \text{constant} = 1 \text{ bit.} \qquad (2)$$

Each time Paul is informed that his RI equals 1 bit, he should find the atom in his RIS detector, if conservation of information applies to Paul's reference frame. Thus, we expect that the number of atoms, N_1, detected by Paul would increase as the photon flux of SS1

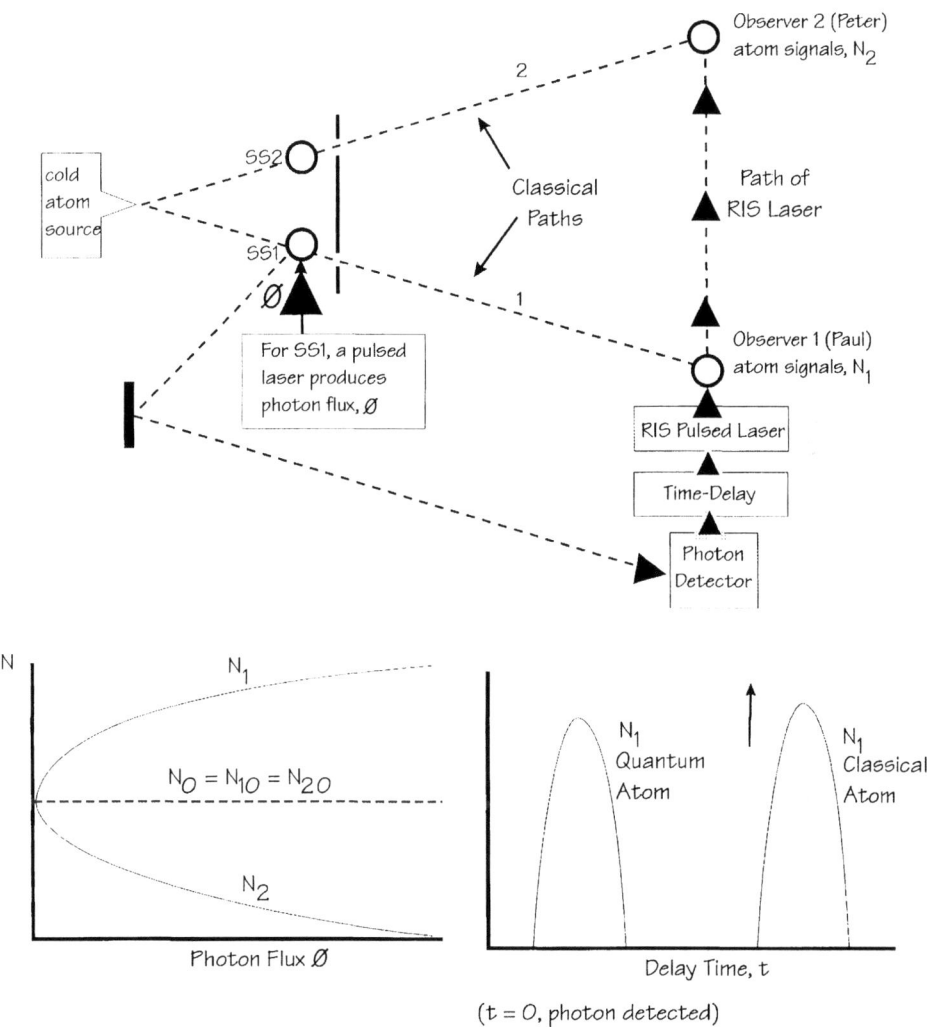

FIGURE 4. A "gedanken" experiment to illustrate collapse of the wave function with relative information.

increases, and that Peter's detections, N_2 will decrease. The use of RI to collapse the wave function has robbed Peter to pay Paul. The function $N(\Psi)$ could be analyzed to obtain the cross section for the photon interaction that collapses the wave function.

But a crucial question is yet to be asked: What is the *time delay*, if any, between the detection of the photon at SS1 and Paul's observation of an atom? If the distances are large, it is necessary to account for the time that is required for light to travel the path from SS1 to Paul's photon detector.

The time delay question will first be examined on the logic of the information conservation concept itself. Imagine that RI and MI form an information *couplet* in which the existence of one of them implies the existence of the other. With this there are four possibilities: (1) RI could arrive immediately with MI trailing, (2) MI could arrive immediately with RI trailing, (3) both MI and RI could be delayed by the same amount, or (4) finally both RI and MI arrive without delay. If information is conserved at any place and at all times, MI and RI must always be synchronistic, eliminating two of these possibilities. The proposed experiment could test: (a) whether the atoms take a long time to appear (depending on the velocity of a cold atom), or (b) if the atoms appear instantaneously. These two possible results are illustrated in Fig. 4. There are two choices: either the atom follows a classical path with classical velocity, or it appears instantly at the detector in phase with RI. An experiment would easily resolve these two cases. It could turn out that the atom arrives in sync with the signal at the photon detector. This case would amount to the teleportation of an atom over any arbitrary distance.

Thus we have concluded that teleportation of atoms may be possible, based on the conservation of information concept applied to the experimental situation where we can legitimately speak of *information entanglement*. Information entanglement means that two locations at any distance can be coupled with information. This idea reminds us of the EPR experiment (gedanken) where the wave function may be concentrated at distant locations. The philosophical questions raised by this proposal were clarified with real experiments based on the work of John Bell. It has been amply demonstrated that quantum mechanics is a complete theory, showing that the Copenhagen interpretation of reality (that reality is created by the measurement) is correct, and that instantaneous "action at a distance is possible." These conclusions are now commonly cited in both the technical and the popular literature [8-10], but perhaps no one has given more insight into quantum mechanics than Heinz Pagels [10].

The properties of entangled quantum states have been used to demonstrate teleportation of a photon [7, 11]. Here, using information entanglement has been considered, along with the conservation of information concept to suggest that massive objects, such as atoms, could be teleported. Though a new method and new objects have been introduced as speculation for teleportation, we must point out that the primary interest of this paper is not teleportation; rather it is the conservation of information concept. With this concept it is possible to use RIS (or other sensitive techniques for the detection of single atoms) to gain a new understanding of the collapse of the wave function in quantum mechanics.

ACKNOWLEDGMENTS

More information can be found on this subject in a book in press [12]. The author wishes to thank C.H. Chen and Darlene Holt at Oak Ridge National Laboratory, as well as Tom Whitaker and Terry Copeland of Atom Sciences for suggestions and support.

REFERENCES

1. Shannon, C. E. and Weaver, W., *The Mathematical Theory of Communications*, Dover Publications, Inc., New York, 1950.

2. Feynman, R. P., Leighton, R.B., Sands, M., The *Feynman Lectures on Physics,* Addison Wesley Publishing Company, Menlo Park, California, 1965, Vol. III.
3. Horgan, J., *Scientific American*, 94-104, (July 1992).
4. Begley, S. *Newsweek*, 67-69,(June 19, 1995).
5. Chiao, R. Y., and Steinberg, A. M., "Tunneling Times and Superluminality" in *Progress in Optics*, edited by E. Wolf, Elsevier, New York, 1997, pp. 345-405.
6. Hurst, G. S., *J. Korean Phys. Soc.*, **33**, 197-206 (1998).
7. Bouwmeester, D., Pan, J. W., Mattle, K., Eibl, M., Weinfurter, H., Zeilinger, A., *Nature*, **390**, 575-579, (1997).
8. De Baere, W., *Adv Electronics and Electron Phys.*, **68**, 245-337, (1986).
9. Stapp, H. P., *Mind, Matter, and Quantum Mechanics*, Springer Verlag, Berlin, 1933.
10. Pagels, H. R., *The Cosmic Code*, Bantum Books, New York, 1982.
11. Bouwmeester, D., Pan, J.W., Daniel, M., Weinfurter, H. and Zeilinger, A. *Phys. Rev. Lett.*, **82**, 1345-1349, (1999).
12. Hurst, G. S., *Quantum Doors*, in press.

Laser Desorption Mass Spectrometry for Biomolecule Detection and Its Applications

C. H. Winston Chen*, L. J. Sammartano[†], N. R. Isola[‡] and S. L. Allman*

Life Sciences Division, Oak Ridge National Laboratory, Oak Ridge, TN 37831-6378
[†]Biology Department, St. Olaf College, 1520 St. Olaf Avenue, Northfield, MN 55057
[‡]Oak Ridge Associated Universities, P. O. Box 117, Oak Ridge, TN 37830

Abstract. During the past few years, we developed and used laser desorption mass spectrometry for biomolecule detections. Matrix-assisted laser desorption/ionization (MALDI) was successfully used to detect DNA fragments with the size larger than 3000 base pairs. It was also successfully used to sequence DNA with both enzymatic and chemical degradation methods to produce DNA ladders. We also developed MALDI with fragmentation for direct DNA sequencing for short DNA probes.

Since laser desorption mass spectrometry for DNA detection has the advantages of fast speed and no need of labeling, it has a great potential for molecular diagnosis for disease and person identification by DNA fingerprinting. We applied laser desorption mass spectrometry to succeed in the diagnosis of cystic fibrosis and several other nerve degenerative diseases such as Huntington's disease. We also succeeded in demonstrating DNA typing for forensic applications.

INTRODUCTION

The conventional method used for most biomolecule measurement is gel electrophoresis. In this process, an electric field is applied to the gel medium to cause the drift of biomolecules that typically exist as charged particles. However, this process in general is slow and needs hours to separate molecules of different sizes. Furthermore, the resolution is not very good due to the limitation caused by the physical size of the electrophoresis instrument. Mass spectrometry has been used for several decades to measure the molecular weight of gas samples by mass-to-charge ratio. Due to the similarity of the principles of using molecular mass for separation between a time-of-flight mass spectrometer (TOFMS) and a gel electrophoresis device, a time-of-flight mass spectrometer is most often be used for biomolecule detection. In 1987 Hillenkamp and his coworkers[1] discovered that large protein molecular ions could be produced without much fragmentation by laser desorption if these biomolecules were mixed with smaller organic compounds which served as a matrix for strong absorption of a laser beam. This process is now called matrix-assisted laser desorption and ionization (MALDI). The typical preparation technique for MALDI is to dissolve biomolecular samples in solution, then prepare another solution containing small organic compounds such as 3-hydroxypicolinic acid. These two solutions are subsequently mixed and a small amount of solution is

CP584, *Resonance Ionization Spectroscopy 2000: 10th Int'l. Symp.*, edited by J. E. Parks and J. P. Young placed on a metal sample plate to dry. The molar ratio of matrix to analyte is typically more than 100 to 1. After the crystallization of the sample, the sample plate is placed in the mass spectrometer for analysis.

Since the discovery of MALDI, many research groups have succeeded in measuring various proteins and large organic compounds by MALDI. MALDI has also been applied to DNA segments.[2] Recently, rapid sequencing of DNA has been considered critically important for biomedical applications. It is natural to consider using TOFMS for DNA sequencing to reduce the sequencing time. Instead of taking several hours by gel electrophoresis, the TOFMS can take less than one millisecond to finish the separation of different sizes of DNA segments. However, the parent DNA ions must be produced with high efficiency and without serious fragmentation. Since the MALDI process has been successfully used in producing parent ions of small DNA segments without serious fragmentation, it could have an important role in rapid DNA sequencing in the future. We recently have succeeded in sequencing DNA by laser desorption mass spectrometry with various approaches.

The discoveries of restriction fragment length polymorphism (RFLP) and polymerase chain reactions (PCR) are two of the most important DNA analysis methodologies in the past few decades. PCR is broadly used in nearly every DNA laboratory to selectively amplify a segment of DNA of interest from genomic DNA. We have also applied laser desorption mass spectrometry for RFLP and PCR product analysis. The success of using mass spectrometry for measurements of RFLP and PCR products paved the way for several important applications which include disease diagnosis and DNA fingerprinting for forensic applications.

RESULTS AND DISCUSSION

DNA Detection by MALDI

With the use of mixtures of organic compounds as matrices, we succeeded in detecting DNA with the size up to 500 base pairs (Fig. 1). However, only single stranded DNA was observed. Recently, we observed the detection of DNA with the size of 3199 base pairs. However, the resolution and reproducibility still needs improvement. Nevertheless, it can be concluded that large DNA fragments can be measured by MALDI. However, the mass resolution for large DNA fragments (>300 nt) is still very poor (M/DM < 50).

In order to study the cause of poor mass resolution, we measured the velocity distributions of desorbed oligomers and found the distribution more or less follows the distribution of matrix.[3] In general, the mass resolution of MALDI spectrum of DNA is relatively poor compared to small organic compounds. This relatively poor resolution is due to (1) high initial velocity distribution, (2) non-homogenous DNA samples distributed on metal plates, (3) the adducts of matrix materials or metal ions to DNAs and (4) fragmentation of DNA molecules due to the laser ablation process. Since DNAs are passively carried by matrix materials, the velocity spreads are expected to be similar to the velocity distribution of the matrix material. When a biomolecule sample mixed with matrix materials begins to dry on the plate, the process of crystal growth makes the sample distribution non-homogeneous. The thickness of the sample can also cause lower resolution. However, a reflectron TOFMS can compensate for the ionization volume to achieve a higher resolution. Since most spectra by MALDI have resolutions of only a few hundred, the limited resolution is primarily from adduct and fragmentation formation. If a biomolecule can be attached by a matrix molecule, or part of a matrix molecule, the resolution then depends on the molecular weight of the attached particle. It is not unusual to observe alkali metal ions attached to DNA molecules.

In addition to the above factors, we discovered that the existence of

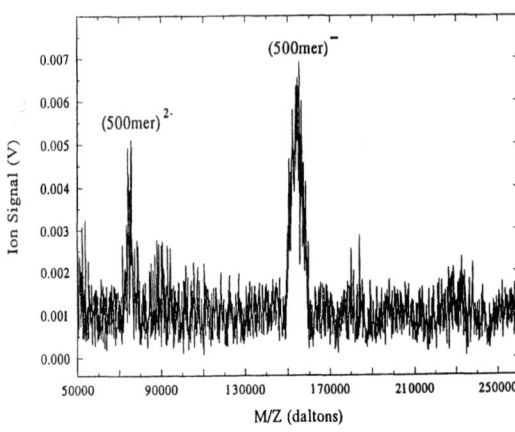

Fig. 1 Negative-ion spectrum of a 500 base pairs DNA amplified from bacteriophage lambda genome. Total amount of DNA loaded was about 2.5 pmol. 2.4 µmol picolinic acid +0.3 µmol 3-HPA was used as a matrix. Laser wavelength was 266 nm and laser fluence was 200 mJ/cm^3. Both single-charged and doubly-charged ions are observed.

metastable negative ions during the MALDI process can also lead to poor mass resolution if a reflectron is used. With the modification of the TOFMS electrical parameters to assure that no long-lasting positive or negative ions can be detected by the TOFMS (Fig. 2), we still observe biomolecule signals. This indicates the existence of metastable states of negative oligomer ions during the MALDI process. The lifetimes of small negative

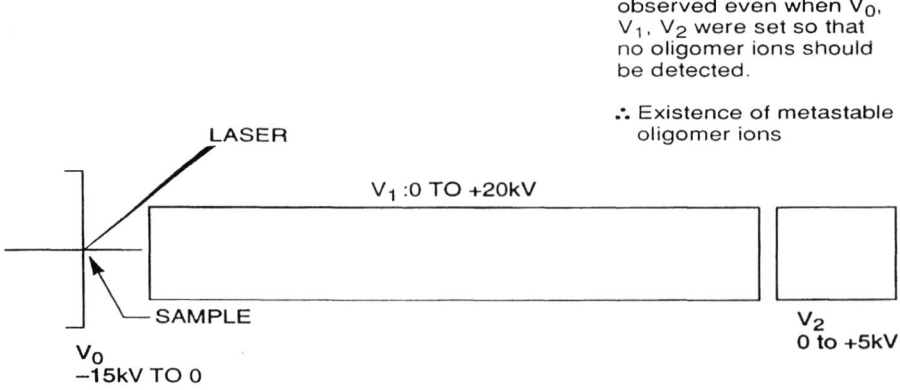

Fig. 2 Oligomer signals were observed even when V_0, V_1, V_2 were set so that no oligomer ions should be detected. This confirms the existence of metastable DNA ions.

oligomer ions were measured to be a few hundred microseconds. The lifetimes become shorter for larger negative oligomer ions compared to small ones. These results indicate that the detection of large negative oligomer ions will be very difficult for Fourier transform, ion-cyclotron, and ion-trap mass spectrometers with MALDI because of the inherent long ion lifetimes required by these instruments.

DNA Sequencing by Mass Spectrometry

It is obvious that a complete sequencing of human genome for each individual can provide very valuable information for gene discovery, disease diagnosis and forensic applications. However, sequencing with gel electrophoresis is a time consuming and labor intensive process. DNA sequencing by mass spectrometry can be orders of magnitude faster than gels.

DNA Sequencing by Mass Spectrometry with DNA Ladders

DNA sequencing has been broadly used for biomedical research and clinical applications during the past two decades. Rapid and reliable DNA sequencing can also be very valuable for forensic applications. With the conventional sequencing approach, different sizes of DNA ladders, which are produced by either Sanger's enzymatic method[4] or Maxam-Gilbert's chemical cleavage method,[5] are separated by gel-electrophoresis to achieve sequencing. With MALDI for sequencing, the speed can be significantly faster than gel electrophoresis. In addition, MALDI sequencing does not require labeling for identification, which saves both time and cost.

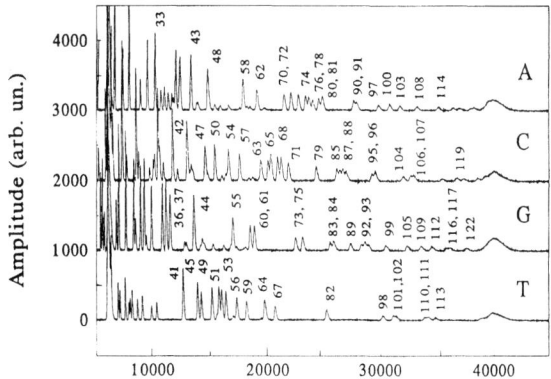

Fig. 3 Negative ion mass spectra of Sanger sequencing ladders produced by dieoxynucleotide chain terminaton of the PCR product of a double-stranded 130 bp template with a 20 nt reverse primer.

During the past few years, mass spectrometry for sequencing short DNA fragments has been pursued. In 1997, we demonstrated a complete sequencing effort for a ss-DNA with 50 bases with no false stops, no serious fragmentation, and no unidentified peaks.[6] Then, we succeeded in sequencing ss-DNA higher than 100 nucleotides by using cycle sequencing to produce DNA ladders.[7] Figure 3 shows the negative-ion mass spectra of DNA ladders from A, C, G and T reactions using double-stranded DNA 130 bp as the template with reverse primer. Forward primers were also used for sequencing with results similar to the data shown in Fig. 3. Since sequencing by using both forward and backward primers can reach 120 mer, complete sequencing of 200 bp DNA can be achieved. Due to the measurement of mass to identify the size of DNAs, there is no concern about a missing band that can occur in the gel electrophoresis method. Thus, redundant sequencing can possibly be eliminated with MALDI DNA sequencing.

With MALDI now able to sequence 200 bp, mass spectrometry is emerging as a useful tool for DNA sequencing and re-sequencing. Mass spectrometry sequence detection utilizes inexpensive unlabeled primers and also has the additional advantages of higher sequencing speed and the ability to obtain sequence information close to the primer.

MALDI for Disease Diagnosis

Disease due to the DNA mutation can be classified into three major categories: (1) deletion or insertion, (2) point mutation and (3) dynamic mutation. We have applied MALDI to all three different cases.

Diagnosis of Disease due to Base Deletion or Point Mutation

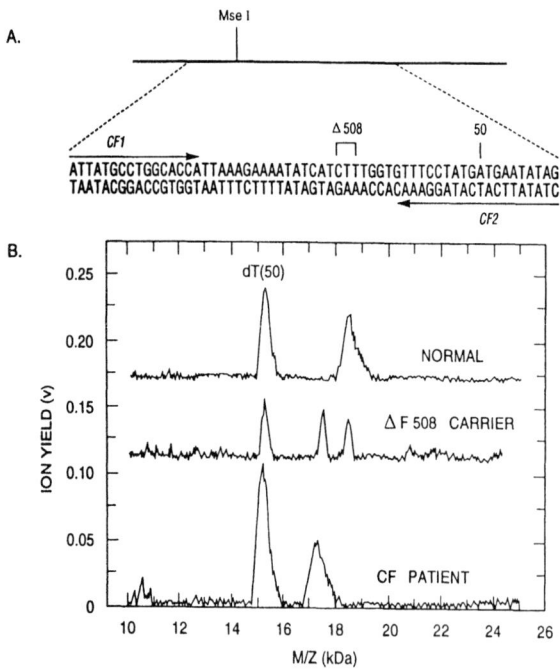

Fig. 4 Mass spectrometry for CF ΔF508 CTT delection detection.

In 1994, we gave the first demonstration of the detection of mutations in the cystic fibrosis (CF) gene in clinical samples.[8] In the North American population, about 70% of CF carriers have a 3-bp deletion in exon 10, resulting in the loss of phenylalanine residue at codon 508 (ΔF508). In our approach, two oligonucleotide primers, CF1 and CF2, were designed to amplify a DNA segment spanning the deletion, thus generating a 59-bp or 56-bp fragment of the normal CF gene and ΔF508 mutation respectively (Fig. 4). Analysis of the PCR amplified products by MALDI resolved the 3-bp difference between the normal and the ΔF508 alleles in the spectra in Fig. 4. Encouraged by these results, we also tried to use allele specific PCR to measure G551D point mutation[9] in cystic fibrosis transmembrane conductance regulator (CFTR).

Dynamic Mutation Detection

There are numerous genetic diseases such as Huntington's disease (HD), dentatorubral-pallidoluysian atrophy (DRPLA), Kennedy's disease, and a number of spinalcerebellar ataxias, which are due to the abnormal trinucleotide expansion. Rapid measurement of the number of these trinucleotide repeats can give accurate identification of normal, heterozygous carrier and homozygous patients. In collaboration with Dr. Karla Matteson and Dr. Nicholas Potter of the University of Tennessee Medical Center, we gave the first demonstration of using laser desorption mass spectrometric technology for rapid diagnosis of Huntington's disease and DRPLA.[10]

It is clear that laser desorption mass spectrometry can potentially be used for diagnosis of many genetic diseases.

DNA Fingerprinting for Forensic Applications

DNA fingerprinting is a technology that identifies individuals based on patterns of DNA markers detected in the genomic DNA. It can be used in forensic analysis to identify suspects or victims. It is particularly valuable at scenes of violent crimes where a body may not be available, or in instances where decomposition or dismemberment excludes the use of standard forensic technique. There are several different approaches available to obtain DNA fingerprints for forensic application. These include restriction fragment length polymorphism (RFLP), short tandem repeats (STR), and single nucleotide polymorphism (SNP).

The foundation of DNA typing for person identification was established by the observation of restriction fragment length polymorphism (RFLP) in which polymorphic DNA loci are determined from restriction enzyme digestion.[11]

Since the confirmation of matching the suspect with the evidence DNA often requires the analysis of several loci, the time needed for DNA profiling can be long and the cost high. At present, all forensic DNA samples for court evidence are analyzed by gel electrophoresis. Visualization of DNA in this procedure requires the use of either radioactive material or dye tagging. MALDI can become an efficient tool for DNA typing for forensic applications.

Mass Spectrometry DNA Detection for Forensic Applications

Each DNA typing involves 1) measurements of DNA fragments associated with genetic markers; 2) match determination for various genetic markers for different samples; and 3) statistical analysis for the type match to determine the possibility of an accidental match. All sequencing or size determination of DNA fragments for forensic applications have been performed with gel electrophoresis. Various approaches for DNA applications in the forensic community include the analysis of VNTR, STR, SNP and DNA sequencing. With the recent progress in development of mass spectrometry for DNA analysis, there is a good potential that mass spectrometry can be used for all the above applications.

When an individual's entire genomic DNA is digested by a restriction endonuclease, hundreds to thousands of restriction fragments are produced. A special class of RFLPs are based on DNA sequences that occur in tandem repeats.[12] Tandem repeats arise when slippage mutation occurs during DNA replication. It has been observed that this type of mutation occurs frequently enough to generate significant variation over many generations. Polymorphism for these markers is so high in humans that only identical twins will have the same patterns. This factor makes the use of tandem repeats analysis extremely valuable to forensic scientists. There is a very large number of STR distributed throughout the human genome. STR analysis can be used for as many loci as needed for reliable identification. PCR is nearly always used in STR analysis. Since PCR can amplify the selected DNA segments by more than six orders of magnitude, only a very small quantity of DNA sample is required. Since the analysis of STR is by measurement of the sizes of DNA products from the PCR process, the size of DNA fragments is often less than 300 bp and the quantity is often in the range of a few picomoles which can be readily detected by MALDI.

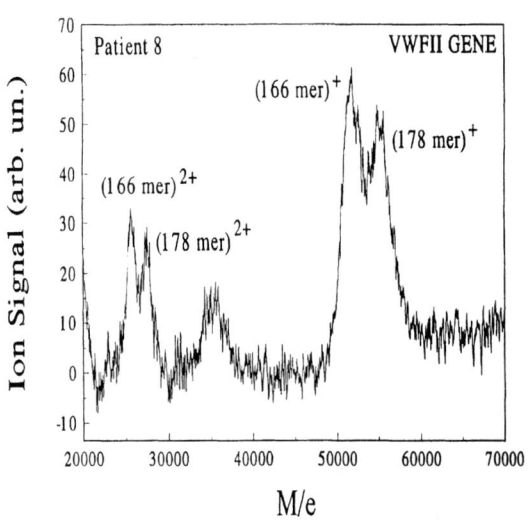

Fig. 5 Negative-ion mass spectrum of PCR product: 164 and 176 bp in the intron 40 of human vWFII gene.

A typical mass spectrum of a DNA sample amplified by PCR from a part of the vWFII gene is shown in Fig. 5. Peaks for PCR products are clearly visible, a difference of 12 bases with three 4-base tandem repeats can also be observed. Both primary and doubly charged ion peaks are observed. The spectrum indicates that one chromosome in this region is 12 bp longer than the same region in the other chromosomes. Thus, the result indicates the sample is from a person with heterozygous vWFII. Several different genes were used for MS detection. Take F13B locus as an example. The maximum difference for the number of 4-base repeats is 5. There are 15 different genotypes for this locus. Since different people have different numbers of repeats in each of these genes, measurements of the number of repeats for several loci can be used for individual identification. Examples of various mass spectra for different people for different loci are shown in Fig. 6. The results were compared with results from gel electrophoresis with good agreement. There are many tetranucleotide repeats that can be used for forensic identification.

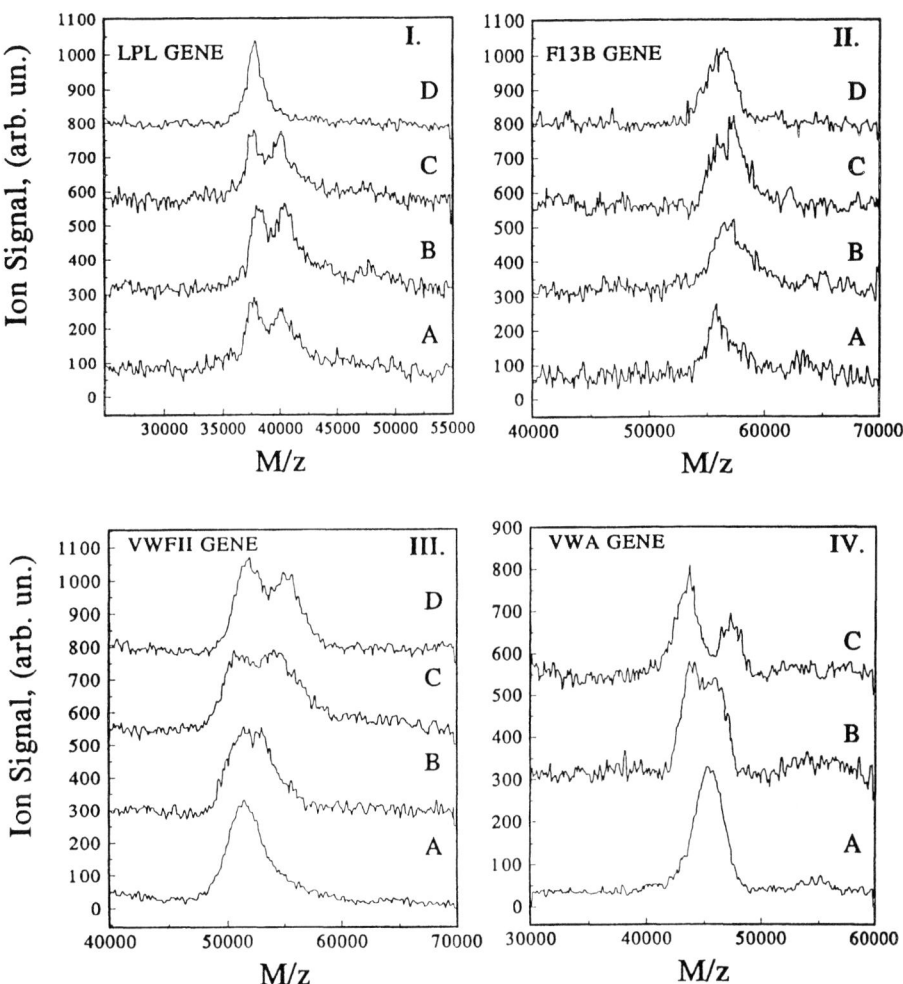

Fig. 6 Negative-ion mass spectra of PCR products of LPL, F13B, vWFII and vWA genes from different persons. Homozygous and heterozygous as well as the number of repeats can be determined.

Gender Determination

Sex determination is often important for identifying suspects and victims. A reliable gender test is PCR amplification of a segment of the X-Y homologous gene amelogenin.[13] A single pair of primers spanning part of the first intron generates 106 bp and 112 bp PCR products from the X and Y chromosomes respectively. MALDI can be subsequently used for the measurements of PCR products. Recently, it was found that the deletion of Y-encoded gene occurs in a small percentage of Y chromosomes. Thus, it is more reliable

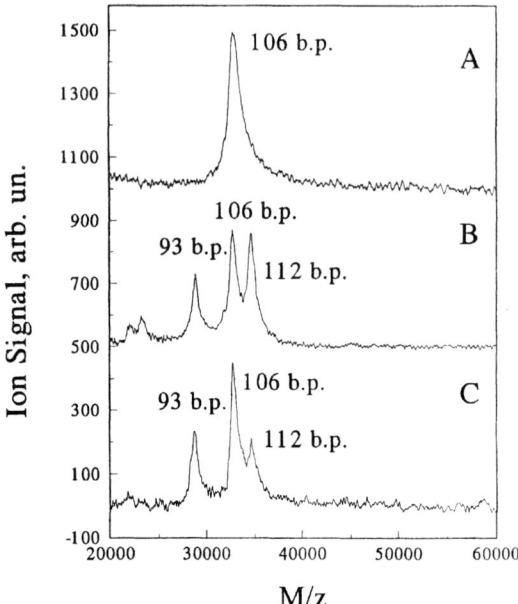

Fig. 7 Negative-ion mass spectra of sex-specific co-amplification (Amel X/Y plus SRY) of loci for (A) female, (B) male, and (C) female plus male (sample ratio: 1 to 1) respectively. Each sample contained 10 ng of DNA. The laser wavelength was 337 nm with a laser fluence of 65 mJ/cm².

to test both amelogenin and the male sex determining gene SRY for gender determination. Male DNA generates three products of 93 bp (SRY), 106 bp (Amelogenin in X chromo-some) and 112 bp (amelogenin in Y chromosome). Co-amplification of female DNA generates only the 106 bp. As shown in Fig. 7, the distinction of genotyping from male and female is quite clear. Figure 7 shows examples of mass spectra of PCR products for a (A) female, (B) male, and (C) mixture of male and female. One of special interests in forensic analysis, especially for sexual assault, is the detection of a trace amount of male-specific DNA in a female sample. Such detection can be performed by demonstrating the presence of male-derived products of 93 bp and 112 bp. We checked MALDI spectra of composite sample containing mostly female species (10 ng) and only a trace amount of male sample (0.01 ng), the trace of the male sample can be detected. This indicates that as little as 0.1 percent of male DNA in female DNA sample can be detected. It also indicates that the detection sensitivity can reach 0.01 ng or less which is about one order of magnitude better than gel electrophoresis.[14]

CONCLUSION

Laser desorption mass spectrometry provides a unique method for nucleic acid detection and sequencing. Applications on disease diagnosis and people identification for forensic applications have been clearly demonstrated. This technology has great potential for routine use in the future.

ACKNOWLEDGEMENT

This work was supported primarily by the National Institute of Justice, grant number 97-LB-VX-A047, National Institute of Health (NIH) and supported in part by the Office of Biological and Environmental Research, U.S. Department of Energy under contract DE-AC05-00OR22725 with UT-Battelle, LLC.

REFERENCES

1. Karas, M., Beckmann, D., Bahr, U., and Hillenkamp, F., *Int. J. Mass Spectrom. Ion Process.*, **78**, 53-68 (1987)
2. Tang, K., Taranenko, N. I., Allman, S. L., Chang, L. Y., and Chen, C. H., *Rapid Comm. Mass Spectrom.*, **8**, 727-730 (1994)
3. Tang, K., Allman, S. L., Jones, R. B., and Chen, C. H., *Organic Mass Spectrom.* **27**, 1389 (1992)
4. Sanger, F., Nicklen, S., and Coulson, A. R., *Proc. Natl. Acad. Sci., USA*, **74**, 5463 (1977)
5. Maxam, A. M., and Gilbert, W., *Proc. Natl. Acad. Sci., USA*, **74**, 560 (1977)
6. Taranenko, N. I., Chung, C. N., Zhu, Y. F., Allman, S. L., Golovlev, V. V., Isola, N. R., Martin, S. A., Haff, L. A., and Chen, C. H., *Rapid Comm. Mass Spectrom.*, **11**, 386-392 (1997)
7. Taranenko, N. I., Allman, S. L., Golovlev, V. V., Taranenko, N. V., Isola, N. R., and Chen, C. H., *Nuc. Acids Res.*, **26**, 2488-2490 (1998)
8. Chang, L. Y., Tang, K., Schell, M., Ringelberg, C., Matteson, K. J., Allman, S. L., and Chen, C. H., *Rapid Comm. Mass Spectrom.*, **9**, 772-774 (1995)
9. Taranenko, N. I., Matteson, K. J., Chung, C. N., Zhu, Y. F., Chang, L. Y., Allman, S. L., Haff, L., Martin, S. A., and Chen, C. H., *Genetic Analysis: Biomol. Eng.*, **13**, 87-94 (1996)
10. Taranenko, N. I., Potter, N. T., Allman, S. L., Golovlev, V. V., and Chen, C. H., *Genetic Analysis: Biomol. Eng.*, **15**, 25-31 (1999)
11. Jeffreys, A. J., Wilson, V., and Thein, S. L., *Nature*, **314**, 67-73 (1985)
12. Nakamura, Y., Leppert, M., O'Connell, P., Wolff, R., Holm, T., Culver, M., Martin, C., Fjuimoto, E., Hoff, M., Kumlin, E., and White, R., *Science*, **235**, 1616-1622 (1987)
13. Ip, N., van de Stadt, I., Loewy, Z., Leary, S., Grzechik, K., and Balazs, I., *Nuc. Acids Res.*, **17**, 4427 (1989)
14. Taranenko, N. I., Potter, N. T., Allman, S. L., Golovlev, V. V., and Chen, C. H., *Anal. Chem.* **71**(18), 3974-3976 (1999)

Isotopic Ratio Measurements of ^{135}Cs / ^{137}Cs using Resonance Ionization Mass Spectrometry

L. Pibida, W. Nörtershäuser[†], J.M.R. Hutchinson, B.A. Bushaw[†]

National Institute of Standards and Technology, Gaithersburg, MD 20899, USA
[†] Pacific Northwest National Laboratory, Richland, WA 99352, USA

Abstract. ^{135}Cs / ^{137}Cs isotopic ratios were measured in the presence of high levels of ^{133}Cs using Resonance Ionization Mass Spectrometry (RIMS). Single- and double-resonance excitation of Cs with high-resolution continuous-wave (cw) lasers along $6s\,^2S_{1/2} \to 6p\,^2P_{3/2}$, $6s\,^2S_{1/2} \to 6p\,^2P_{3/2} \to 7d\,^2D_{5/2}$, and $6s\,^2S_{1/2} \to 6p\,^2P_{3/2} \to 8s\,^2S_{1/2}$ pathways were followed by photoionization with an Ar ion laser. The age of a standard burn-up sample was determined using RIMS and compared to Thermal Ionization Mass Spectrometry (TIMS). The efficiency, selectivity and isobar suppression for TIMS and RIMS has been compared. The overall RIMS efficiency was found to be $\approx 2 \times 10^{-6}$ with a combined selectivity of $\approx 10^{10}$ for both ^{135}Cs and ^{137}Cs against ^{133}Cs using the preferred single-resonance excitation scheme.

INTRODUCTION

The radioisotopes ^{135}Cs and ^{137}Cs are high-yield fission products from ^{233}U, ^{235}U, and ^{239}Pu [1]. The ratio of long-lived ^{135}Cs ($t_{1/2} = 2.3 \times 10^6$ years) to the shorter-lived ^{137}Cs ($t_{1/2} = 30.07$ years) might be used to date radioactive waste. This ratio could also be used to evaluate source conditions during production of radioactive Cs isotopes. While ^{137}Cs is adequately detected by gamma-ray spectroscopy, the pure beta decay and long half-life of ^{135}Cs make its detection by radioactive decay counting impractical. Thus, a sensitive method for measuring both of these radioisotopes, independent of their decay characteristics, would be desirable. Conventional mass spectrometry is a potential solution, but tail interference from large excesses of stable ^{133}Cs and isobaric interferences from barium present serious problems for low-level environmental measurements [2-4].

In this work, RIMS [5] using high-resolution continuous-wave (cw) lasers [6, 7] is applied to measuring Cs isotopic ratios and is compared with conventional TIMS. In RIMS, isobaric interference is inherently avoided because only the target element of interest is ionized [8]. Isotopic selectivity is then provided by the mass spectrometer that analyzes and detects the photoions. High-resolution lasers can add optical isotopic selectivity to that of the mass spectrometer; previous work on Sr [9, 10], Kr [11], and Ca [12] illustrate some of the potential of this technique. For Cs, we have considered single-resonance excitation using optical pumping methods for improving selectivity, compared this to double-resonance excitation, and evaluated both with respect to their analytical performance.

CP584, *Resonance Ionization Spectroscopy 2000: 10th Int'l. Symp.*, edited by J. E. Parks and J. P. Young

EXPERIMENTAL

The RIMS-system has been described in detail previously [9]. It consists of a graphite crucible for sample atomization, the laser system for ionization of the neutral atoms, and a quadrupole mass spectrometer (QMS) to mass analyze and detect the ions. The graphite crucible, of cylindrical geometry with high aspect ratio, acts as self-collimating beam source, resulting in efficient spatial transfer of atoms into the QMS ionization region where the neutral atoms are ionized by the laser beams in perpendicular geometry. The single- and double-resonance excitation of Cs begins with the $6s\,^2S_{1/2}$ ($F = 4$) $\rightarrow$ $6p\,^2P_{3/2}$ ($F' = 3, 4, 5$) transition, excited by an extended cavity diode laser operating at 852 nm. Out of this state, the atoms were either directly ionized with the 488.0 nm line of an Ar ion laser or subjected to a second resonant excitation. The double-resonance schemes involved a second excitation to either the $7d\,^2D_{5/2}$ or $8s\,^2S_{1/2}$ states using a cw Ti:Sapphire (TIS) ring laser operating at 697.5 nm or 794.6 nm, respectively. Ionization was then accomplished with the 514.5 nm line of the Ar ion laser. The mass spectrometer used for TIMS measurements at NIST is a 90° magnetic sector instrument [13] that delivers 7×10^{-3} overall efficiency for Cs detection. It is a single-sector instrument and thus the selectivity for $^{135,\,137}$Cs (relative to ^{133}Cs) is $\approx10^5$, limited by a constant background of scattered ^{133}Cs ions.

Samples containing ^{135}Cs and ^{137}Cs were prepared from two different sources: The first was a NIST ^{137}Cs standard reference material, SRM# 4233B-1, prepared from a nuclear burn-up sample, which had a ^{135}Cs / ^{137}Cs ratio of 1.59717(46), determined by TIMS, at a reference date of August 22, 1977. This standard is known to have an approximate 10^2 excess of stable ^{133}Cs. The second source was derived from a radioactivity standard prepared by the Environmental Monitoring Systems Laboratory (EMSL), source identification number 2437-3, with a known ^{137}Cs activity and ^{133}Cs present at $\approx10^6$ excess, but no information on the ^{135}Cs / ^{137}Cs ratio. For our work, initial stock solutions were limited to "non-radioactive" levels of < 70 Bq/ml and individual sample loads were limited to < 3 Bq ($\approx4\times10^9$ atoms ^{137}Cs).

RESULTS AND DISCUSSION

Initial spectroscopic investigations were performed using the stable isotope ^{133}Cs, for which the hyperfine structure and isotope shifts (IS) in the $6s\,^2S_{1/2} \rightarrow 6p\,^2P_{3/2}$ transition are well known [14, 15]. Note that all ^{133,135,137}Cs isotopes have the same nuclear spin $I = 7/2$ and thus will exhibit similar hyperfine structure. A single-resonance RIMS spectrum for transitions starting in the $F = 4$ level of the ground state, recorded at low laser power of ≈15 µW, is shown in Fig. 1(a). The points represent the experimental data while the solid line is a fit using Voigt profiles. The frequency shifts of -251.0 (3) MHz and -452.5 (4) MHz for the $F = 4 \rightarrow F' = 3, 4$ transitions, relative to the $F = 4 \rightarrow F' = 5$ transition, are in excellent agreement with values given in [14]. In Fig. 1(b), the diode laser power was increased to ≈15 mW and the beam diameter expanded to 4 mm, allowing prepumping of the atomic beam before encountering the ionization laser. Under these conditions, excitation to the $F' = 3, 4$ components results

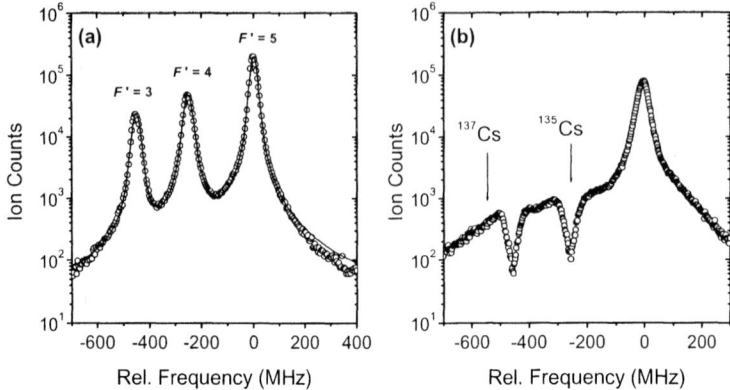

FIGURE 1. RIMS spectra of ^{133}Cs $6s\,^2S_{1/2}$ ($F = 4$) $\rightarrow 6p\,^2P_{3/2}$ ($F' = 3,4,5$) hyperfine structure recorded at (a) low laser intensity and (b) under analytical conditions with higher laser intensity to induce optical pumping and improve optical isotopic selectivity. In (b), the arrows show the position of the desired $F = 4 \rightarrow F' = 5$ transition for the indicated radioisotopes.

in decay into the $F=3$ component of the ground state, from which the atoms are not further excited. This optical pumping is advantageous since the ^{133}Cs ($F = 4 \rightarrow F' = 4$) transition is nearly degenerate with the desired ^{135}Cs ($F = 4 \rightarrow F' = 5$) component. The ^{133}Cs response is reduced by ≈ 3 orders of magnitude when the resonance laser is tuned to either the ^{135}Cs or ^{137}Cs ($F = 4 \rightarrow F' = 5$) positions, as is indicated in Fig. 1(b). This results in a combined (mass spectrometer and laser) selectivity of $\approx 10^{10}$ for both ^{135}Cs and ^{137}Cs against ^{133}Cs. Using this single-resonance approach and discrete samples containing known amounts of ^{133}Cs, we measured an overall efficiency of typically 1×10^{-6} to 3×10^{-6}.

In an attempt to increase the overall efficiency, we considered double-resonance excitations, promoting the $6p\,^2P_{3/2}$ atoms to a second, higher-lying state, followed by photoionization with the 514.5 nm line of the Ar ion laser. Excitation to both the $8s\,^2S_{1/2}$ and $7d\,^2D_{5/2}$ states were investigated, but in both cases the efficiency was found to be somewhat less than the simpler single-resonance scheme. However, we were able to observe previously unknown hyperfine structure in the $7d\,^2D_{5/2}$ state, as is shown in Fig. 2. Here, the first laser was detuned -800 MHz with respect to the $6s\,^2S_{1/2}$ ($F = 4$) $\rightarrow 6p\,^2P_{3/2}$ ($F' = 5$) resonance for ^{133}Cs and the second laser was scanned over a region such that the sum frequency encompassed the $7d\,^2D_{5/2}$ state energy. Thus this is actually a resonantly enhanced two-photon transition (rather than a true double-resonance), which allows the observation of five out of the six hyperfine components of $7d\,^2D_{5/2}$ state. By fitting the positions of the individual hyperfine peaks (shown as dotted lines in Fig. 2) to the Casimir formula, the hyperfine structure constants $A(^{133}$Cs$) = -1.93$ (2) MHz and $B(^{133}$Cs$) = 2.6$ (3) MHz were derived.

Since the double-resonance approach yielded no improvement in efficiency, we returned to the single-resonance scheme to evaluate its capability for measuring the radioisotopes. Figures 3(a) and (b) show experimental measurements of the frequency shifts for the ^{135}Cs and ^{137}Cs ($F = 4 \rightarrow F' = 5$) transitions, respectively. A sample

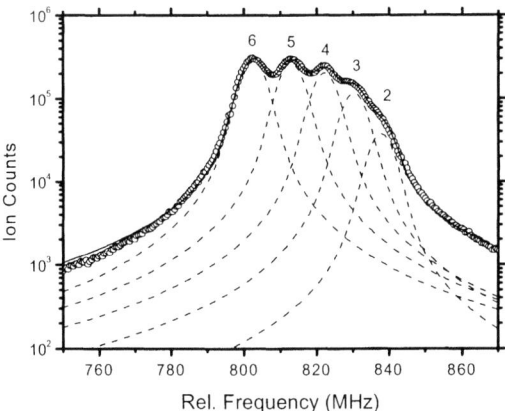

FIGURE 2. Resonantly enhanced two-photon RIMS spectrum of ^{133}Cs $6s\,^2S_{1/2}$ ($F = 4$) $\rightarrow 7d\,^2D_{5/2}$. The transition is enhanced by near resonance with the $6p\,^2P_{3/2}$ intermediate state (see text). Peak labels are the $7d\,^2D_{5/2}$ state F'' hyperfine quantum number.

containing 4×10^9 atoms ^{137}Cs (^{135}Cs content unknown) was prepared from the EMSL stock solution and loaded into the atomization crucible. This sample already contains a 10^6 excess of ^{133}Cs, which was used to record the reference spectra (triangles in Fig. 3) with the mass spectrometer set at mass 133 and the graphite crucible temperature just above the threshold for atomization. Then the mass spectrometer was switched to the desired radioisotope and the laser repeatedly scanned while increasing the crucible temperature to evolve the sample. The individual spectra were summed to give the result shown as circles in Fig. 3. Fitting with Lorentzian profiles yields isotopic frequency shifts (vs. ^{133}Cs) for the $F = 4 \rightarrow F' = 5$ transition of -249.4 (13) and −522.1 (35) MHz for ^{135}Cs and ^{137}Cs, respectively. These are in good agreement with values

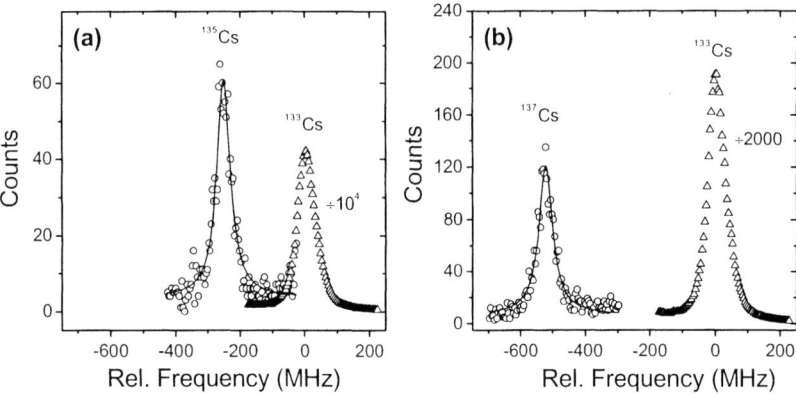

FIGURE 3. Frequency shift measurement for the $6s\,^2S_{1/2}$ ($F = 4$) $\rightarrow 6p\,^2P_{3/2}$ ($F' = 5$) hyperfine component for (a) ^{135}Cs and (b) ^{137}Cs relative to the stable reference isotope ^{133}Cs. Solid lines are Lorentzian fits to the radioisotope data.

109

of −250.03 and −524.74 MHz calculated from hyperfine structure constants and isotope shift data given in [15].

Test measurements to determine $^{135}Cs / {}^{137}Cs$ isotopic ratios and detection limits were performed using both available standards, NIST #4233B-1 and EMSL-2437-3. Figure 4 shows the measurement on a sample prepared from the EMSL standard, containing 0.5 pg ^{137}Cs (^{135}Cs unknown) and 200 µg of ^{133}Cs (4×10^8 excess added to the original solution). The crucible temperature was increased under computer control while switching the QMS and laser frequency every 10 seconds to positions for ^{135}Cs, ^{137}Cs, and a background measurement at mass 139 (with laser still tuned for ^{137}Cs). The background was observed to be independent of mass setting and is attributed to neutral species generated during the atomization process [10]. Thus choosing 139 as the background position is somewhat arbitrary. Integrating ion counts in the three channels (135 : 137 : background) over the evolution of the sample yields 783 : 731 : 186. The corresponding 3σ detection limit and $^{135}Cs / {}^{137}Cs$ ratio are 1.7×10^8 atoms (38 fg) and 1.09(8), respectively. Assuming the background arises from unionized ^{133}Cs, the effective overall selectivity for either ^{135}Cs or ^{137}Cs against ^{133}Cs is 1.2×10^9. This measurement was repeated with 11 sample aliquots and different ^{133}Cs excesses ranging from 10^6 to 4×10^8. All were found to agree within statistical uncertainty and the average $^{135}Cs / {}^{137}Cs$ ratio is 1.038 (14). Similar sets of TIMS measurements were performed, but without extra additions of ^{133}Cs, and the resulting ratio of 1.06 (7) is in good agreement with the RIMS value. Uncertainties for these analytical results are given as the standard error of the mean.

We attempted to estimate the age of the EMSL sample using the fission yields from ^{235}U, ^{233}U, and ^{239}Pu [1]. However, the results are anomalous, predicting an age of 5.0 years for U^{233}, and *negative* 0.7 and 4.6 years for ^{235}U and ^{239}Pu, respectively, while the sample is known to be more than 10 years old. Another EMSL sample has exhibited similar anomalies, which were explained by the thermal neutron flux dependent production of ^{135}Cs [16]. This must be considered when evaluating $^{135}Cs / {}^{137}Cs$ isotopic ratios for dating nuclear waste or environmental contamination. Conversely, if the history of a sample is known, then precise isotope ratio measurements may be used to deduce information about the production conditions.

Sets of RIMS and TIMS measurements were also performed on the NIST sample.

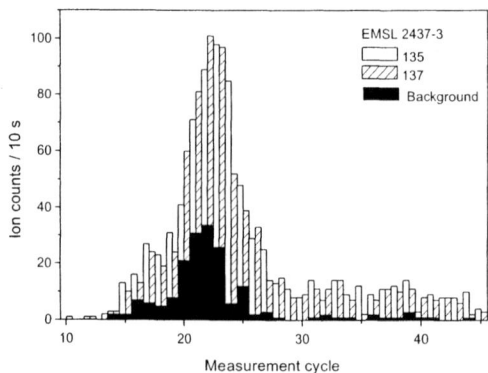

FIGURE 4. RIMS measurement of the $^{135}Cs / {}^{137}Cs$ isotope ratio in the EMSL-2437-3 sample with ^{133}Cs added at 4×10^8 excess. Measurement cycles, which consist of 10 s signal integration on ^{135}Cs, ^{137}Cs, and background (see text), are repeated while the atomization crucible temperature is increased to vaporize the sample.

The average ^{135}Cs / ^{137}Cs ratios were 2.69 (5) by RIMS and 2.66 (6) by TIMS. Both are in excellent agreement with the expected value of 2.6708 (23) calculated from the 1977 reference value. Given the 1977 reference value, the RIMS measurement was able to correctly predict the 22 year elapsed time with an uncertainty of 0.8 years.

CONCLUSIONS

RIMS was tested for measuring ^{135}Cs / ^{137}Cs isotopic ratios. Optical isotopic selectivity of $\approx 10^3$ was observed and combination with a QMS increased overall selectivity to $\approx 10^{10}$. No interference could be observed from barium isobars in the RIMS measurements due to the elemental selectivity of the resonance ionization process. The overall detection efficiency for the RIMS process was $\approx 2 \times 10^{-6}$. Test measurements on samples containing as much as 4×10^8 excess of ^{133}Cs demonstrated detection limits of 1 - 2×10^8 atoms of ^{135}Cs and ^{137}Cs. Isotope ratios for ^{135}Cs / ^{137}Cs were determined both by RIMS and conventional TIMS, and were found to be in excellent agreement. The isotope ratios were able to precisely date a sample whose isotopic composition had been accurately measured two decades previously; however, measurements on another standard with unknown ^{135}Cs content yielded an anomalous ratio that might be attributed to the neutron-flux dependent fission yield of ^{135}Cs.

ACKNOWLEDGMENTS

RIMS measurements performed at PNNL were supported by the U.S. Department of Energy, Office of Science, under contract DE-AC06-76RLO 1830. The Standard Reference Materials Program supported the work performed at NIST.

REFERENCES

1. Walker, F. W., Parrington, J. R., Feiner, F.: *Nuclides and Isotopes*. GE Nuclear Energy 1989.
2. Stoffel(s), J. J., Ells, D. R., Bond, L. A., Freedman, P. A., Tattersall, B. N., Lagergren, C. R., *Int. J. Mass Spectrom. Ion Proc.* **132**, 217 (1994).
3. Lee, T., Ku, T.-L., Lu, H.-L, Chen, J.-C., *Geochim. Cosmochim. Acta* **57**, 3493 (1993).
4. Moreno, J. M. B., Betti, M., Nicolaou, G., *J. Anal. At. Spectrom.* **14**, 875 (1999).
5. Beekman, D. W., Callcott, T.A., Kramer, S. D., Arakawa, E. T., Hurst, G. S., *Int. J. Mass Spectrom. Ion Phys.* 34, 89 (1980).
6. Bushaw, B. A., *Prog. Analyt. Spectrosc.* **12**, 247 (1989).
7. Payne, M. G., Deng, L., Thonnard, N., *Rev. Sci. Instrum.* **65**, 2433 (1994).
8. Hurst, G. S., Payne, M. G., Kramer, S. D., Young, J. P, *Rev. Mod. Phys.* **51**, 767 (1979).
9. Bushaw, B. A., Cannon, B. D., *Spectrochim. Acta B* **52**, 1839 (1997).
10. Bushaw, B. A., Cannon, B. D., *AIP Conf. Proc.* **454**, 177 (1998).
11. Janik, G. R., Bushaw, B. A., Cannon, B. D., *Opt. Lett.* **15**, 266 (1989).
12. Müller, P., Bushaw, B. A., Blaum, K., Nörtershäuser, W., Trautmann, N., Wendt, K., *AIP Conf. Proc.* **454**, 73 (1998).
13. *NBS Technical Note* **277**, July 1966, Editor Shields, W. R., National Bureau of Standards, Washington, DC.
14. Tanner, C. E., Wieman, C., *Phys. Rev. A* **38**, 1616 (1988).
15. Thibault, C. *et al.*, *Nucl. Phys. A* **367**, 1 (1981).
16. Chao, J.-H., Tseng, C.-L., *Nucl. Instr. Meth. Phys. Res. A* **372**, 275 (1996).

RIMS Analysis of Stardust -- Trace, Isotopic Analysis of Individual Micron-Sized SiC Grains*

M. J. Pellin, M. R. Savina, and W. F. Calaway

Materials Science and Chemistry Divisions
Argonne National Laboratory
Argonne, Illinois 60439

Abstract. There exists an important class of analytical problems that requires both sensitivity and discrimination. This class is exemplified by the increasingly stringent demands of the electronic industry for unambiguous quantitative identification of trace impurities in semiconductor materials at high lateral resolution. Recently, particulate analysis, the isotopic and elemental analysis of micron-sized grains, has also begun to interest the analytical community. The difficulty in these two cases arises from the need to make the measurement before consuming the few atoms of the element of interest while discriminating against the vast excess of bulk atoms. Consider trace analysis of one ppm by weight of zirconium in a SiC spherule of 1 μm in diameter. This grain contains approximately 11,000 Zr atoms. For a terrestrial isotopic composition, half of these atoms are ^{90}Zr (the major isotope) while only ~ 300 atoms are in the important (as we shall see later) ^{96}Zr isotope. Analysis is, of course, complicated by the need to discriminate bulk species, some of which (such as Si_3C) have nominally the same mass as the analyte.

    Why would such analyses be important? Some primitive meteorites contain presolar dust grains (such as graphite, silicon carbide, nano-diamonds, and corundum) that have survived the formation of the solar system. It is generally believed that these grains condensed in stellar outflows before being incorporated into meteoritic material of our solar system, and that the elemental and isotopic compositions of these grains preserve a nucleosynthetic record of their parent star. Measurements of the elemental composition and isotopic anomalies in these grains provide information both about stellar nucleosynthesis and about the conditions during circumstellar grain formation. The former can be inferred from the isotopic patterns of heavy element trace impurities such as Mo, Zr, and Sr that, although present at levels of only a few thousand atoms, have been measured for the first time using resonant ionization mass spectrometry (RIMS).

    Currently, isotopic analyses of such grains using the CHARISMA apparatus at Argonne National Laboratory (ANL) are revealing new and important information for the first time about the material that condensed to form our solar system. Grains consistent with formation around thermally pulsing asymptotic giant branch (AGB) stars, such as our sun, make up the majority of grains studied to date. However, more recently several grains have been identified, which are believed to be of supernovae origin. Since each grain represents a unique stellar source, complete elemental and isotopic analysis must be performed on individual grains -- no averaging is possible. Only with the power of RIMS are such analyses possible.

---

*Work supported by the U.S. Department of Energy, BES-Materials Sciences, under Contract W-31-109-ENG-38.

CP584, *Resonance Ionization Spectroscopy 2000: 10th Int'l. Symp.*, edited by J. E. Parks and J. P. Young
2001 American Institute of Physics 0-7354-0024-5

112

INTRODUCTION

There exists an important class of analytical problems that requires both sensitivity and discrimination. This class is exemplified by the increasingly stringent demands of the electronic industry for unambiguous quantitative identification of trace impurities in semiconductor materials at high lateral resolution. Recently, particulate analysis, the isotopic and elemental analysis of micron-sized grains, has also begun to interest the analytical community. The difficulty in these two cases arises from the need to make the measurement before consuming the few atoms of the element of interest while discriminating against the vast excess of bulk atoms. Consider trace analysis of one ppm by weight of barium in a SiC spherule of 1 µm in diameter. This grain contains approximately 11,000 Ba atoms. For a terrestrial isotopic composition, ~70% of these atoms are ^{138}Ba (the major isotope) while only ~ 850 atoms are in the important (as we shall see later) ^{136}Ba isotope. Analysis is, of course, complicated by the need to discriminate bulk, major and minor species, some of which (such as La, Ce) have nominally the same mass as the analyte.

Why would such analyses be important? Some primitive meteorites contain presolar dust grains (such as graphite, silicon carbide, nano-diamonds, and corundum) that have survived the formation of the solar system. It is generally believed that these grains condensed in stellar outflows before being incorporated into meteoritic material of our solar system, and that the elemental and isotopic compositions of these grains preserve a nucleosynthetic record of their parent star.[1-9] Measurements of the elemental composition and isotopic anomalies in these grains provide information both about stellar nucleosynthesis and about the conditions during circumstellar grain formation. The former can be inferred from the isotopic patterns of heavy element trace impurities such as Mo, Zr, Ba, and Sr that, although present at levels of only a few thousand atoms, have been measured for the first time using resonant ionization mass spectrometry (RIMS).[10, 11]

Currently, isotopic analyses of such grains using the CHARISMA apparatus at Argonne National Laboratory (ANL) are revealing new and important information about the material that condensed to form our solar system.[10-24] Grains consistent with formation around thermally pulsing asymptotic giant branch (AGB) stars, such as our sun, make up the majority of grains studied to date.[10, 11, 20, 22] However, more recently several grains have been identified, which are believed to be of supernovae origin.[12-15] Since each grain represents a unique stellar source, complete elemental and isotopic characterization must be performed on individual grains, since averaging together results from various grains looses the detailed information on the parent stars. Only with the power of RIMS are such analyses possible.

EXPERIMENTAL

The RIMS apparatus used here has been described in detail elsewhere[17, 25-27] Samples can be placed into the ultrahigh vacuum instrument through a load lock and subsequently manipulated using a three-axis *in-vacuum* micrometer system based on

Burleigh Inchworm motors. The instrument, shown schematically in Fig. 1, consists of desorption sources, the photoionization lasers, and the time-of-flight (TOF) mass analyzer. Crucial to understanding the time of flight resonance ionization mass spectrometric system is the timing cycle depicted in Figure 2. The experimental cycle begins with a desorption pulse either from a primary ion gun (sputtering) or a microfocus laser source. The purpose of the desorption pulse is to atomize the sample producing to the extent possible a pulse containing only ground state neutral atoms. For most analyte/matrix combination the ground state neutral atom yield is the dominant portion of the desorbing flux, however there do exist cases where this is not the case and the analyst needs to be aware of the particular characteristics of the

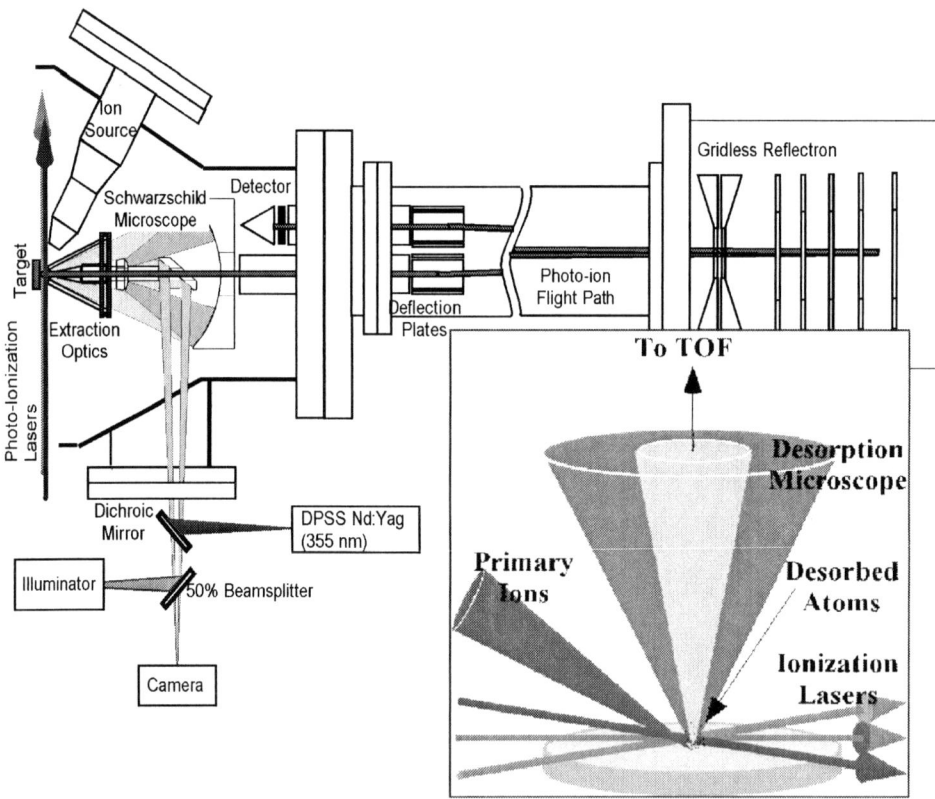

measurement being attempted.

FIGURE 1. Displayed is a schematic diagram of the CHARISMA Resonant Ionization Time of Flight Mass Spectrometer. The inset depicts an expanded and rotated view of the laser interaction region located at the far left of the main diagram. This region includes a desorption microscope built from two spherical mirrors and a metal ion gun capable of sputtering from a 50 nm spot (incident at an angle of 60* from the target normal). The vacuum system allows introduction of many tunable laser beams.

Any one of three possible desorption sources may be used for particulate analysis. Two ion guns are available for surface analysis and sample cleaning. The first is a

Colutron 101 Q ion source that typically can deliver 2 μA of 5 keV Ar$^+$ ions into a sub mm spot onto the target surface. The beam strikes the sample surface at 60° from the target normal. An IonOptika liquid Ga$^+$ ion source is used for microprobe analysis and can deliver >10 nA of 20 keV Ga$^+$ into a submicron target region. The third desorption source, and the one used for the analyses reported here, is the third harmonic light from a diode-pumped, solid-state Nd:YAG laser at 355 nm. (Model DNY from IBL Innovative Berlin Laser AG) This laser radiation (with a typical output of a >1 mj per laser pulse) was attenuated until the incident energy on the target surface was <100 nanojoules per pulse.

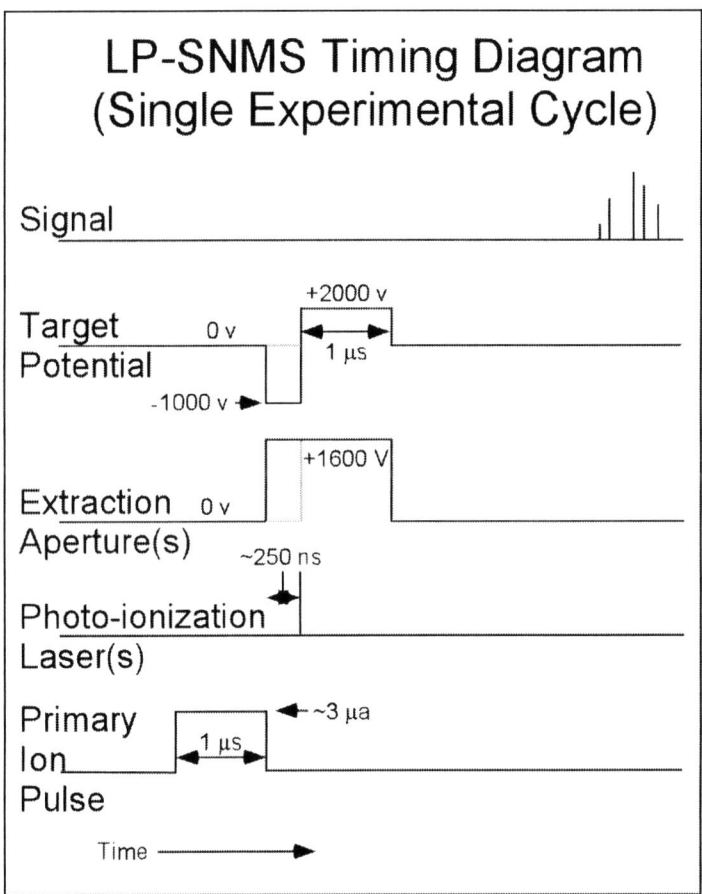

FIGURE 2. Displayed is a timing diagram for the CHARISMA TOF apparatus. Labels refer to various elements in figure 1. The diagram is for a single experimental cycle. Normally this cycle is repeated many times in order to accumulate the necessary signal levels.

Each desorption laser pulse was focused onto the sample surface using an all-reflecting Schwarzschild microscope (shown in the inset of Figure 1).[25] Thus laser desorption from sub micron regions is possible. During the time required for the neutral atoms to leave the sample surface and fill the photoionization volume above the surface (~250 ns for ion sputtered atoms and ~800 ns for laser desorbed atoms) the target potential is reduced in order to sweep out any secondary ions produced in the desorption process. This noise discrimination process is enhanced by also raising the potential of the first extraction aperture.

At the appropriate time the photoionization lasers are fired and the target potential is raised to 2 kV (a potential appropriate for transport of photoions through the TOF mass analyzer). The extraction optics are carefully designed to efficiently extract from the large photoionization volume and yet allow for reasonable mass resolution (m/Δm > 500). The mass resolution is achieved through a combination of pulsed draw-out compression and the use of an ion mirror (reflectron). In combination, these two techniques reduce the mass dispersion caused by the energy spread inherent in the extracted photoions. This energy spread is induced both by the broad energy distribution of desorbed atoms and by the large extraction volume. Currently we are constructing a new TOF with predicted characteristics significantly better than the current design.[26]

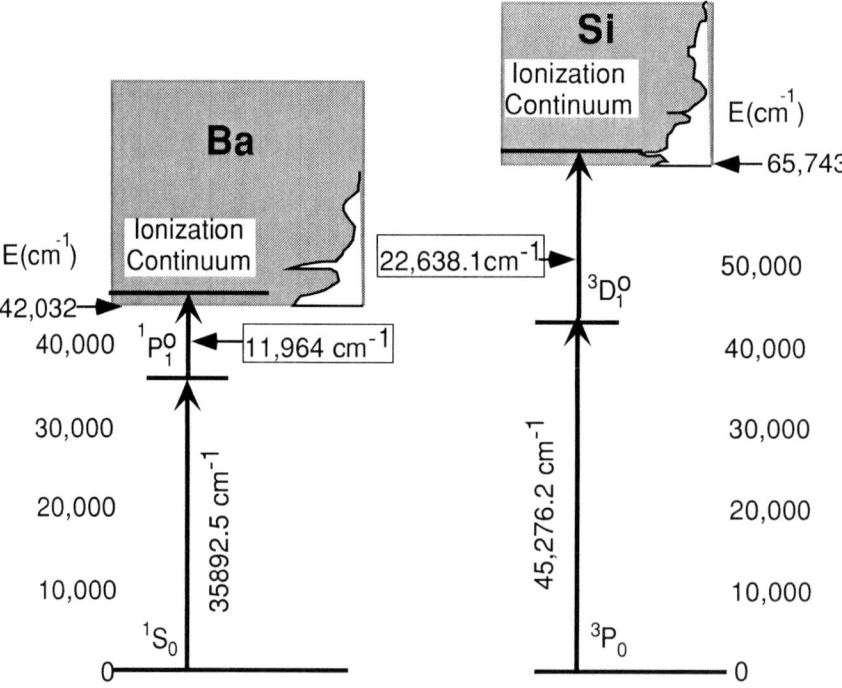

FIGURE 3. Shown above are energy level diagrams for Si and Ba with the photons used to ionize them. In the case of Ba, Ti:Sapphire lasers are employed using a 3,1 scheme – the laser output is first tripled to produce the 35892.5 cm^{-1} resonant photon and then coupled with its fundamental to photoionize Ba. Si uses a 4,2 scheme. The ionization continuum has a structured absorbance for each atom represented by the small absorbance versus energy graph on the top right-hand side of each energy level diagram.

The photoionization lasers used in our apparatus are high repetition rate Nd:YLF laser pumped Ti:Sapphire tunable lasers (2 units from Photonix, Inc.). These lasers are all solid-state broadly tunable lasers capable of ionizing two atoms at once in favorable cases at a repetition rate exceeding 1 kHz. Si and Ba are photoionized using the wavelength scheme depicted in Figure 3.

Each measurement cycle consists of the following sequence: (1) a 355 nm ablation laser pulse from a frequency-tripled Nd:YAG laser is focused on a presolar grain or standard and produces a cloud of neutral atoms, ions and molecules; (2) a voltage pulse is applied to an extraction electrode to electrostatically suppress the ions; (3) two resonant laser beams intersect the cloud of neutral species above the sample, resonantly ionizing Ba or Si with high efficiency and non-resonantly ionizing other species with low efficiency; (4) a positive 2000 V target pulse extracts and accelerates these photoions; (5) the photoions are mass-analyzed in a reflectron-type time-of-flight mass spectrometer with 4 m long flight path; (6) data are collected in an ion counting mode, in which the signal produced by a single photoion is converted to a normalized voltage pulse (500 mV, 5 ns) and digitized in a 500 MHz transient recorder. Subsequent analysis with the lasers slightly detuned identifies the background component of the spectrum.

RESULTS

Trace, isotopic analysis of 1-5 μm SiC grains culled form the Murchison meteorite requires the ability to efficiently photoionize the elements of interest (Si and Ba). Pure Ba and Si are used as terrestrial standards and the atoms are ionized using the scheme displayed in Figure 3 both for standard sample analysis and for meteoritic SiC grain analysis. Using Ti:Sapphire lasers for photoionization inevitably requires that the fundamental laser light tunable in the range 700 nm – 1000 nm be up-converted using non-linear techniques. In the case of Ba, the light is first doubled in a LBO crystal and then tripled in a second LBO crystal. Typically we produce ~100 μj/pulse of light at 35,892.5 cm^{-1}, easily sufficient to saturate the photoionization of Ba. Following resonant excitation the atoms are photoionized using the residual fundamental light from the tripling process. Alternatively, the second Ti:Sapphire laser can be used to produce the desired photons with the advantage of increased laser intensity. Si is photoionized using a 4,2 scheme. That is the fundamental light is first doubled and then doubled again. This fourth harmonic light is coupled with the residual second harmonic light to photoionize Si.

A TOF Mass spectrum for a terrestrial standard Si sample photoionized using the scheme displayed in Figure 3 is displayed in figure 4. Also depicted is the signal detected when the photoionization laser is detuned 0.2 nm from the resonant transition line. Note that the curve has been linearly displaced –0.5 units in order to allow more detailed examination of the differences between it and the resonant laser ionization curve. This second curve is essentially a representation of the noise present in the resonant photoionization of Si from this sample. It is taken with the system completely operational – that is all lasers desorption lasers, etc. running. The only difference is the slight change in laser wavelength. Since only the absorption spectrum of Si shows a sharp change in this wavelength region, this detuning only turns of the signal, not the background. This easy ability to accurately assess the noise in a spectrum represents one of the most powerful aspects of RIMS for trace analysis, for it allows the unambiguous determination of an element of interest and dramatically improves the detection sensitivity of RIMS measurements with simple subtraction techniques.

Currently we are using the detection of Si from the grain to help locate the grain and to establish laser desorption parameters (intensity, focus) for measurement of trace heavy elements such as Ba. Silicon, being a major element, can be measured with very little grain consumption. Indeed, isotopic measurements can be made with the removal of as little as 10 atomic layers from a 1-micron grain. The challenge now is to make high-precision isotopic measurements on the Si isotopes. Increased precision is necessary since the Si isotopes show relatively small changes from grain to grain when compare to the heavy elements. The relatively small Si isotope changes are consistent with the model that most grains (called mainstream) formed around AGB stars. In such stars, the Si isotope distribution will not undergo large changes. Heavy elements are fractionated to a much larger extent since they are being produced in the star itself.

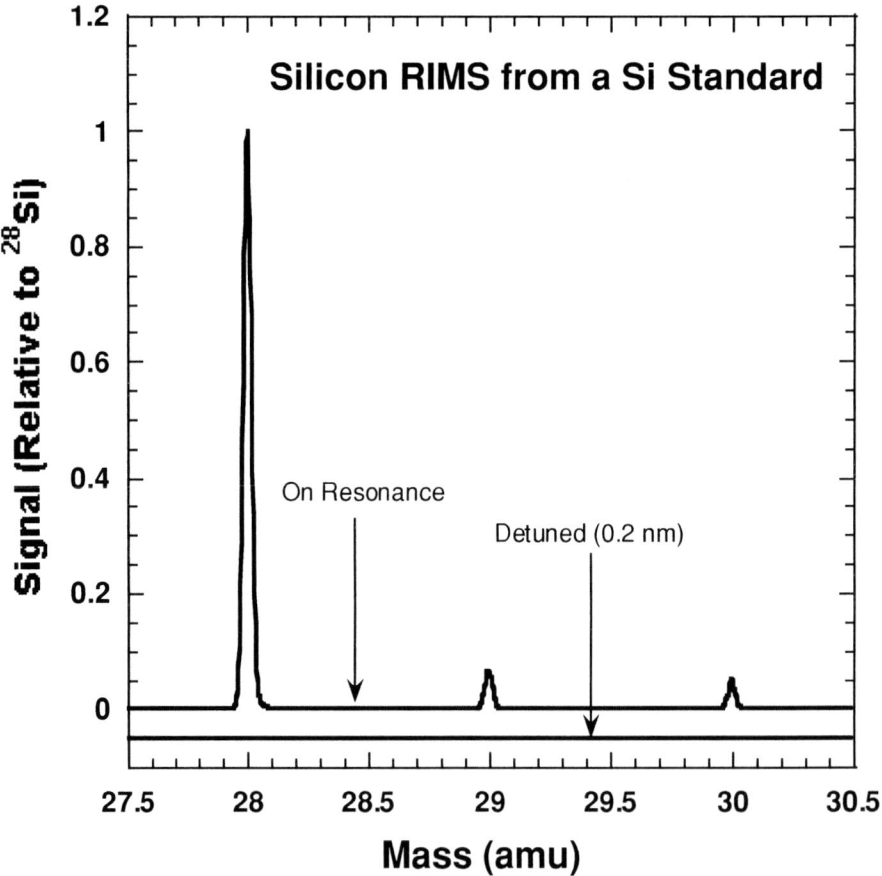

FIGURE 4. Displayed is a mass spectrum of Si isotopes from a Si standard. The upper trace is the signal level for resonant excitation of Si using the $^3P_0 \rightarrow {}^3D_1^{\circ} \rightarrow$ ionization continuum scheme displayed in figure 3. The lower curve is the signal level when the resonant laser is detuned by 0.2 nm.

One should note that the isotope ratios derived from the integrals of spectral peaks, like those in Figure 4, do not precisely represent the terrestrial abundances of ^{28}Si, ^{29}Si, and ^{30}Si. This occurs whether or not laser intensities are sufficient to strongly saturate the relevant transitions and is strongly dependent on the particular resonance ionization scheme chosen and on any wavelength mismatch between the photoionization laser wavelength and the atomic transition. The reasons for this mass fractionation are manifold and have been addressed in many RIS conferences, as well as in the terrestrial photoionization literature.

Instrument mass fractionation is not an unusual problem. All mass spectrometers have mass dependent transmission to some degree. We circumvent this difficulty by comparing all unknown spectra to that of spectra for known terrestrial abundances using the standard "δ" notation.

$$\delta^X Ba = \{(^X Ba/^{136}Ba)_{sample}/(^X Ba/^{136}Ba)_{standard} - 1\} \times 1000 \qquad (1)$$

Using this method, experimental mass specific useful yield effects can be minimized as long as the chosen standard accurately represents these effects for the unknown. Here we have chosen to compare the other isotopes of Ba to the ^{136}Ba isotope in order to emphasize the difference between the nucleosynthetic processing of AGB stars, which have a low neutron density, and the nucleosynthetic processing by supernovae for which the neutron density is extremely high. To understand the significance of the measured Ba isotope ratios, one must first recognize that the seemingly immutable isotope pattern of a Ba on earth is the result of a complex mixing process of all the sources that originally produced it. The reason Ba's isotope pattern is so consistent on earth is a result of the efficient mixing of planetary material within the solar system. Elsewhere in the universe, the Ba isotope pattern varies wildly. In particular, ^{136}Ba, a so-called protected isotope, can only be produced by stepwise single neutron addition and is likely to be enhanced relative to terrestrial abundance if produced in an AGB star. ^{138}Ba on the other hand can be produced in either a low neutron density or high neutron density environment and would be enhanced relative to terrestrial if it condensed in a SiC grain around a supernova star. Finally, it is important to understand the range of δ notation. Examination of equation 1 shows that if a particular isotope is completely depleted in a grain, then it will have $\delta = -1000$. SiC grain compositions that are consistent with terrestrial isotope ratios will have $\delta = 0$. Enhanced amounts of a particular isotope relative to ^{136}Ba will produce $\delta > 0$.

Mass spectra showing the Ba isotopic compositions of a terrestrial standard, an X-grain and a mainstream grain are compared in Fig. 5. The SiC grains are classified with the terms X and mainstream based on the isotopic abundances of the Si, C and N isotopes in the grain and represent an expectation based on these isotopes and modeling studies of stars. Mainstream grains, by far the most abundant type, are likely derived from AGB stars like our sun, while X-grains, representing about 1% of the grain population, are thought to have condensed around supernovae. The two grains displayed in Fig. 5 had their major element isotopic abundances measured with the Washington University ion microprobe, in order to classify their type.

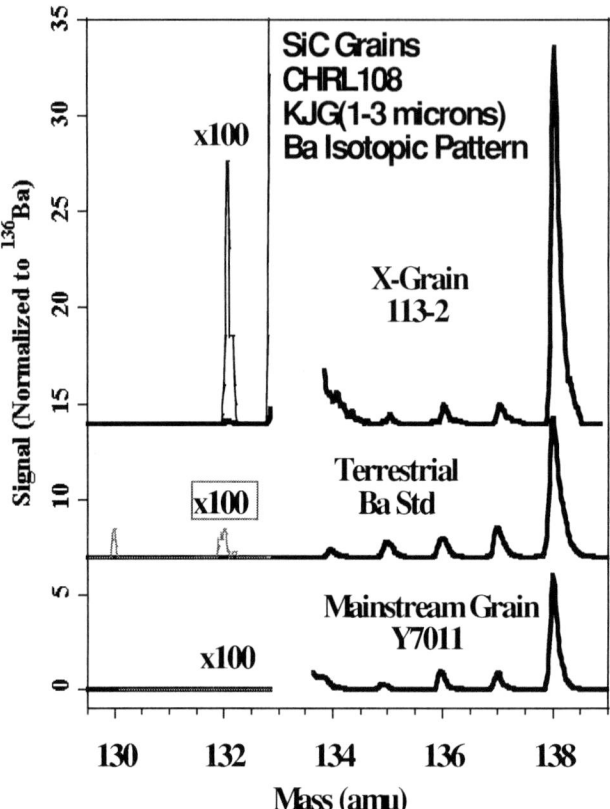

FIGURE 5. The Ba region of mass spectra of a terrestrial Ba standard, a Type X SiC grain and a mainstream SiC grain. Spectra are normalized to the height of ^{136}Ba. The large peak due to implanted Cs at mass 133 has been removed from the SiC spectra for clarity.

Unfortunately, in the Ba spectra, the large amount of Cs implanted during the C, N and Si isotopic measurements gives a large ^{133}Cs$^+$ peak from direct ionization of Cs, and makes ^{134}Ba determination impossible. It is immediately apparent in Fig. 5 that significant differences exist between the terrestrial, mainstream and X-grain patterns for Ba. The most significant difference is the large excesses in ^{138}Ba in the X-grain relative either to the terrestrial standard or to the mainstream SiC grain. While the p-process isotope, ^{132}Ba, relative to solar composition also shows enhancement, this represents very few total counts and is not statistically significant. The isotopic abundances can be found in Table 1 with error limits of 2 σ.

Clearly these two grains have very different isotope ratios. Remember the maximum variation on earth in these units would be <2. Here the difference between the two grains is over *2000* for δ^{138}Ba. Moreover the direction of the change is consistent with a supernova origin for the X-grain. However, the δ^{135}Ba and the δ^{137}Ba are not consistent with a conventional high-density nucleosynthetic mechanism. In conventional r-process nucleosynthesis, both of these isotopes would have enhanced abundances. These discrepancies along with our previously reported measurements of Mo, Zr, and Sr isotopes in these same grains has lead to a suggestion of a new type of nucleosynthetic process based on a compression wave in the supernova explosion.[28] The details of this mechanism are beyond the scope of this paper. It is clear, however, that RIMS studies of these SiC grains is providing new and interesting insights into the origin of the elements.[12, 29]

TABLE 1. Ba isotopic compositions in a SiC grain Type X and a mainstream grain in.

Grain	$\delta^{135}Ba$	$\delta^{137}Ba$	$\delta^{138}Ba$
Mainstream Grain (Y7011)	-554±144	-424±84	-242±54
X-Grain (113-2)	−495±280	−241±210	1890±89

CONCLUSION

Resonance ionization of laser-desorbed neutrals represents a powerful method for the isotopic and elemental analysis of micron sized particulates. Its combination of discrimination and high useful yield can allow unique analyses to be performed. As an example, we have shown that this technique enables isotopic analysis of trace amounts of Ba with concentrations as low as 1 ppm in ancient stardust particles of only a few microns in size.

ACKNOWLEDGMENTS

This work was supported by U. S. DOE, BES-Materials Sciences, under Contract W-31-109-ENG-38. The authors would also like to thank Roy Lewis, Andrew M. Davis, Sachiko Amari, and Robert Clayton for many helpful discussions and for providing the SiC and graphite samples.

REFERENCES

1   S. Amari and E. Zinner, *Nucl. Phys. A* **A621**, 99c-112c (1997).
2   T. J. Bernatowicz, *Astron. Soc. Pac. Conf. Ser.* **122**, 227-251 (1997).
3   T. J. Bernatowicz, *Meteorit. Planet. Sci.* **32**, 607 (1997).
4   L. R. Nittler, *AIP Conf. Proc.* **402**, 59-82 (1997).
5   L. R. Nittler, C. M. O. D. Alexander, X. Gao, et al., *Nucl. Phys. A* **A621**, 113c-116c (1997).
6   E. Zinner, *Meteorit. Planet. Sci.* **33**, 1341 (1998).
7   E. Zinner, *AIP Conf. Proc.* **402**, 3-26 (1997).
8   E. Zinner, *Science (Washington, D. C.)* **271**, 41-42 (1996).
9   E. Zinner, *Astron. Soc. Pac. Conf. Ser.* **99**, 147-161 (1996).
10   G. K. Nicolussi, A. M. Davis, M. J. Pellin, et al., *Lunar and Planetary Science* **28**, 1023-1024 (1997).
11   G. K. Nicolussi, A. M. Davis, M. J. Pellin, et al., *Lunar and Planetary Science* **28**, 1021-1022 (1997).
12   M. J. Pellin, W. F. Calaway, A. M. Davis, et al., *Lunar and Planetary Science* **XXXI**, 1934-1935 (2000).

[13]   M. J. Pellin, A. M. Davis, W. F. Calaway, et al., *Lunar and Planetary Science* **XXXI**, 1917-1918 (2000).
[14]   M. J. Pellin, A. M. Davis, R. S. Lewis, et al., *Lunar and Planetary Science* **XXX**, 1969-1970 (1999).
[15]   M. J. Pellin, G. K. Nicolussi, A. M. Davis, et al., in *Book of Abstracts, 217th ACS National Meeting, Anaheim, Calif., March 21-25*, 1999), p. NUCL-034.
[16]   A. M. Davis, M. J. Pellin, R. S. Lewis, et al., *Lunar and Planetary Science* **XXX**, 1976-1977 (1999).
[17]   M. J. Pellin and G. K. Nicolussi, in *McGraw-Hill Yearbook of Science & Technology 1999*, edited by T. P. Martin (McGraw-Hill, New York, 1998), p. 318-320.
[18]   M. J. Pellin and G. K. Nicolussi, *Proc. SPIE-Int. Soc. Opt. Eng.* **3270**, 148-157 (1998).
[19]   G. K. Nicolussi, M. J. Pellin, A. M. Davis, et al., *Lunar and Planetary Science* **29**, 1415-1416 (1998).
[20]   G. K. Nicolussi, M. J. Pellin, R. S. Lewis, et al., *Geochim. Cosmochim. Acta* **62**, 1093-1104 (1998).
[21]   G. K. Nicolussi, M. J. Pellin, R. S. Lewis, et al., *Astrophys. J.* **504**, 492-499 (1998).
[22]   G. K. Nicolussi, M. J. Pellin, R. S. Lewis, et al., *Phys. Rev. Lett.* **81**, 3583-3586 (1998).
[23]   G. K. Nicolussi, M. J. Pellin, W. F. Calaway, et al., *Anal. Chem.* **69**, 1140-1146 (1997).
[24]   G. K. Nicolussi, A. M. Davis, M. J. Pellin, et al., *Science (Washington, D. C.)* **277**, 1281-1283 (1997).
[25]   Z. Ma, R. N. Thompson, K. R. Lykke, et al., *Rev. Sci. Instrum.* **66**, 3168-3176 (1995).
[26]   I. V. Veryovkin, W. F. Calaway, and M. J. Pellin, *Second. Ion Mass Spectrom., SIMS XII, Proc. Int. Conf., 12th*, 337-340 (2000).
[27]   G. K. Nicolussi, M. J. Pellin, K. R. Lykke, et al., *Surf. Interface Anal.* **24**, 363-370 (1996).
[28]   B. S. Meyer and D. D. Clayton, *Lunar and Planetary Science* **XXXI**, 1458-1459 (2000).
[29]   M. J. Pellin, A. M. Davis, M. R. Savina, et al., *Lunar and Planetary Science* **32**, 2125-2126 (2001).

SESSION V

Laser resonance ionization mass spectrometry for failed fuel detection and location in the experimental fast reactor JOYO.

Hideki HARANO*, Shoichi NOSE*, Kazuhiro ITO*,
Kenichi WATANABE**, Tetsuo IGUCHI**

Oarai Engineering Center, Japan Nuclear Cycle Development Institute
4002 Narita, O-arai, Ibaraki 311-1393 Japan
**Department of Nuclear Engineering, School of Engineering, Nagoya University*
Furo-cho, Chikusa-ku, Nagoya, 464-8603, Japan

Abstract. For the improvement of the failed fuel detection and location technique base on a tagging gas method, we are developing a RIMS prototype system and plan to demonstrate its performance at the experimental fast reactor JOYO. We report the basic design of the system and preliminary experimental results for trace Xe isotopic analysis.

INTRODUCTION

Immediate and precise identification of failed fuel assemblies which should be removed are quite significant in fast reactors for the improvement of their safety and reliability as well as plant availability which has a great influence on the economical efficiency. This identification is called FFDL, which is the initials of failed fuel detection and location. And a tagging gas method, shown in Figure 1, is one of the well-established techniques of FFDL.

The tagging gases consist of stable isotopes of Kr and Xe and they are enclosed in fuel pins with different isotopic composition for each fuel assembly. If fuel failure happens, the tagging gas leaks out and goes through the liquid sodium coolant to the cover-gas plenum. And finally the tagging gas is diluted to the level of 0.5 ppb with Ar cover gas. Then the isotopic composition of this diluted tagging gas is measured to identify the failed assembly.

Conventional mass spectrometry is currently utilized for the isotope analysis after the pretreatment by an activated charcoal cryogenic adsorption method: Kr and Xe concentration and subsequent Ar-gas separation and He-gas substitution for suppression of isobaric interference. Because of this pretreatment, a half day is required for the analysis and large equipments are necessary.

So we have proposed to apply resonance ionization mass spectrometry (RIMS) to this analysis. For its excellent features of high sensitivity and no isobaric interference, we expect that RIMS makes this pretreatment step unnecessary. This permits a reduction of the equipment and speeding up of FFDL.

CP584, *Resonance Ionization Spectroscopy 2000: 10th Int'l. Symp.*, edited by J. E. Parks and J. P. Young
© 2001 American Institute of Physics 0-7354-0024-5/01/$18.00

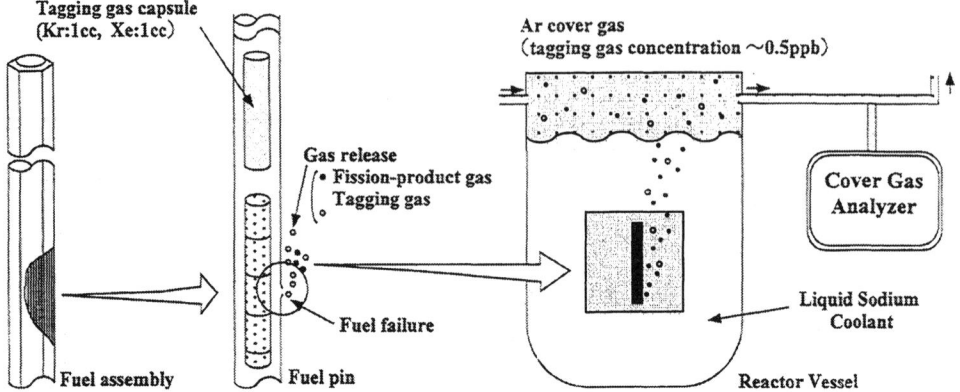

FIGURE 1. Scheme of the tagging gas method.

We are developing the RIMS prototype system for fast reactors for the improvement of FFDL technique based on the tagging gas method. And we plan to demonstrate its performance on the experimental fast reactor JOYO. In this paper, we show the basic design of the RIMS prototype system. And we also report the current status of the system and preliminary experimental results for Xe detection.

REQUIREMENTS

In order to obtain design requirements, basic experiments are performed using a laboratory-scale RIMS device at the nuclear engineering research laboratory (NERL) of the university of Tokyo. The device consists of a narrow-band tunable dye laser system producing 500 μJ pulses of the wavelength of resonance ionization of Kr(216.60nm) and Xe(256.02nm) at 10 Hz, and a reflectron-type time-of-flight mass spectrometer for isotope analysis. The details of the experiments will be reported in ref. [1].

Mass spectra showed very good isotopic ratios with excellent reproducibility. By adding up the accumlated results for 30 minutes, the detection limit was 100 ppb. Since the tagging gas concentration in the cover gas is 0.5 ppb, the detection efficiency must be improved 200 times. We have confirmed that the sensitivity will be enhanced by about one order of magnitude by employing a pulsed supersonic valve for sample gas injection[2]. Then in order to replenish the sensitivity shortage, the laser power must be increased up to 10 mJ at the above UV wavelengths. Fortunately this is obtainable by the use of largest-scale commercial tunable lasers.

And the whole system should have high stability and reliability, However, the sensitivity was a little bit unstable probably due to various instabilities of lasers. So here, the other stable system shown below is combined for sensitivity calibration of RIMS.

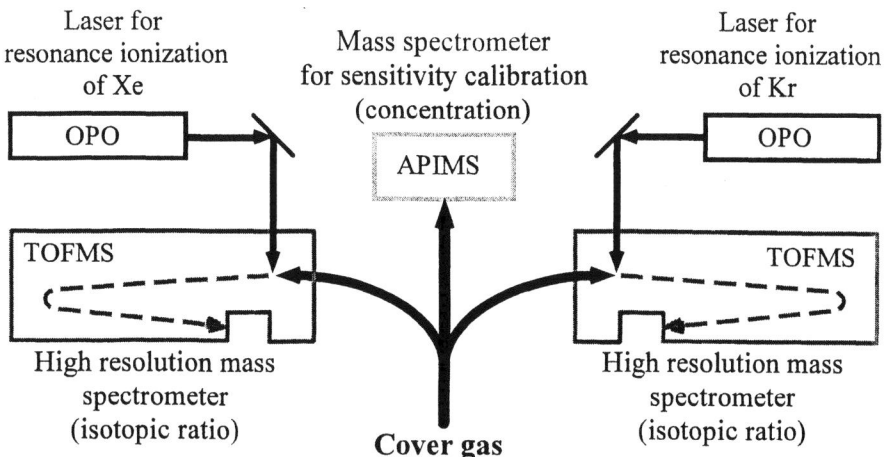

FIGURE 2. RIMS-based cover gas analysis system at the JOYO.

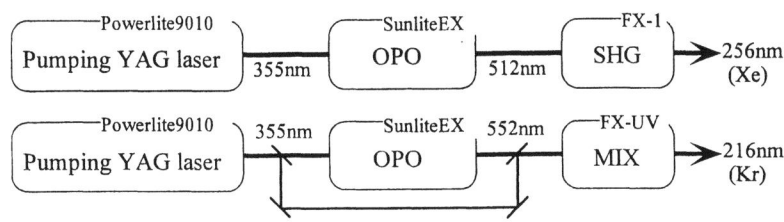

FIGURE 3. Lasers for the RIMS prototype.

PROTOTYPE SYSTEM

Based on the above information, the basic design of the prototype system has been determined as below. The whole system configuration of the prototype is shown in Figure 2.

Lasers

Pulsed OPO (optical parametric oscillator) lasers are utilized for resonance ionization in the prototype. The main reason for choosing the complete solid state systems is dye lasers need combustible solvents such as alcohol, and we do not want to put it near the reactor. The schematic diagrams of the systems are shown in Figure 3. These systems are produced by Continuum, Inc. The OPOs (SunliteEX) consist of oscillators and two stages of amplifiers, pumped by the third harmonics of Nd:YAG lasers. For the generation of the 256 nm for the Xe resonance ionization, the OPO outputs are doubled in frequency by the SHG crystal. The laser beam for 216 nm generation for Krypton is produced by mixing the OPO outputs with the third harmonics of Nd:YAG lasers. The outputs are more than 10 mJ which satisfy the design requirement.

Mass Spectrometers

We chose the reflectron-type time-of-flight mass spectrometer produced by R.M.Jordan Company, Inc. which is almost the same as the system at the univ. of Tokyo. The pulsed supersonic valve is employed as the standard way of sample gas injection. This valve provides a gas jet of 55 microsecond pulsewidth at 10 Hz repetition rate and operates synchronized with laser pulse injection. The ions are detected by a Z-Gap type MCP with high gain. The obtained 10 Hz TOF signals are completely accumulated by a digital signal averager optimized for fast TOF data aquisition (FastFlight, Perkin Elmer, Inc.). And we utilize an oil-free vacuum system consisting of turbo molecular and scroll pumps in a tandem configuration so as to avoid an oil contamination of the reactor exhaust gas system.

Sensitivity Calibration

In order to compensate the RIMS sensitivity fluctuation due to laser instabilities, an atmospheric pressure ionization mass spectrometer (APIMS) was employed for monitoring the absolute concentration of Kr and Xe. The APIMS has very excellent features of high detection sensitivity with remarkable long-term stability. And this is well-established technology with high reliability and widely used in semiconductor processing where ultrahigh purity control is needed. Since the APIMS has unavoidable isobaric interference, only the peaks of the isotopes with no isobaric interference, such as ^{82}Kr and ^{129}Xe, are monitored to obtain the absolute concentration.

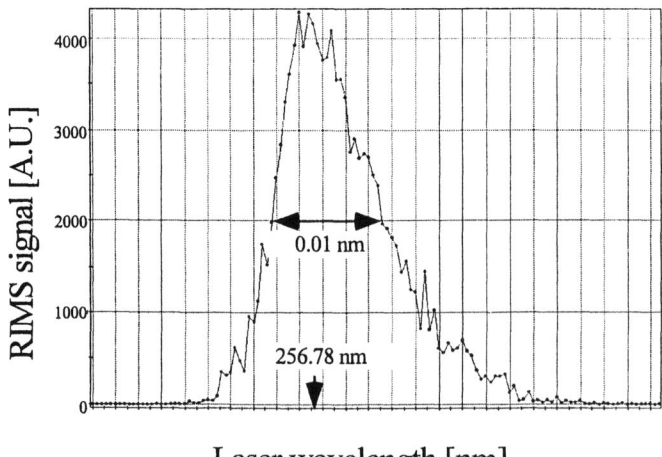

Laser wavelength [nm]

FIGURE 4. RIMS signal response in laser wavelength scanning.

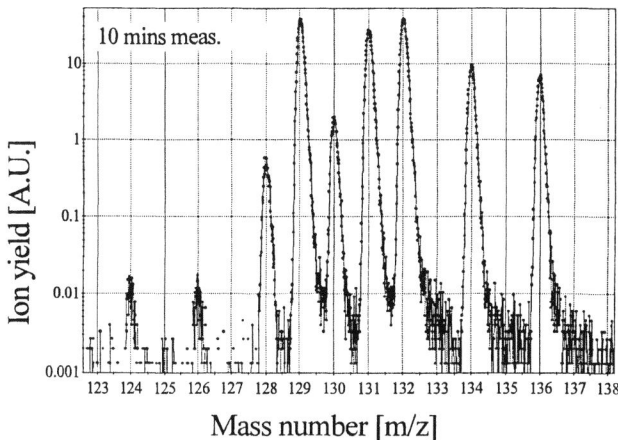

FIGURE 5. Xe mass spectrum obtained by the prototype.

TESTS

The construction and adjustment of the system are in progress. We report the preliminary experimental results for Xe detection. The sample gas is Ar gas containing 30 ppb of Xe of natural isotopic abundance, which is injected by the PSV at 10 Hz into the the ionization area of TOFMS. The laser beam was focused by a fused-silica plano-convex lense of 30 cm focal length. Figure 4 shows the change of RIMS signals when the wavelength scanning of the laser was made. The detection efficiency was peaked at the resonance wavelength of Xe. Figure 5 shows the excellent mass spectrum obtained at the resonance wavelength.

SUMMARY

We are developing the RIMS-FFDL prototype system for fast reactors at the experimental fast reactor JOYO. The basic design of the whole system was determined based on the experiments at the university of Tokyo. We are constructing and tuning up the system and the excellent mass spectrum was obtained for Ar gas containing 30 ppb Xe in preliminary experiments.

REFERENCES

1. Watanabe,K., Ogita,T., Uritani,A., Iguchi,T. and Harano,H., "Development of Failed Fuel Detection and Location Technique using Resonance Ionization Mass Spectrometry," submitted to Journal of Nuclear Science and Technology.
2. Watanabe,K., Iguchi,T., Ogita,T., Watanabe,T., Uritani,A., Kawarabayashi,J., Harano,H. and Nose,S., "Development of Failed Detection and Location Technique using Resonance Ionization Mass Spectrometry: Improvement of the Detection Limit with Pulsed Supersonic Valve",Proc.14 th W.S. on Radiation Detectors and Their Uses, KEK Proc., 00-8(2000).

Which Isotope Effects Can Be Observed In Heavy Rydberg Atoms Using Resonance Ionization Mass Spectroscopy?

Peter Bisling, Karen Ong, Bernd Neidhart, and Claus Weitkamp

GKSS–Forschungszentrum Geesthacht GmbH, Postfach 11 60, D–21494 Geesthacht, Germany

Abstract. Relative optical isotope shifts are observed in mercury at different energy levels by multiphoton resonance excitation combined with mass spectrometric detection. The sign of isotope shifts changes from negative to positive for states of higher energy. It is negative up to the total excitation energy of $Hg[6s(^2S)8s\,^3S_1]$. It changes to positive for higher energy levels as measured for example in $Hg[6s(^2S)nd\,^3D_2]$ with $7 \le n \le 60$.

INTRODUCTION

Any element with more than one isotope has energy levels that are shifted relative to each other because of two different isotope effects. Firstly, differences in nuclear mass induce an isotope shift (IS), usually with a slightly higher energy of the heavier isotope, which is then defined as positive. Secondly, differences in nuclear volume of heavy atoms mostly cause an opposite effect which is consequently called negative. Therefore elemental isotope analysis profits a lot from resonance ionization mass spectrometry (RIMS), particularly when interfering background from complex matrices cannot be eliminated by standard sample preparation methods. Previous work on RIMS of heavy Rydberg atoms is continued here [1] in order to investigate some implications of isotopic resonance effects for isotope analysis.

RIMS experiments are described in which various excitation schemes for mercury isotopes are tested. Three–color experiments are used for the measurement of ISs at bound electronic states. Two–color laser radiation is used for the excitation of the intermediate and high–lying bound Rydberg states. They are detected by pulsed field ionization (PFI), which is much more efficient than direct photoionization because PFI cross sections are strongly enhanced. New data on ISs at electronic and Rydberg levels are presented and investigated by varying the bandwidth of the exciting radiation.

EXPERIMENTAL

The experimental setup has been described recently [2]. Briefly, a vacuum apparatus with three differentially pumped pressure stages houses a reflectron time–of–flight mass spectrometer (TOF). Its typical mass resolution of $m/\Delta m \approx 2000$ at full width at half maximum is adjusted in each measurement to be sufficient for the given require-

CP584, *Resonance Ionization Spectroscopy 2000: 10th Int'l. Symp.*, edited by J. E. Parks and J. P. Young

ment of isotope selective mass detection. A mixture of Hg vapor with He at room temperature is expanded from a total stagnation pressure of 2800 hPa into the vacuum through a pulsed nozzle valve.

Two pulsed laser systems provide the radiation for resonance excitation in the modified Wiley–McLaren ion source of the TOF. The first system is composed of a frequency–tripled Nd:YAG pump laser and a tunable, frequency–doubled dye laser, whose wavelength is set to $\lambda_{air,1} = 253.65$ nm. The bandwidth of its fundamental line is about $\Delta \tilde{v}_{L1} = 0.2$ cm^{-1} ($\Delta v_{L1} = 6$ GHz) and the laser pulse energy is between 3 and 7 μJ. This radiation excites the $Hg[6s(^2S)6p\,^3P_1]$ resonance (1), which is the first step in all four excitation schemes A – D of this work.

$$Hg(6s^2\,^1S_0) \xrightarrow{\;253.65\text{ nm}\;} Hg[6s(^2S)\,6p\,^3P_1] \longrightarrow \left\{ \begin{array}{l} A \\ B \\ C \\ D. \end{array} \right. \tag{1}$$

The second laser system comprises a XeCl excimer laser and a further dye laser; its wavelength is tuned for double–resonance excitation of the mercury atoms. The characteristics of this radiation are similar to those of the first excitation step. The subsequent excitation steps to three different electronic states in Hg (schemes A, B, and C) are described by Eqn. (2).

$$\left. \begin{array}{l} A \xrightarrow{\;289.36\text{ nm}\;} Hg[6s(^2S)\,8s\,^3S_1] \\[4pt] B \xrightarrow{\;285.69\text{ nm}\;} Hg[6s(^2S)\,8s\,^1S_0] \\[4pt] C \xrightarrow{\;265.20\text{ nm}\;} Hg[6s(^2S)\,7d\,^3D_2] \end{array} \right\} \xrightarrow{\;532\text{ nm}\;} Hg^+ + e^-. \tag{2}$$

Because of moderate laser pulse energy these excitation schemes need the radiation of the second harmonic of the Nd:YAG laser at 532 nm as a third color in order to provide appreciable continuum photoionization efficiency for mass spectrometric detection. Using these excitation schemes, the laser pulses are adjusted to cross the ion source simultaneously in the time domain.

Alternately, high–lying bound Rydberg states (scheme D) are excited according to Eqn. (3)

$$D \xrightarrow[\substack{223.9\text{ nm} \\ 223.4\text{ nm}}]{} Hg[6s(^2S)\,nd\,^3D_2] \xrightarrow{\;PFI\;} Hg^+ + e^-. \tag{3}$$

Their excitation is delayed between 70 and 90 ns with respect to the first excitation step. If the wavelength of the second laser is now scanned between $\lambda_{air,2} = 223.9$ and 223.4 nm at reduced bandwidth of the fundamental ($\Delta \tilde{v}_{L2} < 0.04$ cm^{-1} corresponding to $\Delta v_{L2} < 1$ GHz), the $nd\,^3D_2$ series is excited in Hg isotopes for $33 < n < 75$ at field–

free conditions in the ion source. After a further delay between 1 and 2 μs with respect to the second laser pulse, these Rydberg states are ionized by pulsed electrical fields which are applied to the electrodes of the ion source.

RESULTS

As is well known, the isotope components of the Hg[$6s(^2S)6p\,^3P_1$] intermediate show a large negative IS [3]. This is depicted in Figure 1, where the IS relative to ^{200}Hg is plotted versus nucleon and neutron number A and N measured, e.g., with a two–color photoionization scheme [2]. The ISs of the odd isotopes are determined by the center of gravity (CoG) of the hyperfine structure (HFS) components. A staggering in the optical shifts between even and odd isotopes is obvious, suggesting a larger nucleus for an odd than for an even number of neutrons. In summary, the nuclear volume effect dominates, and thus the field inside the nucleus is not a pure Coulomb field.

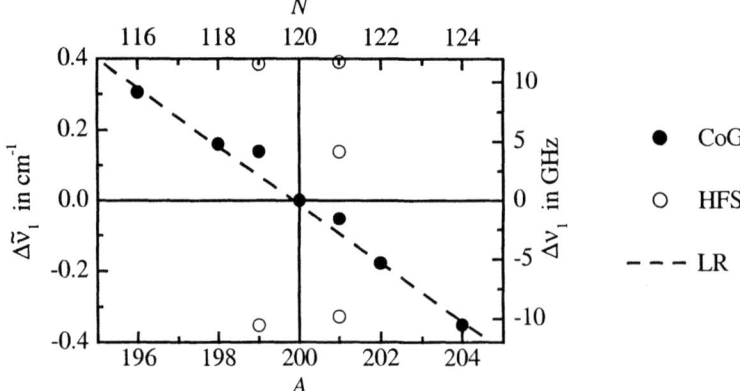

FIGURE 1. Negative IS $\Delta\tilde{\nu}_1$ for the $6s(^2S)6p\,^3P_1$ level [2, 3] relative to ^{200}Hg versus nucleon and neutron number (A and N). The hyperfine structure (HFS) splitting of odd–mass isotopes is represented by its center of gravity (CoG). The dashed line of the linear regression (LR) through the IS of even–mass isotopes emphasizes an odd–even staggering.

The ISs of this work for states of higher energy, i.e. $8s\,^3S_1$, $8s\,^1S_0$, and $7d\,^3D_2$ from excitation schemes A, B, and C, are shown in Figure 2. Here all Hg[$6s(^2S)6p\,^3P_1$] isotopes are more or less coherently excited due to the laser line bandwidth and atomic line power broadening. These shifts are about an order of magnitude smaller and there is a change of sign from negative to positive, which starts at the energy of the $6s(^2S)8s\,^1S_0$ state. In order to carefully confirm this observation, the positive IS is measured at various levels of the $6s(^2S)nd\,^3D_2$ Rydberg series (Figure 3), which allows enhanced detection by delayed PFI in spite of soft radiation conditions. Table 1 summarizes the results on ISs. It is obvious at levels of higher excitation energy that the addition of up to three pairs of neutrons influences the ISs less by the volume effect than by the mass effect and that the core field for the corresponding electronic states behaves more like a Coulomb field.

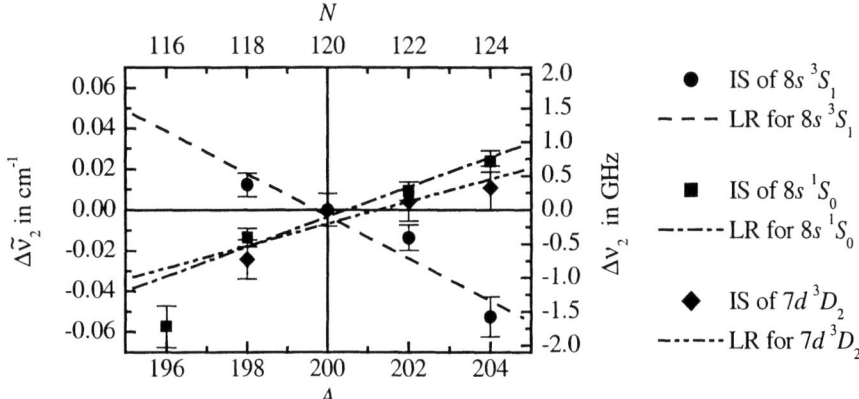

FIGURE 2. Negative and positive ISs $\Delta \tilde{\nu}_2$ for electronic states of even–mass mercury isotopes relative to ^{200}Hg versus nucleon and neutron number (A and N). Data result from double–resonance schemes A, B, and C by three–color RIMS with $\Delta \tilde{\nu}_{L1, L2} < 0.2$ cm^{-1} of the fundamental lines of both dye lasers.

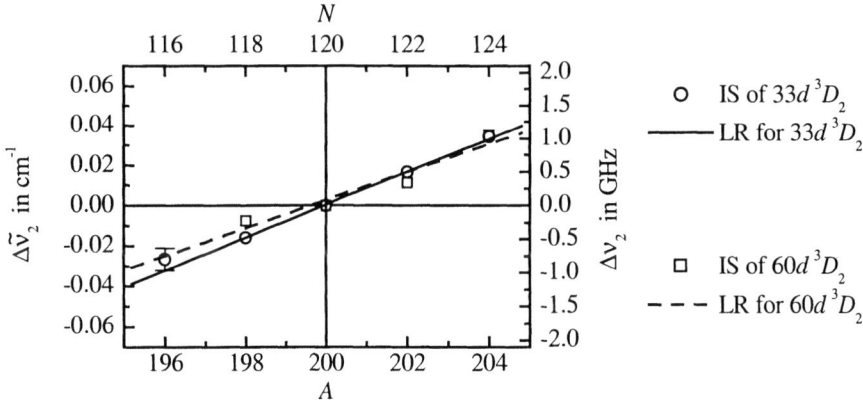

FIGURE 3. Positive ISs $\Delta \tilde{\nu}_2$ for Rydberg states of even–mass mercury isotopes relative to ^{200}Hg versus nucleon and neutron number (A and N). Data result from double–resonance scheme D by two–color RIMS with $\Delta \tilde{\nu}_{L1} < 0.2$ cm^{-1} and $\Delta \tilde{\nu}_{L2} < 0.04$ cm^{-1}.

TABLE 1. Isotope Shifts of Even Mercury Isotopes.

Atomic State	(Scheme)	Wavenumber in cm^{-1}	Isotope Shift
$6p\,^3P_1$		39412.3	Negative[a]
$8s\,^3S_1$	(A)	73961.3	Negative
$8s\,^1S_0$	(B)	74404.6	Positive[b]
$7d\,^3D_2$	(C)	77107.9	Positive
$33d\,^3D_2$	(D)	84061.4	Positive
$41d\,^3D_2$	(D)	84108.5	Positive
$60d\,^3D_2$	(D)	84150.4	Positive

[a]Ref. [2, 3]
[b]See Also Ref. [4].

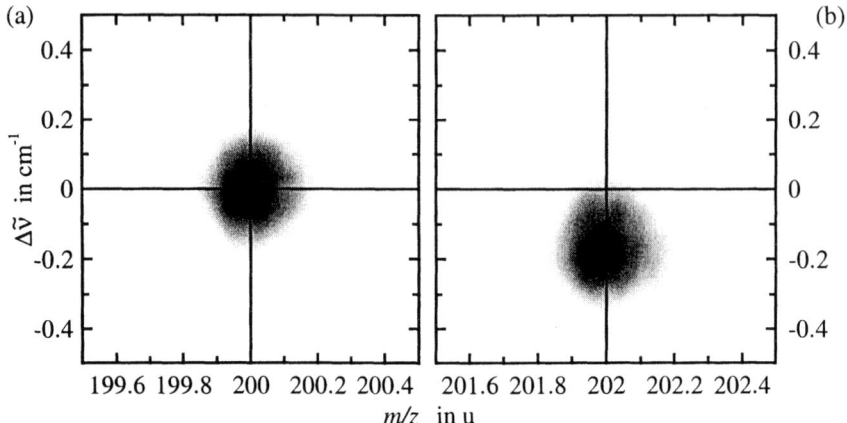

FIGURE 4. Negative IS $\Delta\tilde{\nu}$ for ^{202}Hg[$2s(^2S)60d\,^3D_2$] (b) relative to ^{200}Hg[$2s(^2S)60d\,^3D_2$] (a) at 84150.40(3) cm^{-1}. Data result from two delayed PFI measurements using the double–resonance scheme D with $\Delta\tilde{\nu}_{L1,L2} < 0.04$ cm^{-1} for both the $6s(^2S)6p\,^3P_1$ intermediate and Rydberg excitations.

These results allow optimizing some experimental conditions such as to get favorable for analytical applications like isotope analyses. It is easy to reduce the laser bandwidth for the excitation according to Eqn. (1) to a value low enough to allow selecting only one isotope, e.g. ^{200}Hg or ^{202}Hg, because $\Delta\tilde{\nu}_1^{200\,202} \equiv \Delta\tilde{\nu}_1(202) - \Delta\tilde{\nu}_1(200) = -0.177(2)$ cm^{-1} at $6s(^2S)6p\,^3P_1$. Although this shift might be slightly reduced by the second excitation, i.e. by $\Delta\tilde{\nu}_2^{200\,202} \equiv \Delta\tilde{\nu}_2(202) - \Delta\tilde{\nu}_2(200) = 0.013(4)$ cm^{-1} at $6s(^2S)60d\,^3D_2$ in Figure 3, only one of all seven isotopes is detected. Two separate measurements of ^{200}Hg or ^{202}Hg single isotope detection are shown in Figure 4 (a) and (b). Here both lasers are adjusted to narrow–band radiation for the double–resonance excitation of scheme D. Any of the other Hg isotopes or background ions disappear completely in the TOF. Assuming that the contributions to the IS of each excitation step are separately known from the previous measurements, a common relative wavenumber scale is applied to Figure 4 (a) and (b) with $\Delta\tilde{\nu}^{200\,202} \equiv \Delta\tilde{\nu}_1^{200\,202} + \Delta\tilde{\nu}_2^{200\,202} = -0.164(6)$ cm^{-1}. The conclusion is that these and related phenomena are useful for the design of analytic isotope determinations with RIMS.

REFERENCES

1. Bisling, P., Dederichs, J., Neidhart, B., and Weitkamp, C., "High–Lying Bound Rydberg States of Excited Hg($6s6p\,^3P_1$) Atoms from Two–Color Resonance Ionization Mass Spectroscopy" in *Resonance Ionization Spectroscopy*, edited by J. C. Vickerman et al., AIP Conference Proceedings 454, New York: American Institute of Physics, 1998, pp. 33–36.
2. Bisling P., Dederichs, J., Neidhart, B., and Weitkamp, C., *Fresenius J. Anal. Chem.* **364**, 79–86 (1999).
3. Schweitzer, Jr., W. G., *J. Opt. Soc. Am.* **51**, 692–693 (1961).
4. Dyer, P., Baldwin, G. C., Kittrell, C., Imre, D. G., and Abramson, E., *Appl. Phys. Letters* **42**, 311–313 (1983).

Optimization of Trace Element Analysis using Resonant Laser Ablation

K. Watanabe and T. Iguchi

Department of Nuclear Engineering, Graduate School of Engineering, Nagoya University
Furo-cho, Chikusa-ku, Nagoya 464-8603, Japan

Abstract. Resonant Laser Ablation (RLA), which combines Laser Ablation (LA) and Resonance Ionization Spectroscopy (RIS) simultaneously with a single laser, can be used as a simple analytical technique of trace elements with high sensitivity and elemental (isotopic) selectivity for solid samples. However, the elemental selectivity falls off in the higher laser power range because of an increase of an ion yield produced non-resonantly in laser ablation process. To enhance an ion yield produced through the resonant ionization process, the incident laser is split into two beams for LA and RIS, respectively. In the present study, we have developed a simple theoretical model to simulate the RLA process and checked the validity of the model by analyzing some experiments to detect a trace of Al. The dependence of the elemental selectivity and the detection limit on the incident laser power is mainly discussed to optimize the trace element analytical technique with RLA.

INTRODUCTION

Resonance Ionization Spectroscopy (RIS) has recently got into the spotlight as an innovative technology for ultra high sensitive trace element analysis with elemental (isotopic) selectivity and is being developed into more extensive application in various engineering fields with improvement of tunable laser performance. Resonance Ionization Mass Spectroscopy (RIMS), which combines RIS with mass spectroscopy, is used as useful mass spectroscopy without isobaric interference because RIS can make the elemental selective ionization. Resonant Laser Ablation (RLA), which combines Laser Ablation (LA) and RIS simultaneously with a single laser, is a simple analytical technique of solid samples based on resonance ionization process, on which the possibility of many kinds of application has been so far reported [1-4].

In the present study, we are trying to apply this RLA to the detection of trace long-lived radionuclides and/or stable nuclides produced in nuclear reactor materials from the viewpoint of new neutron dosimetry technique as an alternative of conventional activation analysis. Other mass spectrometry techniques, such as ICP-MS and AMS, are also promising candidates for this application, but they have difficulty on isobaric interference, which can be easily avoided in RIMS.

RLA, however, has a problem that the elemental selectivity falls off in the higher laser power range because of an increase of an ion yield produced non-resonantly in LA process. To enhance an ion yield produced through the resonant ionization process, it is useful that the incident laser is split into two beams for LA and RIS, respectively.

CP584, *Resonance Ionization Spectroscopy 2000: 10th Int'l. Symp.*, edited by J. E. Parks and J. P. Young

The elemental selectivity depends on the ratio of neutral atoms to ions produced in LA process. To consider the LA laser power dependence of the elemental selectivity, we have developed a simple theoretical model on the neutral atom and ion yields in RLA process.

In this paper, we focus on the optimized detection of a trace of Al with RLA, practically aiming at the detection of a long-lived radionuclides [26]Al produced by [27]Al(n,2n) reaction in high energy neutron fields, which would be affected by the isobaric interference of [26]Mg. Through comparison between experiment and calculation, we mainly discuss the dependence of the elemental selectivity and the detection limit on the incident laser power to optimize the trace element analytical technique based on RLA.

OUTLINE OF THEORETICAL MODEL

This theoretical model is coupled the vaporization and ionization process on the sample surface [5]. In the vaporization process of this model, the particle flux balance is calculated between the vapor and solid phases of sample materials through the ablated sample surface. We can defined the net number of vaporized particles through the ablated sample surface, $f_{net}(T_s, N_v)$, as follows:

$$f_{net}(T_s, N_v) = f_{out}(T_s) - f_{in}(T_s, N_v), \qquad (1)$$

where $f_{out}(T_s)$ and $f_{in}(T_s, N_v)$ are the particle fluxes emitted out from and flowed into the sample surface, respectively, as a function of T_s; the temperature of the sample surface and N_v; the vapor density near the sample surface. On the sample surface temperature T_s, we solve the equation of heat conduction. On the vapor density near the sample surface N_v, we treat the motion of vapor through the equations for gas dynamics.

The ionization probability in the net particle flux consists of two components from resonant and non-resonant (or thermal) ionization process.

EXPERIMENTAL SETUP

Figure 1 shows a schematic of our experimental system, which consists of one tunable dye laser pumped by a pulsed Nd: YAG laser and reflectron TOF mass spectrometer with a field-free drift region approximately 150 cm long and a microchannel plate (MCP) ion detector. The RIS laser was delayed to the LA laser for 50 ns because it needs a finite time for diffusion of the atoms vaporized by LA. In basic experiments to investigate a property of LA process, Nd: YAG laser was used. The bandwidth of the dye laser was 0.14 cm^{-1}, the laser pulse duration 10 ns and the repetition rate 10 Hz, respectively. The laser beams were focused at an ionization area with a diameter of 100 μm. The typical laser output energy was 500 μJ/pulse ($\approx 6 \times 10^8$ W/cm^2) for RIS and 50 μJ/pulse ($\approx 6 \times 10^7$ W/cm^2) for LA.

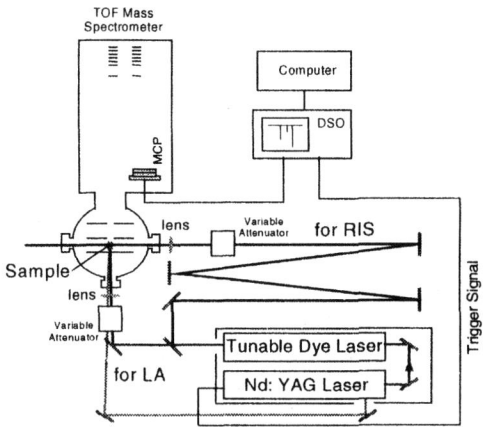

FIGURE 1. A schematic of our experimental system

As a sample, kovar (Ni: 29 %, Co: 17 %, Mg: <0.1 %, Al: <0.1 %, Fe: Balance) rod was used. The ionization schemes for a variety of elements used in this paper are listed in Table 1. As their schemes, two-photon excitation followed by one photon ionization at the same wavelength, that is, the "2+1" transition was adopted because it is quite simple and suitable for one tunable laser.

TABLE 1. RIS schemes

Element	Wavelength [nm]	Lower Level	Energy [cm⁻¹]	Excited Level	Energy [cm⁻¹]
Fe	499.036	$3d^7(^4F)4s\ a^5F_5$	6928.266	$3d^7(^4F)5s\ e^5F_5$	47005.508
	531.833	$3d^7(^4F)4s\ a^5F_3$	7728.056	$3d^6(^5D)4s(^6D)5s\ e^5D_2$	45333.874
Ni	489.564	$3d^9(^2D)4s\ ^1D_2$	3409.937	$3d^9(^2D_{3/2})5s\ ^2[3/2]$	44262.599
Co	484.270	$3d^8(^3F)4s\ b^4F_{9/2}$	3482.82	$3d^8(^3F)5s\ e^4F_{9/2}$	44782.13
Al	496.623	$3s^2(^1S)3p\ ^2P^o_{1/2}$	0	$3s^2(^1S)5p\ ^2P^o_{1/2}$	40271.978

The elemental selectivity depends on the ratio of neutral atoms to ions produced in LA process. First, to understand a basic property of LA process, the LA laser power dependence of the neutral atom and ion yield was measured by using Nd: YAG laser for LA and compared with the theoretical calculation. The neutral atoms produced in LA process are ionized through RIS process by using a tunable dye laser.

Secondly, to evaluate the performance of the trace element analysis with RLA, that is, using a single dye laser shared for LA and RIS, we measured the dependence of the elemental selectivity and the detection limit for Al in a kovar sample on the laser power for LA and compared with the calculation.

RESULTS AND DISCUSSION

The elemental selectivity depends on the ratio of neutral atoms to ions produced in LA process. The elemental selective ionization is carried out only in RIS process. The ions produced non-resonantly in LA process may occur isobaric interference. Therefore, it needs to keep low the ion yield produced in LA process to obtain the high elemental selectivity. Figure 2 shows the laser power dependence of the neutral atom and ion yields of Fe, Co and Ni produced from the kovar sample in LA process. The lines represent the calculation results. From Fig. 2, it is found that the present model calculation gives quite good agreement with the experiment on the laser power dependence of the neutral atom and ion yields, except for the Fe ions. The threshold laser power to produce neutral atoms in LA process is determined by thermal properties of a sample, for example, the boiling point, the latent heat, the thermal conductivity, and so on. Therefore, all elements have the same threshold laser power of the neutral atom yield. On the other hand, the threshold laser power of the ion yield depends on an ionization potential and is, therefore, higher than that of the atom yield. It is considered reasonable that the optimal laser power range of LA for the trace element analysis with an isobar should be from the threshold laser power of the neutral atom yield to that of the ion yield occurring isobaric interference, which can be well predicted by the present model calculation. On the discrepancy of the Fe ion results, the ablated Fe neutral atoms may be ionized by Nd: YAG laser resonantly because the wavelength of second harmonics of Nd: YAG laser (532.1 nm) is near that of the Fe two-photon excitation from $(3d^7(4F)4s)a^5F_3$ to $(3d^6(5D)4s(6D)5s)e^5D_2$ (531.833 nm).

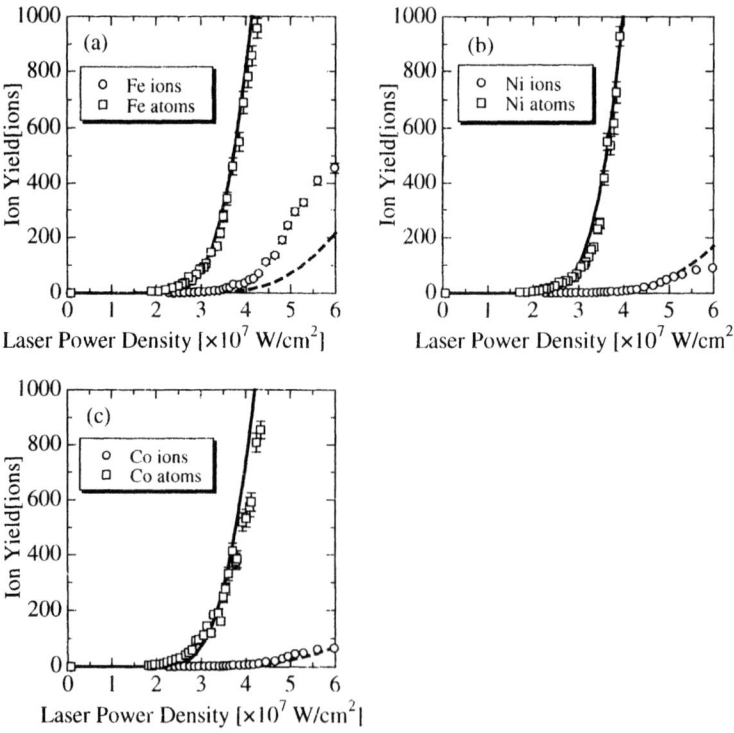

FIGURE 2. The LA laser power dependence of the neutral atom and ion yields of Fe (a), Ni (b) and Co (c) produced in LA process.

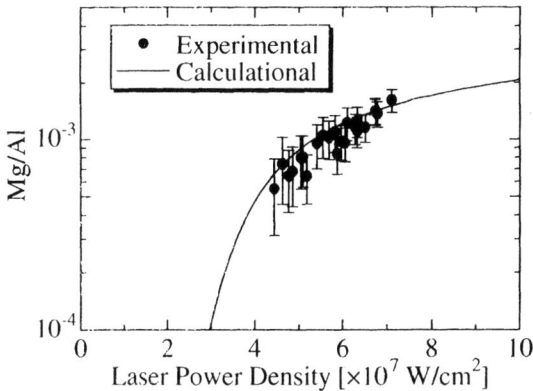

FIGURE 3. The LA laser power dependence of the elemental selectivity.

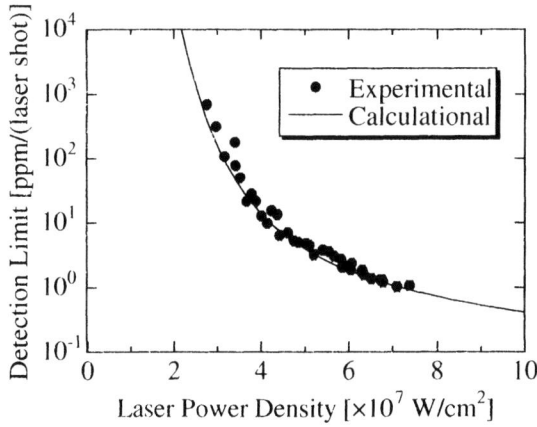

FIGURE 4. The LA laser power dependence of the detection limit.

The RLA experiment, where the single dye laser is used for both LA and RIS, was made to estimate the elemental selectivity and the detection limit for Al in a kovar sample. The ratio of Mg to Al ions should be small so that ^{26}Mg can not give significant isobaric interference to the ^{26}Al measurement. Fig. 3 shows the laser power dependence of the elemental selectivity, which is defined as the ratio of Mg to Al ions, when the laser wavelength is tuned to the Al resonant "2+1" transition at 496.510 nm. It is found that the Al ion yield exceeds 10^3 times larger than the Mg ion yield below the laser power density of 7×10^7 W/cm^2. The LA laser power dependence of the detection limit, which is defined as the concentration that can be detected an ion by MCP, is shown in Fig. 4. Because the detection limit is approximately 1 ppm per one laser shot at the LA laser power density of 7×10^7 W/cm^2, the Al content in a sample with 10 ppb can be detected by accumulating 100 laser shots (i.e. 10 sec.). From the theoretical calculation, it is found that the detection limit could be improved by increasing the LA laser power if the elemental selectivity is not important. The theoretical calculation also gives good agreement with the experiment on the elemental selectivity and the detection limit, which leads to the possibility of finding the optimal laser power for the trace element analysis with RLA by the present model calculation.

CONCLUSION

In this study, we have developed a simple theoretical model to simulate the RLA process and verified the model through analyzing some experiments to detect a trace of Al. From the comparison results between experiment and calculation, it is concluded that the present model is reasonable for estimating the LA laser power dependence of the neutral atom and ion yields, and also useful for finding the optimal laser power for the elemental selectivity and the detection limit in the trace element analysis with RLA.

As the future work, we will try to detect the trace long-lived radionuclides and/or stable nuclides produced by nuclear reactions in neutron irradiated materials as an example of application of RLA.

REFERENCES

1. Wang, L., Borthwick, I. S., Jennings, R., Mccombes, P. T., Ledingham, K. W. D., Singhal, R. P. and Mclean, C. J., *Inst. Phys. Conf. Ser.* **114** ,455-458 (1991).
2. Wang, L., Borthwick, I. S., Jennings, R., Mccombes, P. T., Ledingham, K. W. D., Singhal, R. P. and Mclean, C. J., *Appl. Phys.* **B 53**, 34-38 (1991).
3. Eiden, G. C., et al., *Microchem. J.* **50**, 289-300 (1994).
4. Gill, C. G., et al., *Spectrochimica Acta* **B 51**, 851-862 (1996).
5. Watanabe, K. and Iguchi, T., *Appl. Phys.* **A 69**, S845-S848 (1999).

High-Resolution, Three-Step Resonance Ionization Mass Spectrometry of Gadolinium

K. Blaum[*], B.A. Bushaw[†], W. Nörtershäuser[‡], K. Wendt[*]

[*]Institut für Physik, Universität Mainz, D-55099 Mainz, Germany
[†]Pacific Northwest National Laboratory, PO Box 999, Richland, WA 99352
[‡]Gesellschaft für Schwerionenforschung mbH, D-64291 Darmstadt, Germany

Abstract. High-resolution resonance ionization mass spectrometry has been used to measure triple-resonance autoionization (AI) spectra of gadolinium. AI resonances as narrow as 10 MHz have been observed and isotope shifts and hyperfine structure have been measured in selected AI states. The strongest AI state observed at 49663.576 cm^{-1} with a photoionization cross section of >3.6x10^{-15} cm^2 was found to have an overall detection efficiency of >3x10^{-5}, allowing application to a number of ultratrace determination problems. Analytical measurements with a diode-laser-based system have been successfully performed on bio-medical tissue samples.

INTRODUCTION

Ultratrace determination of the rare earth element gadolinium (Gd) has a variety of applications in a number of fields [1,2]: In medicine, the chelate Gd-DTPA is used as a primary contrast medium for magnetic resonance imaging, but little is known about the physiological properties of the highly toxic Gd, such as retention and redistribution kinetics in the human body [3]. In astrophysics, the ratio of the minor stable isotopes ^{152}Gd/^{154}Gd in presolar grains from meteorites can be used to determine the temperature during the formation of these isotopes [4]. Furthermore an efficient excitation scheme employing multi-step RIS might be used for isotope enrichment of ^{155,157}Gd, which are used as burnable poison in nuclear reactors [5]. Elemental and isotopic selectivities of >10^6 and overall efficiencies of ~10^{-4} are needed to attain sensitivities required for the analytical applications. Thus we have investigated triple-resonance excitation to autoionizing (AI) states following the scheme in Fig. 1a.

Knowledge of isotope shifts (IS) and hyperfine structure (HFS) in all excitation steps is needed to determine quantitative isotope ratios and precise abundances employing isotope dilution techniques. The IS and HFS in the first and second step have been studied previously, with the first step investigated using a 1+1′ ionization scheme for six resonance lines in the wavelength range of 422-429 nm [2]. For the analysis of the complex hyperfine spectrum in the second step and to assign the previously unknown electronic configuration of the $6s\,8s\,^9D_6$ level, we used the technique of "intermediate-level hyperfine state selection" described in [6], which has also been applied to determine the HFS and the J values of some of the AI states reported in this paper.

CP584, *Resonance Ionization Spectroscopy 2000: 10th Int'l. Symp.*, edited by J. E. Parks and J. P. Young

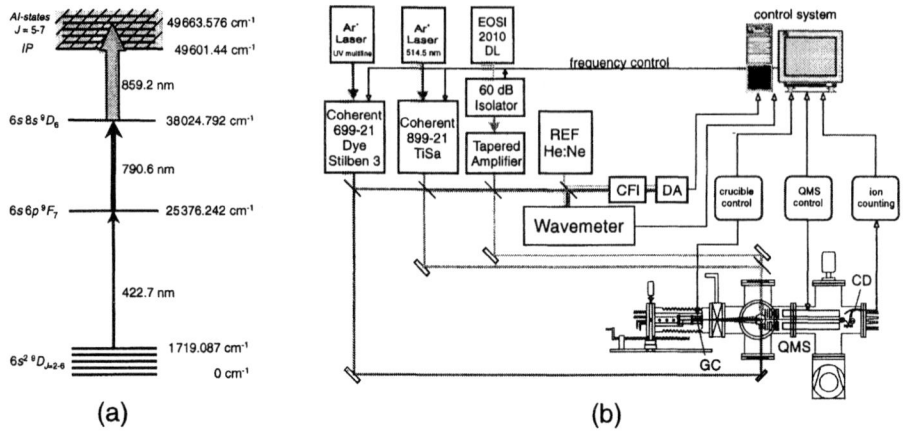

(a)	(b)

FIGURE 1. (a) The three-step RIS scheme for Gd. (b) Schematic diagram of the experimental setup including vacuum chamber with graphite crucible (GC), quadrupole mass spectrometer system (QMS), and channeltron detector (CD) as well as the laser system for the triple-resonance measurements together with the confocal Fabry-Perot interferometer (CFI) and the detector assembly (DA).

EXPERIMENTAL

The experimental apparatus (Fig. 1b) used for spectroscopic measurements at Pacific Northwest National Laboratory has been described previously [2,7], so only a brief description is given here. Resonance ionization is performed on a collimated thermal atomic beam of Gd atoms in the ionization region of a quadrupole mass spectrometer (QMS). The excitation ladder (Fig. 1a) uses a dye laser (Coherent 699-21) for the first step at 422.7 nm, a titanium-sapphire laser (Coherent 899-21) for the second step at 790.6 nm and a tapered diode amplifier (SDL 8630) seeded by a single-frequency diode laser (EOSI 2010) for photoionization. Resulting photoions are extracted with conventional ion optics and mass analyzed with a commercial QMS. Ions passing through the mass filter are detected with an off-axis continuous dynode electron multiplier. The apparatus used at the University of Mainz is similar, except that diode lasers are used for all steps, with the 422.7 nm light obtained by frequency doubling in an external enhancement cavity.

RESULTS AND DISCUSSION

Scanning the third laser from 847.4 – 865.8 nm, we observe about 150 AI states, as shown in Fig. 2a; many with linewidths (FWHM) smaller than 100 MHz. Detailed IS and HFS analysis of selected states (numbered 1 - 8) will be treated in a forthcoming publication. Excitation to the AI state at 49663.576 cm^{-1} (number 3, shown in Fig. 2b) was found to have the highest overall detection efficiency at $>3 \times 10^{-5}$ and the resonance exhibited both Lorentzian and Gaussian linewidth components of 9(1) MHz. The third-step photoionization cross-section was determined to be $>3.6 \times 10^{-15}$ cm^2, nearly 5 orders of magnitude stronger than ionization into the nearby continuum. The broader AI resonances on either side of the narrow resonance are well fit by Fano profiles [8], as indicated by the solid curves in Fig 2b.

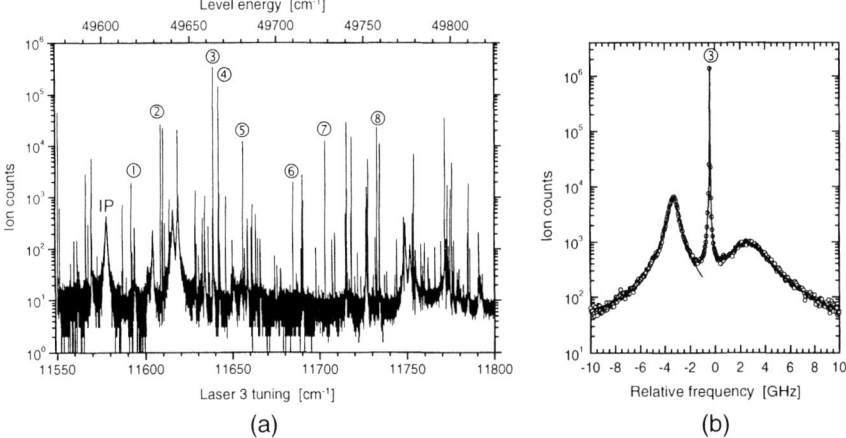

(a)   (b)

FIGURE 2. (a) Observed spectrum of autoionizing resonances of Gadolinium. The tuning range of the ionization laser was 250 cm⁻¹ (7.5 THz) with a step size of 130 MHz. IP is the ionization potential. (b) Strongest autoionizing resonance in the spectrum (a) on an expanded scale.

Both IS and HFS in this strong AI state were measured using the known IS and HFS in the first two excitation steps. Figure 3a shows a composite spectrum of all stable, even isotopes of Gd in the $6s^2\ ^9D_6 \rightarrow$ AI (49663.576 cm⁻¹, $J = 7$) transition on a logarithmic scale. The resonances were recorded with the mass spectrometer set to the proper mass, the first- and second-step laser fixed on the respective $6s^2\ ^9D_6 \rightarrow 6s\ 6p$ $^9F_7$ and $6s\ 6p\ ^9F_7 \rightarrow 6s\ 8s\ ^9D_6$ isotopic resonances, as determined by IS and HFS data given in [2] and [6], and then the ionization laser was scanned across the three-photon resonance in 3-MHz steps. The sum detuning for all three steps has been used for the frequency axis and thus the peak separation reflects the transition isotope shift (TIS) for the three-photon transition starting in the $6s^2\ ^9D_6$ fine structure component of the ground state and terminating in the AI state.

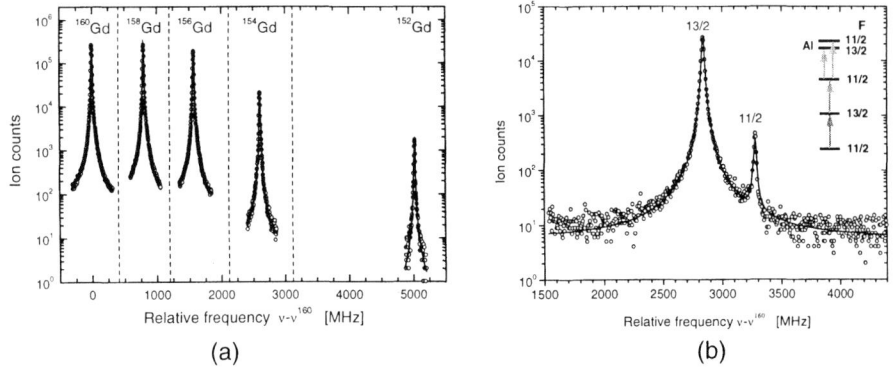

(a)   (b)

FIGURE 3. Experimental spectra (circles) and fitted Voigt profiles (lines) used to determine isotope shifts for the even Gd isotopes (a) as well as hyperfine structure for ¹⁵⁵,¹⁵⁷Gd (b, ¹⁵⁵Gd shown) in the $6s^2$ $^9D_6 \rightarrow$ AI (49663.576 cm⁻¹, $J = 7$) transition. The inlay in (b) shows the principle of the intermediate level hyperfine state selection method: tuning the first- and second-step lasers to specific hyperfine-transitions allows only a limited and defined set of hyperfine levels to be observed in the third excitation.

143

For analysis of complex hyperfine spectra, and to determine the J value for AI states, we used intermediate-level hyperfine state selection as described in [6]. Fig. 3b shows an example. Resulting IS values for all stable Gd isotopes are listed in Table 1.

TABLE 1. Measured isotope shifts in the $6s^2\,^9D_6 \to$ AI (49663.576 cm^{-1}, $J = 7$) transition of Gd I. All values in MHz. Uncertainties are given as the standard error of the mean for three measurements.

^{160}Gd	^{158}Gd	^{157}Gd (cg)	^{156}Gd	^{155}Gd (cg)	^{154}Gd	^{152}Gd
0.0	798.3(3)	1508.8(6)	1567.5(4)	2146.7(15)	2605.9(3)	5020.1(3)

The 49663.576 cm^{-1} AI resonance was employed for further analytical work using the Mainz RIMS spectrometer [1], which performed the triple-resonance excitation completely with diode lasers. Using this system, we have demonstrated isobaric and isotopic selectivities of $>10^7$ and detection limits of $<4\times10^9$ Gd atoms, corresponding to <1 pg Gd, in prepared samples as well as in bio-medical tissue samples.

SUMMARY

High-resolution triple-resonance ionization mass spectrometry has been used for spectroscopic studies on gadolinium, with precision IS and HFS measurements in the $6s^2\,^9D_6 \to$ AI (49663.576 cm^{-1}, $J = 7$) transition. An overall efficiency approaching 10^{-4} can be reached. First analytical measurements have been successfully performed on mouse tissue samples [9].

ACKNOWLEDGMENTS

Spectroscopic measurements reported in this work were performed at Pacific Northwest National Laboratory with support from the U.S. Department of Energy, Office of Science under Contract No. DE-AC06-76RLO 1830.

REFERENCES

1. Blaum, K., Bushaw, B.A., Geppert, C., Müller, P., Nörtershäuser, W., Schmitt, A., Trautmann, N., and Wendt, K., "Diode-Laser-Based Resonance Ionization Mass Spectrometry of Gadolinium", in *Resonance Ionization Spectroscopy*, edited by J.C. Vickerman et al., AIP Conference Proceedings 454, New York: American Institute of Physics, 1998, pp. 275-278.
2. Blaum, K., Bushaw, B.A., Diel, S., Geppert, Ch., Kuschnick, A., Müller, P., Nörtershäuser, W., Schmitt, A., and Wendt, W., *Eur. Phys. J. D* **11**, 37-44 (2000).
3. Mathur-De Vré, R., and Lemort, M., *Br. J. Radiol.* **68**, 225-247 (1995).
4. Käppeler, F., Gallino, R., Busso, M., Picchio, G., and Raiteri, C.M., *Astrophys. J.* **354**, 630-643 (1990).
5. Santala, M.I.K., Daavittila, A.S., Lauranto, H.M., and Salomaa, R.R.E., *Appl. Phys. B: Lasers Opt.* **64**, 339-347 (1997).
6. Nörtershäuser, W., Bushaw, B.A., and Blaum, K., *Phys. Rev. A* **62**, 022506 1-4 (2000).
7. Nörtershäuser, W., Trautmann, N., Wendt, K., and Bushaw, B.A., *Spectrochim. Acta B* **53**, 709-721 (1998).
8. Fano, U., *Phys. Rev.* **124**, 1866-1878 (1961).
9. Geppert, C., Blaum, K., Diel, S., Müller, P., Schreiber, W., and Wendt, K., "Gadolinium Trace Determination in Biomedical Samples by Diode Laser based multi-step RIMS", in *Resonance Ionization Spectroscopy* 2000, contribution to this issue.

Effects of dc Electric Fields on Multiphoton Ionization of Rubidium Atoms at Low and High Densities

Nathan I. Hammer[*] and Robert N. Compton[*†]

Departments of Chemistry[*] and Physics[†]
The University of Tennessee
Knoxville, TN 37996

Abstract. Multiphoton ionization (MPI) of rubidium atoms at both low (atomic beam) and high (heat pipe) densities is studied using a tunable OPO laser. At high Rb densities ionization of the laser excited *ns*, *np*, and *nd* states occurs both through photoionization and collisional ionization. Excitation of the *np* states is found to be induced by the external electric field at both low and high densities. In addition, *np* signal is also seen at very low (E→ 0) fields in the heat pipe, providing evidence for collision mixing as well as field mixing. At low densities, signal for the high *np* states initially increases with applied field, but soon saturates (i.e. becomes field independent) while the signal for high *nd* states decreases with increasing field. At low Rb densities strong resonance features are observed in the energy region between the zero field limit (IP) and the field ionization limit. These features, as well as the field ionization threshold, are found to be dependent upon the angle between the laser polarization and the direction of the applied dc field. Evidence for tunneling through the barrier created by the $-e^2/r - eEr$ potential is also presented for *ns* and *nd* states.

INTRODUCTION

The spectroscopy of atoms in high Rydberg states has been a topic of intense study for a number of years.[1,2] Recently, the growing use of Rydberg charge-exchange reactions for the creation of negative ions has reinitiated an interest in efficiently obtaining these states of high principal quantum number, *n*. The ground and excited states of an atom are termed Rydberg states if the energy levels can be described as a quasi-hydrogenic "one-electron" atom and the energy levels relative to the ground state follow the Rydberg formula

$$E_{n,\ell} = IP_A - \frac{R_A}{n^{*2}} \tag{1}$$

where IP_A represents the ionization potential of the atom, R_A is the Rydberg constant for the atom and n^* is the effective principal quantum number ($n^* = n - \mu_\ell$, with μ_ℓ

CP584, *Resonance Ionization Spectroscopy 2000: 10th Int'l. Symp.*, edited by J. E. Parks and J. P. Young

being the ℓ-dependent quantum defect and n relates to the hydrogen atom). The various n and ℓ states can be obtained either by a single or multi-photon process. For example, the *np* states of rubidium can be easily obtained with one photon of one color and the then the *ns* and *nd* states can be accessed from that state through the use of another photon of different color. Alternately, the *ns* and *nd* states can be obtained directly via a two-photon one-color process. Figure 1 shows the *ns*, *np*, and *nd* energy levels for rubidium and some possible excitation schemes for populating high n_ℓ Rydberg states. The advent of tunable OPO lasers with resolutions on the order of 0.02 cm^{-1} now allows easy access to very high Rydberg states of atoms through any of the above excitation mechanisms.

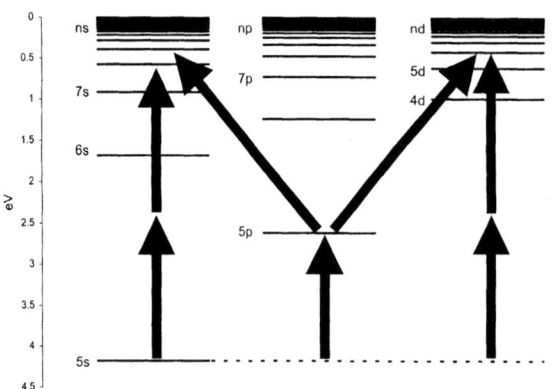

Figure 1. Excitation scheme for producing one and two-photon excited *ns*, *np*, and *nd* states in rubidium

It has been known for many years that the presence of a dc electric field effectively lowers the ionization potential of an atom. In addition, resonance features were reported in the region above the zero field threshold.[3] Considerable experimental and theoretical interest has been devoted to these above-threshold resonances. Field induced resonances have also been seen in the region between the zero field threshold (IP) and the field ionization limit as discussed below.[4,5]

The potential energy for a Rydberg electron in the presence of an electric field is

$$V(r) = -\frac{e^2}{r} - eEr \qquad (2)$$

where e is the charge of an electron, r is the distance of the electron from the core, and E is the magnitude of the external electric field. As the field is increased the ionization potential is lowered as high Rydberg states can be field-ionized. Figure 2 shows the electric field modified potential energy diagram for rubidium. The ionization potential is lowered to the effective ionization potential by the amount

$$V_{lowering} = 2e\sqrt{eE} \qquad \text{where} \qquad V_{max} = V(R) - V_{lowering} \qquad (3)$$

Ryderg states with energy above this new effective potential escape over the barrier and those below V_{max} can only tunnel through the potential barrier. Thus ions result from either three-photon ionization below the field ionization threshold, tunneling at the top of the barrier, or field ionization.

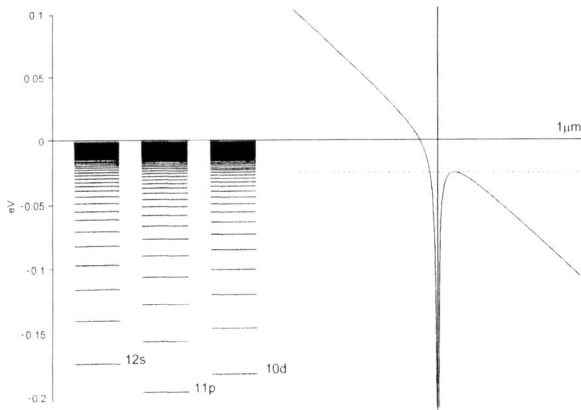

Figure 2. Electric Field modified potential energy diagram for rubidium showing field ionization of high Rydberg states

The effects of a dc electric field on multiphoton ionization of an alkali atom (cesium) have been reported previously.[5,6] In addition to the well-known field-ionization effects, a dc field was seen to induce access to the dipole forbidden two-photon np series. Previously, Zhang, et al.[7] observed ionization through two-photon excited np states of sodium at high densities (heat pipe). The origin of this forbidden excitation was uncertain and is one of the subjects of this study.

EXPERIMENTAL

The output of a tunable OPO laser (Continuum Sunlite) pumped by the third harmonic ($\lambda = 355$ nm) of the fundamental ($\lambda = 1064$ nm) of a Nd:YAG laser (Continuum Powerlite) was transmitted though a heat-pipe oven containing rubidium vapor for high density studies and into a rubidium atomic beam apparatus for studies at low density. The heat pipe was 25 cm long with a 2 cm internal diameter. The rubidium pressure in the heat pipe and the density of the atomic beam were controlled by the temperature ($20°C \leqslant T \leqslant 200°C$) of each. The Continuum Sunlite OPO has a narrow linewidth (approximately 0.07 cm^{-1}), which is ideally suited to study the high Rydberg states of alkalis through multiphoton excitation/ionization. The absolute wavelength was determined to within 0.001 nm with a Burleigh Wavemeter model WA-4500-0. The ionization signal from the heat pipe was collected on a biased insulated collector wire which passed through the active region of the heat pipe. The ions or electrons in the atomic beam apparatus were detected using a dual-microchannel plate detector. A voltage applied to a metal pusher plate was used to accelerate the ions. In some studies the pusher voltage was continuous, but in others it was pulsed with a 1 μs delay and a 5 μs duration. Thus n-photon excitation could be studied with or without the influence of an electric field. The ionization signal from both the heat pipe and atomic beam setup was observed as either a negative or positive signal, depending on whether electrons or positive ions were being collected. The positive or negative voltage pulses were displayed on a digital oscilloscope, averaged by a boxcar integrator (Stanford Research Systems model SR250), and recorded by a data acquisition computer program.

RESULTS AND DISCUSSION

Figure 3 shows a typical 2+1 multiphoton ionization Rydberg spectrum for rubidium. This particular spectrum was taken in the atomic beam apparatus at an electric field of 2300 V/m and shows many of the possible Rydberg states (n⩾14). One can see the prominent 2+1 MPI nd transitions, smaller ns and forbidden np transitions near the ionization potential, and the field-induced ionization signal. The two-photon zero field IP is 593.63 nm. The ns transitions are small because of their small ionization cross-sections. Signal from most of the nd transitions is saturated. The nd signal is seen to fall off roughly as n^{*-7} as seen previously and is accounted for by a decreasing excitation probability as well as decreasing photoionization cross section with n^{*}.

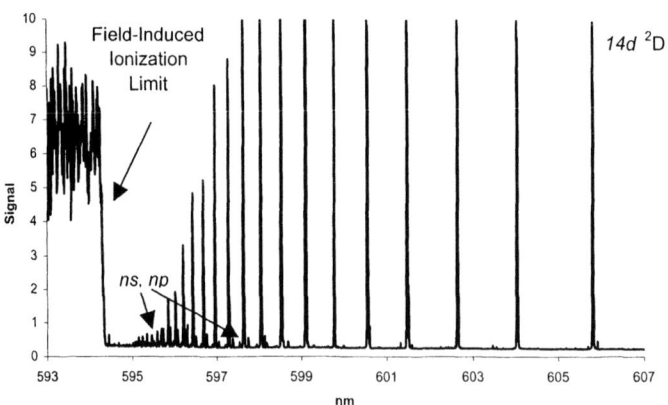

Figure 3. Typical rubidium Rydberg spectrum

High Density

In the atomic beam experiment the lower nd transitions are most prominent relative to ns or np states due to their larger photoionization cross-sections. However, in the heat pipe collisional ionization also plays a role and as a result higher ns and nd Rydberg states can be readily seen. It was also observed that the symmetry forbidden *np* states were prominent at high n and that increasing the electric field in the heat pipe or atomic beam experiment increased the *np* signal. Shown in Figure 4 is a typical spectrum obtained in the heat pipe showing the *ns*, *np*, and *nd* states. Figure 5 shows the ratio of the signal intensities of the *41p* and the *41s* states as the voltage on the collection wire (and hence the electric field in the heat pipe) is increased. The absolute intensity of the *ns* transition is not affected by the presence of an electric field and therefore such a ratio was used in order to compensate for any variations in laser power or alkali vapor pressure. It appears that the increase in *np* signal (relative to a nearby *ns* state) is linear with respect to the applied voltage with a non-zero intercept. The non-zero intercept may be due to collisionally induced population of *np* states.

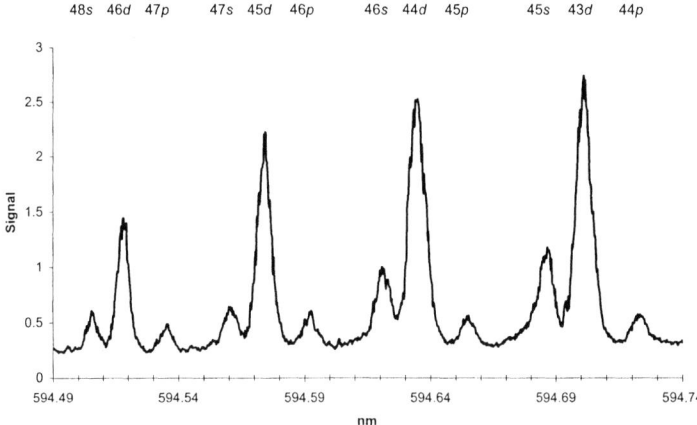

Figure 4. High-n Rydberg states of rubidium obtained in a heat pipe with a small electric field (~10 V on wire)

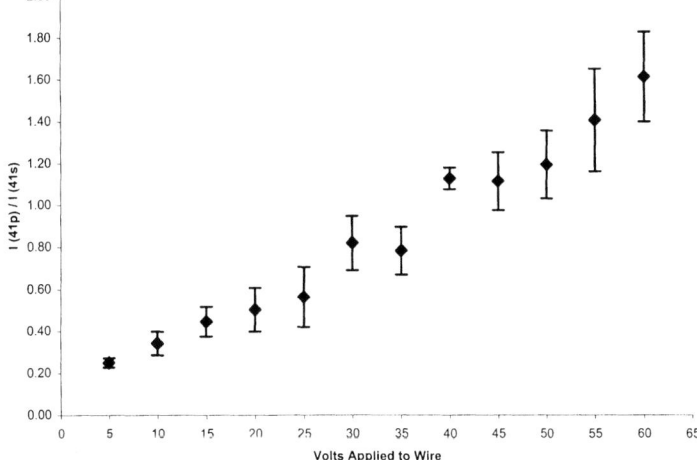

Figure 5. Ratio of intensities of the *41p* and the *41s* states of rubidium as a function of applied voltage

149

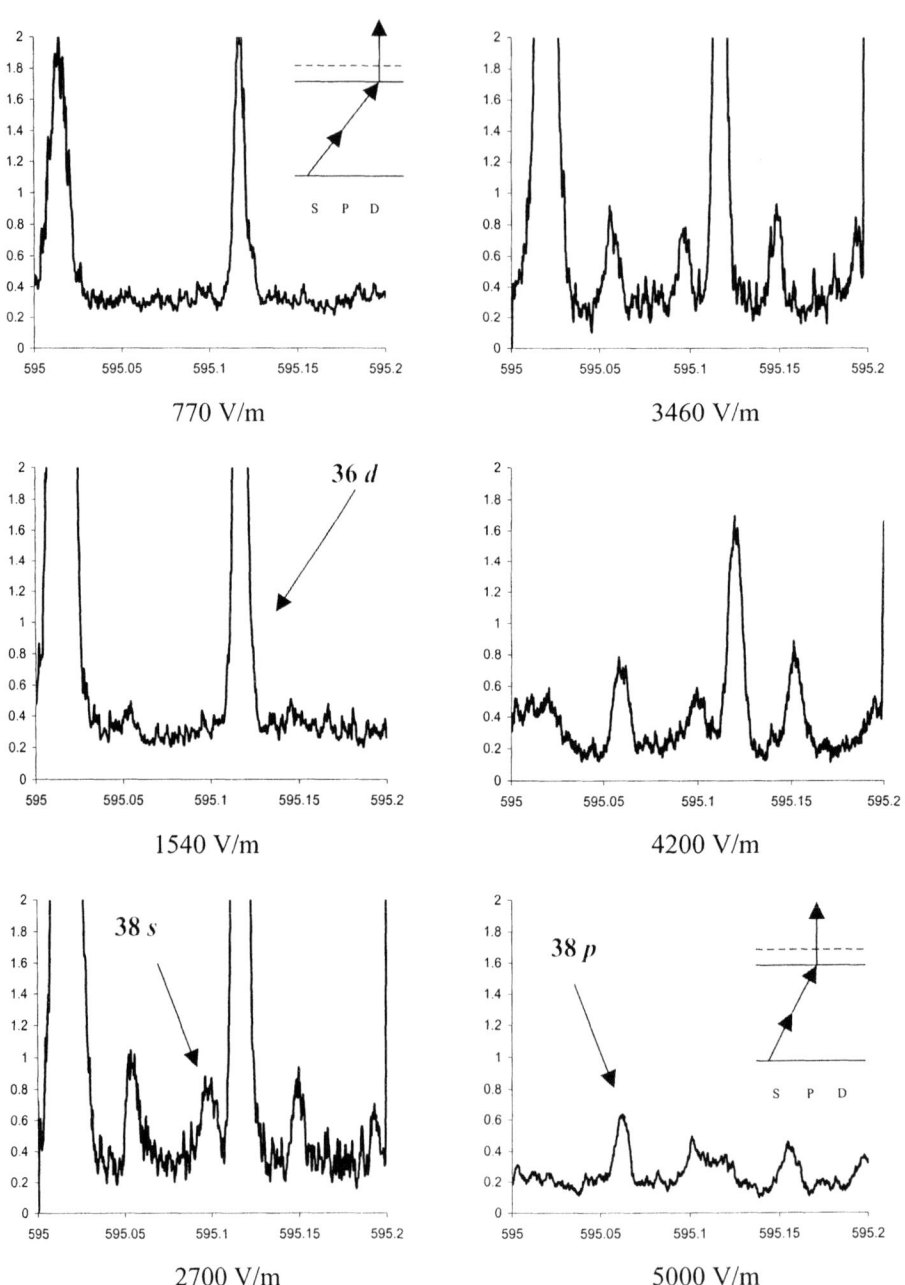

Figure 6. Variation of high *s*, *p*, and *d* Rydberg states with increasing electric field

Low Density

Results for the atomic beam experiment were similar to those obtained in the heat pipe except that collisional ionization was absent. Shown in Figure 6 are 6 MPI spectra taken at various increasing electric field strengths. The forbidden *np* transitions were observed only through the application of an electric field. The intensity of the *np* transitions increases with applied electric field, but then levels off and becomes unaffected by further increases in field strength. Interestingly, while the *np* signal remains constant with ever-increasing field the *nd* signal decreases. Presumably, the magnetic sublevels of the *nd* states are spreading out and becoming more diffuse.

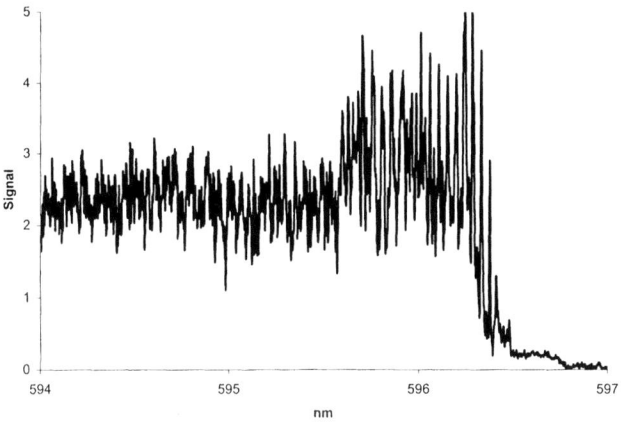

Figure 7. Above threshold resonances with a field of ~70000 V/m and polarization of the laser perpendicular to the field

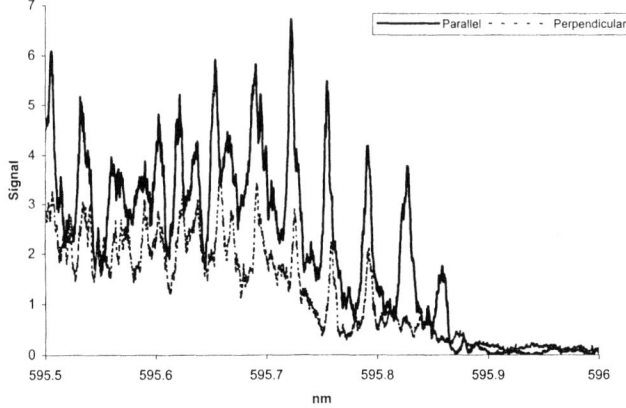

Figure 8. Above threshold resonances with a field of ~45000 V/m with laser polarizations perpendicular and parallel to the electric field

Distinct resonances were observed in the region between the two-photon IP and the field induced ionization limit when a dc draw-out field was applied. Figures 3, 7, and 8 show such resonances. Similar resonances have been seen in potassium,[3] rubidium,[4]

and cesium.[5] It was found that these resonances also depend upon the direction of polarization of the laser relative to the electric field, in a manner similar to that reported earlier for cesium. These features are clearly field-induced as shown by measurements under pulsed field conditions, i.e. field ionization occurs from well-derived n_ℓ states produced under zero field conditions. Figure 9 shows the falling off of the *nd* series as the field-induced limit is approached. Interesting features appear in the region near the field ionization threshold. Note that the transition from the three-photon ionization to two-photon excitation/field ionization is not totally abrupt. The *nd* level just below the field ionization limit appears to be due to both tunneling and three-photon ionization. More dramatic is the onset of *ns* levels in this region. Ionization of *ns* states is primarily through field ionization, however, it would appear that transmission through the barrier is greater for *ns* states than *nd* states. This might be accounted for by the absence of a centrifugal barrier for *ns* states. Figure 10 shows very high field-ionized Rydberg states as the zero-field limit (593.63 nm) is approached.

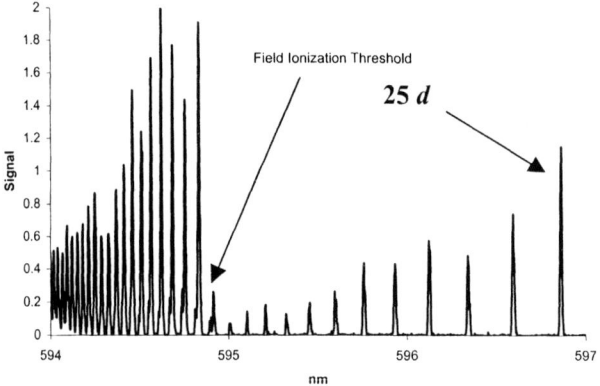

Figure 9. Three-photon and field-ionized *ns* and *nd* Rydberg states

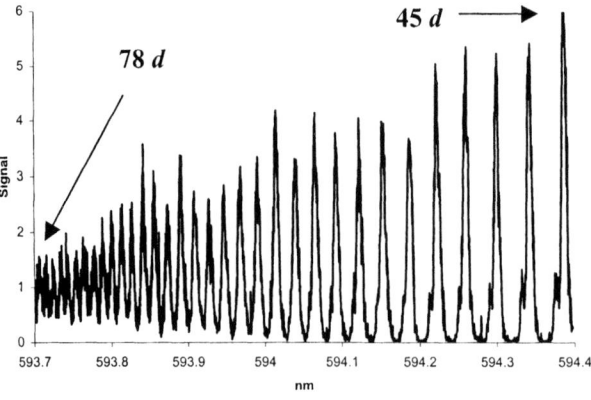

Figure 10. Very high field-ionized *ns* and *nd* Rydberg states

CONCLUSIONS

The present study demonstrates that resonantly enhanced 2+1 multiphoton ionization of rubidium atoms through the two-photon forbidden np Rydberg states occurs primarily because of state mixing induced by the presence of an external electric field. At high rubidium pressures (heat pipe) there is also evidence for collisional mixing of forbidden and allowed states leading to ionization. The non-zero intercept in Figure 5 leads one to believe that there is some collisionally-induced state mixing. Similar results were reported in our earlier studies of sodium,[7] and are fully explained by the conclusions reported herein.

In the heat pipe experiments, collisions primarily account for the ionization of the resonantly excited n_ℓ states. This is apparent due to the presence of strong ns ionization signal at high densities. Signal resulting from ns states is very weak at low densities. MPI via the 2+1 ns intermediate state is small because the photoionization cross-section for the ns states is very low compared to that of the nd states. On this basis, one would also conclude that the ionization signal for the forbidden np states would also be collisionally-induced.

Data from the pulsed electric field ionization studies of high Rydberg ns and nd states are seen to fit the field ionization model very well. Evidence is also presented for tunneling through the top of the Coulomb barrier. A higher tunneling rate is observed for the ns states than the nd states. In the dc field ionization studies, rich structure is seen in the energy region between the field ionization and zero field ionization thresholds. The above threshold resonances observed here are very similar to others seen previously for potassium,[3] rubidium,[4] and cesium.[5] Results from all of these experiments await a proper theoretical interpretation.

ACKNOWLEDGMENTS

We would like to thank Dr. Shannon Mahurin and Ashley Greer for their initial contributions to this work and the National Science Foundation (CHE-9981945) for financial support.

REFERENCES

1. Gallagher, T., *Rydberg Atoms*. Cambridge University Press, 1983.
2. Stebbings, R. F., and Dunning, F. B., *Rydberg States of Atoms and Molecules*, Cambridge Press, 1983.
3. Freeman, R. R., Economou, N. P., and Bjorklund, G. C., *Phys. Rev. Lett.* **41**, 1463 (1978).
4. Sato, Y., Teraoka, Y., and Murakami, J., "Electric Field Induced Resonances of Potassium Atoms as Observed by Multiphoton Ionizations," in *Oji International Seminar On Highly States of Atoms and Molecules*, edited by S. S. Kano and M. Matsuzawa, Fuji-Yoshida, Japan, 1986, pp. 17-21.
5. Klots, C. E., and Compton, R. N., "Effects of dc Electric Fields on Multiphoton Ionization Spectra in Cesium," in *Multiphoton Processes*, edited by P. Lambropoulos and S. J. Smith, Springer-Verlag, Berlin, 1984, pp. 58-66.

6. Klots, C. E., and Compton, R. N., *Phys. Rev. A* **31**, 525 (1985).
7. Zhang, J., Lambropoulos, P., Zei, D., Compton, R. N., and Stockdale, J. A. D., *Z. Phys. D.* **23**, 219 (1992).

Progress in ^{41}Ca ultratrace determination by diode-laser-based RIMS

*P. Müller, ‡B. A. Bushaw, *K. Blaum, *S. Diel,
*Ch. Geppert, †N. Trautmann, *K. Wendt

*Institut für Physik, †Institut für Kernchemie,
Johannes Gutenberg-Universität, D-55099 Mainz, Germany
‡ Pacific Northwest National Laboratoray, Richland, WA 99352, USA

Abstract. We report on progress in development and application of ^{41}Ca ultratrace determination by diode-laser-based RIMS. Applications include biomedical isotope-tracer studies of human calcium kinetics, cosmochemical investigations of meteorites, environmental dosimetry and radiodating. Depending on the application, ^{41}Ca needs to be determined at isotopic abundance in the range of 10^{-9} to 10^{-15} relative to the major stable isotope ^{40}Ca. We use either a double- or triple-resonance excitation scheme and subsequent non-resonant photo-ionization of calcium atoms in a collimated atomic beam. All resonant steps are excited with narrow bandwidth extended cavity diode lasers, non-resonant photo-ionization is attained with either an argon ion laser or a CO_2 laser. The resulting photo-ions are detected with a quadrupole mass spectrometer. With double-resonance excitation, the optical isotopic selectivity for ^{41}Ca against ^{40}Ca is 2×10^4, while the triple-resonance scheme provides optical selectivity of more than 10^9. By adding the third resonant step, overall detection efficiency increases from 1×10^{-6} to 5×10^{-5} and the detection limit for relative ^{41}Ca abundance improves from 5×10^{-10} to 2×10^{-13}. Both schemes have been applied to various sample types and accuracy and reproducibility of the resulting ^{41}Ca/^{40}Ca isotope ratios have been determined to be better than 5%.

INTRODUCTION

In recent years we have been developing a method for sensitive and selective ultratrace determination of the long-lived radioisotope ^{41}Ca ($t_{1/2} = 1.04 \times 10^5$ a) using diode-laser-based, high-resolution resonance ionization mass spectrometry [1]. Targeted applications include using ^{41}Ca as biomedical tracer isotope for in vivo studies of human calcium kinetics [2], cosmochemical investigations of the ^{41}Ca concentration in meteorites [3] including fundamental research on the production cross section for ^{41}Ca in corresponding nuclear reactions [4], and the determination of the specific ^{41}Ca activity in concrete or other calcium-rich material for integrated neutron dosimetry [5,6]. In all these applications ^{41}Ca must be determined at abundances ranging from 10^{-9} to below 10^{-12}, relative to the predominant stable isotope ^{40}Ca. Detecting ^{41}Ca concentrations at ambient levels, which are expected to be on the order of 10^{-15}, may also permit radiodating of geological and anthropological samples [7]. Thus, extremely high isotopic selectivity against ^{40}Ca, as well as high suppression of isobaric atomic and molecular interferences, predominantly ^{41}K and ^{40}CaH, is

CP584, *Resonance Ionization Spectroscopy 2000: 10th Int'l. Symp.*, edited by J. E. Parks and J. P. Young
© 2001 American Institute of Physics 0-7354-0024-5/01/$18.00

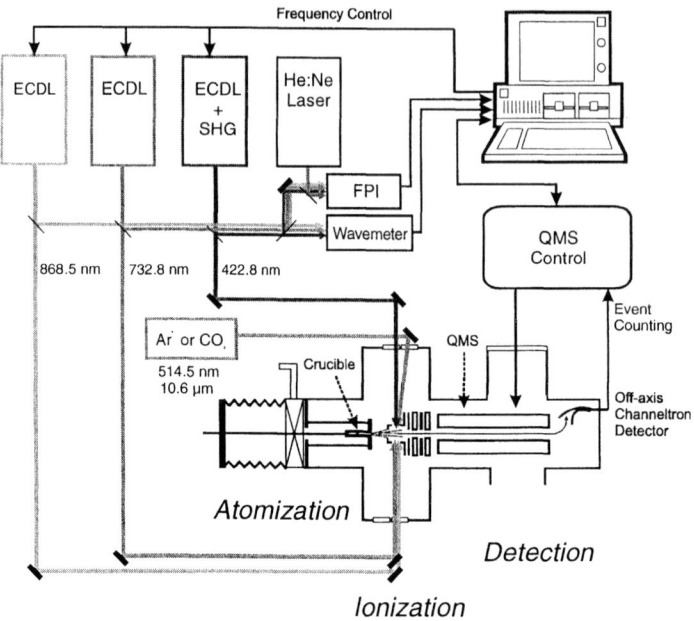

FIGURE 1. Experimental setup for high-resolution RIMS.

required. For practical ultratrace measurements, these characteristics must also be combined with high overall efficiency. Until now, these specifications could only be reached by large-scale accelerator mass spectrometry. The high isobaric selectivity of our approach using resonance ionization mass spectrometry with narrow-band extended-cavity diode lasers (ECDL) is based on resonant optical excitation and subsequent ionization of the desired species in a collimated atomic beam, which also yields high isotopic selectivity, due to the resolution of isotope shifts in the successive optical transitions. In the absence of uncorrelated background events, this optical isotopic selectivity may be multiplied with the mass selection obtained with a commercial quadrupole mass spectrometer (QMS) used in detection of the photo-ions.

EXPERIMENTAL SETUP

The experimental detection process can be split into three different parts: atomization of the sample, resonant excitation and photo-ionization of the desired species, and mass selective detection of the photo-ions. A diagram of the apparatus is shown in Fig. 1. Atomization of the sample is performed in a resistively heated graphite crucible at temperatures of up to 2000 °C. The narrow crucible channel acts as a self-collimating atomic beam source with a full-angle divergence of ~10°, which provides good transport of atoms to the laser ionization region as well as reduction of Doppler broadening. At a distance of 2 cm from the crucible orifice, the atomic beam is perpendicularly overlapped with up to four laser beams. These begin with resonant excitation along the $4s^2\,^1S_0 \rightarrow 4s4p\,^1P_1 \rightarrow 4s4d\,^1D_2$ pathway. In double-resonance experiments the atoms undergo subsequent non-resonant photo-ionization by 514.5 nm light from an argon-ion laser, as shown in Fig. 2 A. For triple-resonance measurements, the atoms are further excited to the $4s15f\,^1F_3$ Rydberg state, and finally photo-ionized by the 10.6 μm radiation of a CO_2-laser, as shown in Fig. 2 B. Laser light for the first resonant step is produced by frequency doubling the 845.6 nm output of an ECDL in a $KNbO_3$ crystal, which is placed in an external resonance enhancement cavity. The second and third step are directly excited by the 732.8 nm and 868.5 nm light from two additional ECDLs. The blue laser beam and the two red beams are counter-propagated to further reduce Doppler broadening by partial cancellation of photon momenta. The optical isotopic selectivity of the resonant laser excitation is derived from the narrow

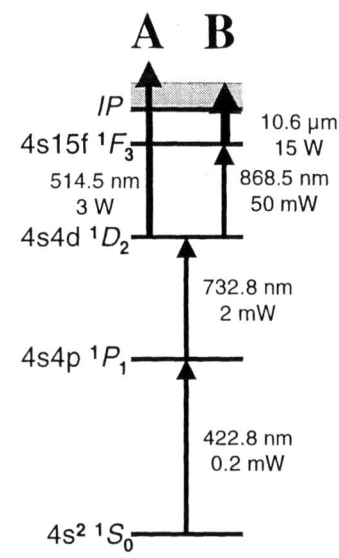

FIGURE 2. Excitation scheme and respective laser wavelengths and powers for double-resonance (A) and triple-resonance (B) excitation of calcium

line widths of the optical resonances in combination with the isotope shifts [8,9]. For ^{41}Ca detection, the selectivity relative to the predominant stable isotope ^{40}Ca is most important and has been determined to be 2×10^4 for double resonance excitation [10] and more than 10^9 for triple resonance excitation [11]. Precise frequency control is necessary for exact and reproducible tuning of all diode lasers to the center of their respective resonances for each isotope measured. Therefore, the frequencies of the diode lasers are computer controlled via their transmission patterns through a scanning Fabry-Perot interferometer in comparison to those of a stabilized helium neon laser. This setup provides frequency stabilization against long-term drifts (< 3 MHz/day) and offset tunings to an arbitrary frequency with an accuracy of < 300 kHz. The ionization volume is located inside a standard crossed-beam ion source of a commercial QMS and is held at a positive potential to reject any surface ions created at the atomization crucible. The photo-ions are focussed into the QMS, which is set to transmit only the mass corresponding to the isotope ionized by the lasers, with ions of neighboring mass suppressed by at least 8 orders of magnitude [12]. Transmitted ions are counted by a channeltron detector.

MEASUREMENT PROCEDURE

In our RIMS measurements, we determine isotope ratios by switching both the QMS and the lasers between the selected target isotopes. The stable isotope ^{43}Ca is used as the best reference for the determination of the relative abundance of ^{41}Ca. Both odd-massed calcium isotopes have a nuclear spin of $I = 7/2$, and a corresponding reduction of ionization efficiency due to hyperfine structure is practically the same for both isotopes. It has been shown that the reduction in efficiency for odd-massed isotopes can be essentially avoided by optical pumping [13], which enables direct ratio measurements between even- and odd-massed isotopes. However, this requires additional polarization optics, laser power, and careful sizing and alignment of the laser beams, and thus it has not been employed here. The natural ^{43}Ca isotopic abundance of 1.336×10^{-3} does not vary significantly for all sources of solar origin and reduces the dynamic range required for ^{41}Ca determination. Prior to a measurement, all lasers are centered on their respective resonance positions for ^{40}Ca and laser beam overlap and ion optics setting are optimized for maximum signal intensity. Then the lasers and the QMS are switched cyclically between the following three settings:

- ^{43}Ca reference: lasers tuned on-resonance for ^{43}Ca, QMS set for mass 43
- ^{41}Ca signal: lasers tuned on-resonance for ^{41}Ca, QMS set for mass 41,
- background: one ore more lasers tuned off-resonance for ^{41}Ca, but to a position with comparable ionization efficiency for ^{40}Ca, QMS set for mass 41.

At each setting the ion counts are integrated for several seconds while the crucible temperature is slowly increased to achieve and maintain a stable ^{43}Ca$^+$ count rate of about 1 MHz, until the sample is exhausted or the ^{41}Ca signal reaches a level that is sufficient for the desired statistical uncertainty. For analyzing the data, the integrated counts at the background position are subtracted from the ^{41}Ca signal, and the result is divided by the dead-time corrected ^{43}Ca counts. Multiplication with the natural ^{43}Ca abundance yields the ^{41}Ca abundance relative to the total calcium. Background on mass 43 can be neglected. A final small correction for mass fractionating effects is made using calibration data obtained from samples with known ratios of the even-massed isotopes.

ANALYTICAL RESULTS

A number of sample types have been measured with double- and triple-resonance excitation schemes, two of these will be discussed here. To test the linearity, reproducibility and accuracy of the ^{41}Ca determination, synthetically prepared samples have been investigated with double-resonance excitation. The primary sample was produced by neutron irradiation of highly enriched ^{40}Ca in a research nuclear reactor. From this, a series of isotopically diluted solutions were produced and measured. The results are listed in Table 1a) and show a very good linearity over three orders of magnitude. The small deviation in the 10^{-4} dilution can most probably be attributed to uncertainties in the dilution process itself. The precision of the results approaches the 1% level, while the accuracy error is estimated to be below 5%, limited by uncertainties in the neutron flux and ^{41}Ca production cross section.

We also have addressed one application in environmental research that involves a series of measurements on concrete samples taken from the building of a nuclear research reactor undergoing decommissioning [6]. The task was to measure the specific [41]Ca activity and to confirm that it was below 100 Bq/g, which would allow unclassified release of the material. Because of the unfavorable decay properties of [41]Ca, standard radiometric techniques are not feasible for these measurements. Three different concrete samples were to be investigated: a blank sample from outside the reactor building as reference, a sample from the wall of the biologic shield and a sample from the floor below the reactor core. The samples were first split into a number of aliquots. Each aliquot was chemically treated to extract the calcium and to determine the overall calcium concentration. To calibrate the results, a number of aliquots were initially spiked with a known amount of [41]Ca. RIMS results obtained with double-resonance measurements for spiked aliquots are listed in Table 1b) and show a good agreement with the amount of spike added. For the blank sample this demonstrates the accuracy of the measurements, for the other two samples this implies an original [41]Ca content in the material much smaller than the spike. Similarly, double-resonance measurements on aliquots without spike showed no significant results above the relative abundance detection limit of $\sim 5 \times 10^{-10}$, which is determined by the background at mass 41. In double-resonance measurements, this background arises from $^{40}CaH^+$ ions that are produced by collisions of $^{40}Ca^+$ with hydrogen from residual gas molecules inside the vacuum chamber, e.g., water or molecular hydrogen, and occurs at an intensity of about $\sim 2 \times 10^{-5}$ relative to production of $^{40}Ca^+$ ions. Hence, the interference from this background is determined by the optical $^{41}Ca/^{40}Ca$ selectivity, and occurs at the 10^{-9} level. The much higher optical selectivity of triple-resonance excitation reduces this background below 2×10^{-14} and other sources of background become limiting. Correspondingly, for triple-resonance measurements on the aliquots

TABLE 1. Analytical results for [41]Ca ultratrace determination for various sample types with double- and triple-resonance excitation scheme

Sample	rel. [41]Ca abundance	Comment
a) Irradiated Ca		
10^{-2} dilution	$4.23(3) \times 10^{-7}$	
10^{-3} dilution	$4.16(4) \times 10^{-8}$	
10^{-4} dilution	$4.55(9) \times 10^{-9}$	
b) Reactor Concrete with Spike		Spike added
Blank	$2.72(24) \times 10^{-9}$	$2.96(23) \times 10^{-9}$
Wall	$7.14(21) \times 10^{-9}$	$7.54(58) \times 10^{-9}$
Floor	$8.33(36) \times 10^{-9}$	$7.00(54) \times 10^{-9}$
c) Reactor Concrete without Spike		Spec. [41]Ca activity
Blank	$<6 \times 10^{-11}$	<20 mBq/g
Wall	$3.7(7) \times 10^{-10}$	$61(12)$ mBq/g
Floor	$2.5(6) \times 10^{-10}$	$25(6)$ mBq/g

without spike, as listed in Table 1c), a significant level of ^{41}Ca for the wall and floor sample could be determined, while no signal was found for the blank sample. The corresponding specific ^{41}Ca activity is also given in Table 1c) and lies well below the limit for unclassified release of the material. Additionally, the relative ^{41}Ca abundance allows calculation of the neutron fluence at the sampling position during the operating time of the reactor and is therefore valuable information for neutron dosimetry.

Further triple-resonance measurements have shown that the detection limit of about 5×10^{-11} is dominated by pressure background, which results from neutral species traversing the QMS and being ionized in the detector region by either collisional ionization or field emitted electrons. This neutral background can be substantially reduced by placing the QMS perpendicular to the direction of the atomic beam, but with the loss of a factor of about 5 in efficiency, due to unfavorable ion optics. However, preliminary measurements using this perpendicular geometry have demonstrated a detection limit as low as 2×10^{-13}. This improvement enables measurements on urine samples for human calcium kinetics studies and samples for $Fe(p,x)^{41}Ca$ cross section determination, which are currently under investigation.

CONCLUSION AND OUTLOOK

Thus far, the performance of ^{41}Ca ultratrace determination obtained by diode-laser based RIMS matches the requirements of all targeted applications, as discussed above, except for the most demanding application of radiodating. Analytical measurements on various sample types have been successfully performed. To further improve detection limits, the primary task is to increase overall efficiency. This will be addressed in a number of steps: adaptation of the ion optics to the perpendicular geometry, increase of the laser power in the third resonant step for saturation of this transition, application of optical pumping in the first step, and improved furnace design for better atomization and beam collimation. Altogether, these improvements may increase the overall efficiency by at least another factor of 100 and ultimately enable the ultratrace detection of ^{41}Ca at ambient levels.

REFERENCES

1. Bushaw, B. A., et al., in *RIS-1996*, AIP Conf. Proc. **338**, 115 (1996)
2. Freeman, S. P. H. T., et al., *Nucl. Instr. Meth. Phys. Res.* **B 123**, 266 (1997)
3. Nishuuzumi, K., Arnold, J. R., Fink, D., Klein, J., Middleton, R., *Meteoritics* **26**, 379 (1991)
4. Fink,D., et al., *Nucl. Instr. Meth. Phys. Res.* **B 29**, 275 (1987)
5. Rühm, W., et al., *Rad. Envir. Biophys.* **37**, 293 (1998)
6. Müller, P., et al., *Radiochimica Acta*, **88**;8, 487 (2000)
7. Henning, W., et al., *Science* **236**, 725 (1987)
8. Nörtershäuser, W.Trautmann, N., Wendt, K., Bushaw, B. A., *Spectrochim. Acta* **B 53**, 709 (1998)
9. Müller, P., Bushaw, B. A., Nörtershäuser, W., Wendt, K., *Eur. Phys. J.* **D 12**, 33, (2000)
10. Bushaw, B. A., Nörtershäuser, W. , Wendt, K., *Spectrochimica Acta* **B 54**, 321 (1999)
11. Nörtershäuser, W., Müller, P., Wendt, K., Bushaw, B. A., *Appl. Opt.*, **39**, 5590 (2000)
12. Blaum, K., et al., *Int. J. Mass Spectr. Ion Process.* **181**, 67 (1998)
13. Müller, P., et al., in *RIS-1998*, AIP Conf. Proc. **454**, 73 (1998)

SESSION VI

High Efficiency Four-wave Mixing at Low Power Densities and Low Concentrations: The Role of Coherence

M.G. Payne[1], L. Deng[2], and X.J. Wang[1]
[1]Department of Physics, Georgia Southern University,
Statesboro, GA 30460-8031
[2]Electron and Optical Physics Division
National Institute of Standards and Technology,
Gaithersburg, MD 20899

Abstract

Two transform limited nanosecond lasers are used in a delayed pulse mode to prepare a highly coherent mixture of the ground state and the upper state of a two-photon transition. A picosecond laser co-propagates through the prepared medium at a delayed time and generates a four-wave mixing field. With a carefully chosen time delay, the wave mixing process is phase matched for all detunings and concentrations, and all of the third laser photons can be converted to the mixing field, independent of the intensity of the short-pulse laser. The condition for optimum conversion efficiency does not usually correspond to maximum atomic coherence. In the case of a near resonance short-pulse laser, the condition for phase matching leads to efficient conversion, with the mixing field propagating at the speed of light in vacuum with no distortion.

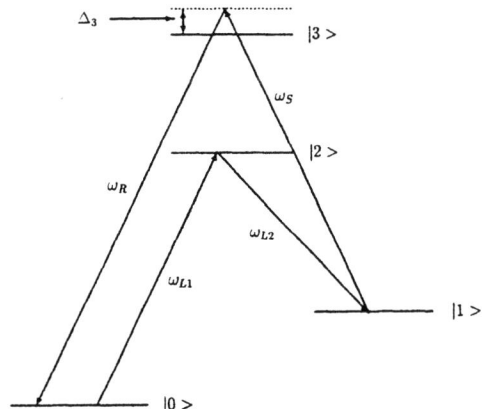

Figure 1: Energy level diagram showing relevant laser couplings. Transform limited bandwidth nanosecond lasers are tuned to resonance between $|0> \rightarrow |2>$ and $|2> \rightarrow |1>$. The laser 1 (ω_{L1}) is time delayed by about a half pulse length from the laser 2 (ω_{L2}). At power densities such that both transitions are strongly saturated the state $|2>$ is a "dark state" which accumulates no population. With proper restrictions on power densities for the first two lasers and on the concentration, both laser pulses propagate at a group velocity very close to vacuum speed of light with almost no attenuation and no distortion. After a predetermined time delay, a short pulse length third laser (ω_S) coupling $|1> \rightarrow |3>$ propagates along the same beam path as the first two lasers leading to the generation of a four-wave mixing field at $\omega_R = \omega_{L1} - \omega_{L2} + \omega_S$.

CP584, *Resonance Ionization Spectroscopy 2000: 10th Int'l. Symp.*, edited by J. E. Parks and J. P. Young
© 2001 American Institute of Physics 0-7354-0024-5/01/$18.00

I. Introduction

Many studies have been published concerning the use of laser induced transparency to achieve highly efficient four-wave mixing.[1-5] Here we describe a scheme which is similar to one used by Harris[4] in an experiment where around 40% efficiency was achieved in four-wave mixing. However, we show that with our means of phase matching, the criteria for high efficiency is not determined by making the the the coherence between $|0>$ and $|1>$ a maximum. Instead, it depends on a relation involving a combination of the relative populations of states $|0>$ and $|1>$, the oscillator strengths of the $|0>\rightarrow|3>$ and the $|1>\rightarrow|3>$ transitions, and the product of the density and the path length of the laser beams in the medium. This criteria is a combination of a phase match condition and another condition to achieve constructive interference between two terms in the expression for the four-wave mixing field. In addition, when the detuning from the three-photon resonance is small we show that the four-wave mixing field propagates through the medium at the speed of light in vacuum without suffering any pulse distortion, even though the medium is highly dispersive.

II. Solution of the model

We assume the two transform limited bandwidth lasers at angular frequencies ω_{L1} and ω_{L2} are sufficiently powerful to strongly saturate the $|0>\rightarrow|2>$ and the $|2>\rightarrow|1>$ transitions, respectively (see Fig.1). We also assume that the concentration of the atomic vapor is low enough to allow the laser pulses to propagate through the medium without distortion at a speed that is very close to the velocity of light in vacuum. When the laser at ω_{L1} is time delayed by about a full width at half maximum of the pulses, the above assumptions imply the following criteria

$$\frac{\kappa_{02}c}{|\Omega_{02}|^2 + |\Omega_{12}|^2} << 1,$$

$$\frac{\kappa_{12}c}{|\Omega_{02}|^2 + |\Omega_{12}|^2} << 1.$$

where $\kappa_{02} = (2\pi\omega_{L1}N|D_{02}|^2)/(\hbar c) = 0.0265NF_{02}$, and $\kappa_{12} = (2\pi\omega_{L2}N|D_{12}|^2)/(\hbar c) = 0.0265NF_{12}$, with N being concentration in cm^{-3}, and F_{02} and F_{12} being absorption oscillator strengths for the respective transitions. In the criteria given above, $\Omega_{02} = D_{02}E_{L10}/(2\hbar)$ and $\Omega_{12} = D_{12}E_{L20}/(2\hbar)$ are half of the Rabi frequency for the $|0>\rightarrow|2>$ and $|1>\rightarrow|2>$ transitions. E_{L10} and E_{L20} are the amplitudes of the first and second laser fields. With strong transitions, $|\Omega_{02}|$ can be as large as $|\Omega_{02}| \simeq 2 \times 10^8\sqrt{I}$, where I is the power density in W/cm^2. Taking a typical case of 0.001 J per pulse in a 2 ns pulse for laser 1 and 2, and for a beam diameter of 1mm, we could have $|\Omega_{02}|$ or $|\Omega_{12}|$ with peak values above or around 10^{12}/s. With such a value, the quantities on the left hand side of the criteria can be less than 0.02 for concentrations up to about 5×10^{13}/cm^3. In the next part of this section we will see how these conditions come about.

In the pulse-delay mode used here, when laser 1 begins to build in intensity, laser 2 is already very bright, resulting in a large Aulter-Townes splitting in the upper state. Thus, at early time in the pulse of laser 1 the splitting keeps the population in the ground state until laser 1 has become of comparable intensity with laser 2. By the time laser 1 has become much more intense than laser 2, the population has all been transferred to state $|1>$. At this stage, however, the coupling between states $|1>$ and $|2>$ has become too weak to overcome the large splitting that now exists in state $|2>$ due to laser 1. Thus, with laser 1 being time delayed there is never a time when the "gap" between the eigenvalues is relatively small that any population can be transferred to spoil the adiabatic approximation. Without the time delay between the two lasers, the adiabatic approximation would not work at all in this double resonant situation.

The conditions stated above are necessary to insure that these laser fields propagate with no distortion at a velocity nearly equal to the speed of light in vacuum. They also insure that the two polarizations in the equations for the propagation of these laser fields,

$$\frac{\partial E_{L10}}{\partial z} + \frac{1}{c}\frac{\partial E_{L10}}{\partial t} = 4\pi i \frac{\omega_{L1}}{c} P_{L10}^+(z,t),$$

$$\frac{\partial E_{L20}}{\partial z} + \frac{1}{c}\frac{\partial E_{L20}}{\partial t} = 4\pi i \frac{\omega_{L2}}{c} P_{L20}^+(z,t),$$

are sufficiently small to avoid significant changes in propagation velocity from c. If one uses a purely adiabatic theory, the polarizations at ω_{L1} and ω_{L2} are zero when one neglects spontaneous decay during the pulse lengths. However, this is inconsistent with the fact that the number of ω_{L1} photons absorbed is equal to the number of atoms in the beam, while the number of ω_{L2} photons increases by the same number when state $|1>$ lies below state $|2>$. To overcome this inconsistency one must go beyond the usual adiabatic treatment and to look for the next correction in determining the amplitude of state $|2>$. Such an extended treatment gives

$$4i\pi\frac{\omega_{L1}}{c}P_{L10}^+ = -\kappa_{02}\frac{|\Omega_{12}|^2}{[|\Omega_{02}|^2 + |\Omega_{12}|^2]^2}\left[\frac{\partial E_{L10}}{\partial t} - \frac{E_{L10}}{E_{L20}}\frac{\partial E_{L20}}{\partial t}\right],$$

$$4i\pi\frac{\omega_{L2}}{c}P_{L20}^+ = -\kappa_{12}\frac{|\Omega_{02}|^2}{[|\Omega_{02}|^2 + |\Omega_{12}|^2]^2}\left[\frac{\partial E_{L20}}{\partial t} - \frac{E_{L20}}{E_{L10}}\frac{\partial E_{L10}}{\partial t}\right],$$

Our stated conditions insure that both the coupling between the two fields and the terms that modify propagation velocity are very small so that the two pulses propagate without distortion at a speed close the speed of light in vacuum. As an example, let us assume a gaussian pulse shape for lasers 1 and 2 $E_{L10} = E_{L1a}e^{-([t-(3/2)\tau-z/c]/\tau)^2}$, and $E_{L20} = E_{L2a}e^{-([t-z/c]/\tau)^2}$, where E_{L1a} and E_{L2a} are the time independent parts of the field amplitudes for laser 1 and 2, and the time delay of laser 1 is chosen to be $(3/2)\tau$. With this pulse shape, the populations of states $|0>$ and $|1>$ are shown as a function of t/τ in Figure 2. By choosing the proper time at the front (z=0) of the gas cell, the ratio of the two populations can be made to take on any chosen value. Based on adiabatic theory, which requires $\tau\sqrt{|\Omega_{02}|^2 + |\Omega_{12}|^2} \gg 1$ at any time when population transfer could occur. The populations of the levels are given as

$$\rho_{11} = \frac{|\Omega_{20}|^2}{|\Omega_{12}|^2 + |\Omega_{20}|^2},$$

$$\rho_{22} = 0,$$

$$\rho_{00} = \frac{|\Omega_{12}|^2}{|\Omega_{12}|^2 + |\Omega_{20}|^2}.$$

With this time delay (i.e. $(3/2)\tau$) the time relative to the peak in laser 1 intensity at which a fraction F of the atoms are excited is $t_d/\tau = 3/4 + (1/6)\ln((|\Omega_{02a}|^2/|\Omega_{02a}|^2)((1 - F)/F))$. The short pulse laser is timed so that its peak in intensity at the entrance window coincides with this time, if $\rho_{11} = F$ is the value required for phase matching.

What we have discussed so far is the first stage of the problem treated here, i.e. the coherent preparation of the initial condition. Next, we fire a much shorter pulse laser at ω_S at a delayed time t_d. The relative populations of the two levels at the time when the short-pulse laser is fired is solely dependent upon the time delay between the first two lasers and the ratio of the amplitudes of the Rabi frequencies. For small time delays, the population of state $|1>$ is $\rho_{11} \simeq 0$, while for large time delays the population of state $|1>$ is $\rho_{11} \simeq 1$. Ideally, the state $|1>$ should be a low lying metastable state that cannot radiate to any lower state in order to keep ρ_{11} from being depleted by amplified spontaneous emission (ASE). We assume that the short pulse laser is sufficiently weak, so that the populations of levels $|0>$ and $|1>$ are not modified significantly by the four-wave mixing, and that its pulse length is so short that their populations and coherence do not change appreciable due to their adiabatic tracking of lasers 1 and 2. This requires that the number of photons in the generated four-wave mixing field be small compared with the number of excited atoms in state $|1>$ which lies in the laser beams. Obviously the intensity of the laser at ω_S must be quite low. In the situation we are describing here, the off-diagonal elements of the density matrix are given accurately by[6], so that $\rho_{02} = \rho_{12} = 0$ and

$$\rho_{10} = -\frac{\Omega_{02}\Omega_{21}e^{i(k_{L1}-k_{L2})z}}{|\Omega_{12}|^2 + |\Omega_{20}|^2}. \tag{1.}$$

With the assumption that the intensity of laser 3 is such that the population of states $|0>$ and $|1>$ remain nearly unchanged during the short-pulse laser duration, the amplitude of state $|3>$ can be found by solving

$$\frac{da_3}{dt} = i\Omega_{30}e^{i(\omega_3-\omega_0-\omega_R)t}e^{ik_Rz}a_0$$
$$+ i\Omega_{31}e^{i(\omega_3-\omega_1-\omega_S)t}e^{i(k_S+k_{L1}-k_{L2})z}a_1, \tag{2.}$$

where $\Omega_{03} = D_{03}E_{R0}/(2\hbar)$ and $\Omega_{13} = D_{13}E_{S0}/(2\hbar)$ are half Rabi frequencies between $|0> \rightarrow |3>$ and $|1> \rightarrow |3>$, with E_{R0} and E_{S0} being the amplitudes of the four-wave mixing field and the short pulse laser. Here, we have taken the state vector to be

$$|\Psi(z,t)> = a_0e^{-i\omega_0t}|0> + a_2e^{-i\omega_2t}|2>$$
$$+ a_1e^{-i\omega_1t}|1> + a_3e^{-i\omega_3t}|3>, \tag{3.}$$

and the position dependent phase factor from the otherwise time independent a_1 has been combined with the position dependent phase from laser 3. The time dependent Schrödinger equation for a_3 must be solved simultaneously with

$$\frac{\partial E_{R0}}{\partial z} + \frac{1}{c}\frac{\partial E_{R0}}{\partial t} = 4\pi ik_R P^+_{\omega_{R0}},$$
$$\frac{\partial E_{S0}}{\partial z} + \frac{1}{c}\frac{\partial E_{S0}}{\partial t} = 4\pi ik_S P^+_{\omega_{S0}}, \tag{4.}$$

where $P_{\omega_{R0}}$ and $P_{\omega_{S0}}$ are the amplitudes of the polarization of the vapor at the frequencies of the short pulse laser and four-wave mixing field, respectively. They are given by

$$P^+_{\omega_R0} = D_{03}Nb_3a_0^*,$$
$$P^+_{\omega_S0} = D_{13}Nb_3a_1^*, \tag{5.}$$

where $b_3 = e^{i\Delta_3t}e^{-ik_Rz}a_3$, with $\Delta_3 = \omega_R - \omega_3 + \omega_0$. Within our model $|\rho_{01}|^2 = \rho_{00}\rho_{11}$. This assumes that collisional relaxation is very unlikely to occur for a significant fraction of the atoms in the laser beam during the pulse length of lasers 1 and 2. Also, we assume that processes like ASE out of level $|1>$ do not affect the populations or the phase coherence of any of the states involved in our model. This makes the choice of these states of critical importance. We will discuss this further at a later point.

We Fourier transform both sides of Equations (2), (4), and (5) with respect to time, and after combining the results

$$\frac{\partial R}{\partial z} - i\frac{\omega}{c}R = -i\frac{\kappa_{03}\rho_{00}}{\Delta_3+\omega}R - i\frac{\rho_{10}\sqrt{\kappa_{03}\kappa_{31}(\omega_R/\omega_S)}}{\Delta_3+\omega}S,$$
$$\frac{\partial S}{\partial z} - i\frac{\omega}{c}S = -i\frac{\rho_{01}\sqrt{\kappa_{03}\kappa_{31}(\omega_S/\omega_R)}}{\Delta_3+\omega}R - i\frac{\kappa_{31}\rho_{11}}{\Delta_3+\omega}S, \tag{6.}$$

where $\kappa_{03} = 2\pi|D_{03}|^2\omega_R N/(\hbar c)$ and $\kappa_{13} = 2\pi|D_{13}|^2\omega_S N/(\hbar c)$. $S = S(z,\omega)$ and $R(z,\omega)$ are the Fourier transforms of E_{S0} and E_{R0}, respectively. In carrying out the transform we have treated a_0 and a_1 as time independent over the pulse length of the laser at ω_S. In the following, we will assume Gaussian pulse profile for the short pulse laser and let $t_r = t - z/c$.

The above equations can be easily solved and the inverse Fourier transform carried out. We obtain, for frequency ratio times the square of the four-wave mixing field,

$$\frac{\omega_S E_{R0}^2}{\omega_R} = \frac{R_{max}}{4\pi} \left| \int_{-\infty}^{\infty} dw\, S(0,w) e^{-iw(t_r)/\tau} \left(1 - e^{-iv(w)z}\right) \right|^2, \tag{7.}$$

where

$$R_{max} = \frac{1}{\left[\sqrt{\frac{\rho_{11}\kappa_{13}}{\rho_{00}\kappa_{03}}} + \sqrt{\frac{\rho_{00}\kappa_{03}}{\rho_{11}\kappa_{13}}}\right]^2}, \tag{8.}$$

and

$$v(w) = \frac{\rho_{11}\kappa_{13}\tau + \rho_{00}\kappa_{03}\tau}{\Delta_3\tau + w}. \tag{9.}$$

Above, we have introduced the dimensionless variable $w = \omega\tau$, where τ is a measure of the width in time of E_{s0}, being of the order of magnitude of the half-width at half maximum. In terms of w we expect that $S(0,\omega) = \tilde{S}(0,w)$ becomes small once $|w| \gg 1$.[7]

The requirements $|\Delta_2\tau| > 20$ and $|v(0)L| \leq \pi$ assures that only the first two terms in following expansion

$$v(w) = \left(\frac{\kappa_{03}\rho_{00} + \kappa_{13}\rho_{11}}{\Delta_3}\right)\left(1 - \frac{\omega}{\Delta_3} + \left(\frac{\omega}{\Delta_3}\right)^2 + \cdots\right)$$

need to be kept. With this approximation, a much simpler analytic form for Eq. (7) can be derived. For most smooth pulse shapes the requirements $|\Delta_3\tau| > 20$ and $|v(0)L| \leq \pi$ are sufficient to make this approximation accurate. In this limit, most of the contribution to the integral comes while $|\Delta_3\tau| \gg |w|$, so that the above approximation can be truncated after only a few terms. This is because $\tilde{S}(0,w)$ is very small once $|w| > 4$, or so, for most smooth pulse shapes. We find

$$\frac{\omega_S I_R(z,t_r)}{\omega_R I_S(0,t_r)} = R_{max}\left|1 - e^{-iv(0)z}\frac{E_S(0,t - z/v_R)}{E_S(0,t - z/c)}\right|^2, \tag{10.}$$

where v_R is a proper group velocity for the four-wave mixing field. This group velocity satisfies

$$\left|\frac{z}{v_R} - \frac{z}{c}\right| = \frac{(\kappa_{31}\rho_{11} + \kappa_{03}\rho_{00})z}{\Delta_3^2}.$$

As long as the quantity on the left in the last equation is much less in absolute value than τ there will be little slippage between the peaks of the two contributions in Eq. (7). This says that we must have

$$\left|\frac{(\kappa_{31}\rho_{11} + \kappa_{03}\rho_{00})z}{\Delta_3^2}\right| = \left|\frac{v(0)z}{\Delta_3}\right| \ll \tau,$$

or, $|v(0)z/(\Delta_3\tau)|$, so that if $|v(0)z| \leq \pi$ and $|\Delta_3\tau| > 20$, there will be no significant separation of the peaks. The same condition is also (for most pulse shapes) sufficient to make keeping only the first two terms in the series for $v(w)$ sufficient for the accurate evaluation of the inverse Fourier transform. Thus, when $|\Delta_3\tau| > 20$ and $|V(0)z| \ll |\Delta_3\tau|$, the ratio of $E_S(0, t - z/v_R)/E_S(0, t - z/c) \simeq 1$, and Eq. (10) can be written as

$$\frac{\omega_S I_R(z, t_r)}{\omega_R I_S(0, t_r)} = 2R_{max}\left[1 - \cos(v(0)z)\right]. \tag{11.}$$

The nature of this solution is obvious. We get the largest values for the ratio of four-wave mixing photon flux to that of the incident short pulse laser when $\kappa_{03}\rho_{00} = \kappa_{13}\rho_{11}$, resulting in $R_{max} = 1/4$. Later, we show that this is the requirement for phase-matching for any Δ_3. The part of the solution in square brackets (Eq. (11)) represents the interference between a wave that propagates without distortion at the velocity of light in vacuum and another term which can be strongly distorted due to dispersion when $\Delta_3\tau$ is small. For larger detuning from $|3>$ this term also propagates without distortion at a group velocity, as described above. These two terms interfere constructively when $|v(0)L| = \pi$, provided that the two pulses still overlap spatially when the pulses exit on the other side of the cell. Under this circumstance, if the phase matching condition is met the initial flux of short-pulse photons in at ω_S is the same as the flux of photons out at the four-wave mixing frequency.

If the conditions $|\Delta_3\tau| > 20$ and $|v(0)L| \leq \pi$ were not well satisfied, the two waves can physically separate and the waves will spread as they propagate. In this situation a conversion efficiency of 25% to 50% is typical if the phase matching condition which maximizes R_{max} is satisfied. When our approximation is valid the peaks of the two parts separate very slowly and the peak height of the second wave stays close to that of the first wave. Thus, if $|\Delta kL| = \pi$, due to allowing the wave to propagate a distance such that the two terms interfere constructively, we get a conversion efficiency for the light in the short-pulse laser of unity when $R_{max} = 1/4$. At this depth of penetration, all of the incident short pulse photons have been converted to four-wave mixing photons. This is the ultimate in conversion efficiency.

List of References

1. K.J. Boller, A. Imamoglu, and S.E. Harris, Phys. Rev. Lett. **66**, 2593 (1991).
2. G.Z. Zhang, K. Hakuta, and B.P. Stoicheff, Phys. Rev. Lett. **71**, 3099 (1993).
3. A. Kasapi, M. Jain, G.Y. Yin, and S.E. Harris, Phys. Rev. Lett. **74**, 2447 (1995).
4. M. Jain, H. Xia, G.Y. Yin, A.J. Merriam, and S.E. Harris, Phys. Rev. Lett. **77**, 4326 (1996). For more recent experimental works related to a double-Λ scheme see for example, A.J. Merriam et al. Phys. Rev. Lett. **84**, 5308 (2000).
5. M.D. Lukin, P.R. Hemmer, and M.O. Scully, Advances in Atomic, Molecular, and Optical Physics, Vol 42, 347 (2000).
6. J. Oreg, F.T. Hioe, and J.H. Eberly, Phys. Rev. A**29**, 690 (1984). See also: J.H. Eberly, R. Rahman, and R. Grobe, Phys. Rev. Lett. **76**, 3687 (1996).
7. The question of how much $|w$ should be greater than 1 is dependent on the functional form of E_{S0}.
8. M.G. Payne, W.R. Garrett, R.C. Hart, and I. Datskou, Phys. Rev. A**42**, 2756 (1990).

Probing Coherent and Incoherent Intermediate States With RIS: A Laser-Induced Catalyst

W.R. Garrett[1], L. Deng[2], M.G. Payne[3] and Y. Zhu[3]

[1]*Department of Physics, University of Tennessee, Knoxville, TN,* [2]*NIST, Washington, DC*
[3]*Department of Physics, Georgia Southern University, Statesboro, GA*

Abstract. Since its inception, a primary focus of resonance ionization spectroscopy has been sensitive detection of atomic species in one form on another. But RIS has found many applications in spectroscopy where the nature of a resonant transition was of interest, and RIS provided a very useful tool for measuring a given feature. A large set of examples from this category can be found in studies of quantum interferences as they affect multiphoton-resonant transitions in gases and vapors. In many instances RIS has been used to show various effects that can be produced when two different optical paths simultaneously link a resonant transition in a gaseous system. We note that RIS has been used to show dramatic effects of wave mixing interferences on certain multi-photon driven resonant excitation processes. These include: (1) Complete destructive interference from a FWM field can reduce to zero the odd-photon-mediated excitation probability of a dipole allowed transition [1-4]; (2) For dual beam pumping the resonant excitation probability is restored but at a shifted frequency, [2]; (3) Separate resonant excitation profiles of a given transition can be produced under multi-beam pumping, all in conformance with theoretical predictions [3]; Quantum interference can suppress an ac Stark shift [4].

The combination of quantum interference behavior and RIS can be used to demonstrate coherent control of an excitation process in a new and novel way. We describe the results of a study in which three photon induced excitation of an excited state of Kr occurs only through a laser induced collisional energy transfer (LICET) from Xe [5]. The experiment serves as an example of coherent control of a primitive reaction. Or it can be viewed as a laser induced catalyst for a transition. In a broader context we have demonstrated a method of separating coherent from incoherent contributions to a multi-photon excitation involving a resonant intermediate state [6].

We note that the following phenomenon is now well known in the nonlinear optics and multi-photon processes communities. When a one-photon allowed atomic or molecular transition in a gaseous medium is resonantly driven by three-photon absorption (involving one, two, or three coherent beams, possibly of different colors), a nonlinear polarization is created at the sum frequency of the driving fields. This nonlinear electric polarization is the source for creation of a four-wave-mixing (FWM) field in the resonant medium. When a particular condition among a certain combination of physical and experimental parameters is satisfied, the generated FWM field evolves in amplitude and phase such that a strong interference is created between the transition probability amplitude due to three-photon excitation from the laser sources and that due to excitation by one FWM photon. That is, the nonlinear response

CP584, *Resonance Ionization Spectroscopy 2000: 10th Int'l. Symp.*, edited by J. E. Parks and J. P. Young
© 2001 American Institute of Physics 0-7354-0024-5/01/$18.00

of the target gas automatically creates two interfering pathways within the resonant medium. Moreover, the interference between the two transition amplitudes can be varied continuously from totally destructive to totally constructive by changes in either laser beam crossing angles (in multi-beam experiments), gas concentration, laser frequencies, or some combination of these experimental variables. (This evolution from destructive to constructive interference as a function of laser detuning from three-photon resonance is the mechanism responsible for the interference-based frequency shifts that are produced in the class of transitions under consideration here [14]). This well established phenomenon occurs in any circumstance where the product of oscillator strength F_{0n} for the transition, the path length z in the gas, the concentration N_0 of the resonant species, and the effective laser bandwidth Γ_L, satisfy the condition [13],

$$\pi z N_0 \, e^2 \, F_{0n}/mc >> \Gamma_L , \qquad (1)$$

where e and m are the electronic charge and mass respectively.

Thus when one or more laser fields are three-photon-resonant in a target system in which the conditions of Eq. (1) are satisfied, a strongly-influential four-wave mixing field at the sum frequency of the driving field(s) is generated within the target medium. The transition amplitudes from the two paths created by the laser field(s) and the wave-mixing field can be totally destructive, rendering a zero transition probability, or they can be constructive, rendering a maximum in the transition probability. The very robust effect of the interference persists over wide ranges of concentrations, laser bandwidths, admixtures of other gases, etc. and is controllable through beam geometry, target gas concentration, buffer gas concentration and pump frequency, but it is independent of laser intensity [6].

For a path length of ≈ 1 cm with single mode laser sources (narrow bandwidths) for a strong transition (e.g. large oscillator strength in an atomic system), the concentration must be above $\approx 10^{-3}$ Torr. For a weak transition (most molecules), a minimum concentration of 1 to 10 Torr would be required. But this can be an advantage rather than a disadvantage. The phenomenon persists to pressures of more than 10^3 Torr.

With this background we demonstrate an example of a "primitive" reaction and its control by the wave-mixing interference process. This requires two steps. First, we choose a transition that can exhibit the interference phenomenon under three-photon excitation. We elect two-color three-photon excitation of the $4d[1/2]_1$ state of Kr. This transition is also one-photon allowed; consequently three-photon excitation, as depicted in Fig. 1(b), is controlled by the wave-mixing mechanism through generation of ω_m when condition (1) is satisfied (Concentrations above ≈ 0.1 Torr for a 1 cm path). We do not repeat the analysis here, but note that a proper theoretical treatment of three-photon resonant excitation by two photons at angular frequency ω_1 and one photon at ω_2 must also include a nonlinear polarization of the medium that is unavoidably produced at the sum frequency $2\omega_1 + \omega_2$. This polarization generates a new electromagnetic wave, the four-wave-mixing field, at $\omega_m = 2\omega_1 + \omega_2$. This field

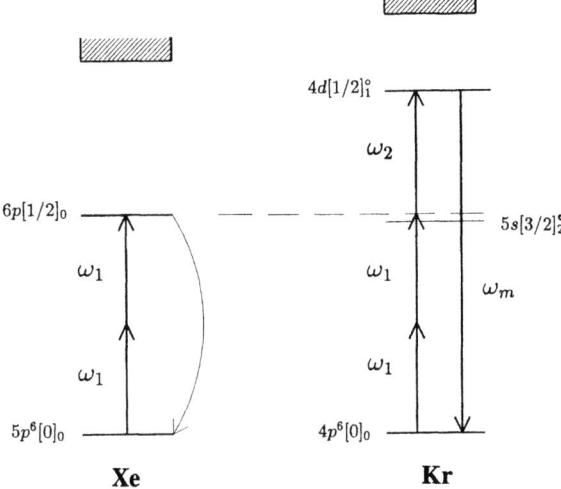

FIGURE 1. Energy levels and excitation scheme used in the Xe-Kr LICET experiment. The curved arrow in the diagram represents the return of Xe* to the ground state during a collision with Kr simultaneous with absorption of an ω_2 photon.

must be determined by solving Maxwell's equation with the nonlinear polarization as a source. But the polarization is proportional to the probability amplitude of the upper state. Thus one must solve the Schrodinger equation along with Maxwell's equation and determine the field and the probability amplitude of the upper state self consistently. Note that the transition amplitude then has two contributing terms, one due to three laser photons and one due to the internally generated wave-mixing field. When the laser photons are all co-directional, the nonlinear response is such that the transition amplitude from the FWM field evolves 180° out of phase and equal in magnitude to that from the laser field. That is, complete destructive interference is produced in the medium and no excitation occurs after propagating the first few absorption lengths into the medium. But if two laser beams are crossed at an angle greater than ≈0.1 radian, then the excitation can be changed from a minimum to a maximum. The resonant excitation profile then has normal line shape and amplitude, but with an interference-induced, intensity independent frequency shift [15,16]. The degree of the destructive interference can be varied from minimum to maximum under the control of the experimentalist through a wide latitude in choices of beam crossing angle, target concentration, relative wavelengths of two-color schemes, and resonant detuning.

Thus, the transition probability for the chosen Kr $4p^6[0]_0 \rightarrow 4d[1/2]_1$ transition, will be reduced to a value too small to observe if pumped with $\omega_r = 2\omega_1 + \omega_2$ photons in overlapped co-directional beams (ω_r is the resonant transition frequency). But if the two laser beams are counter-propagated, excitation will occur with normal probability,

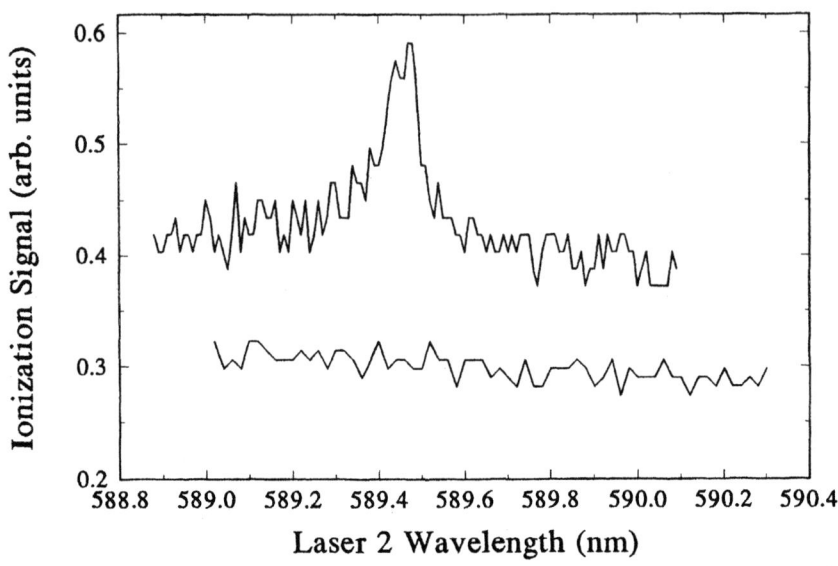

FIGURE 2. Resonant three-photon excitation profiles in Kr as one laser is held fixed at 249.6 nm while the second is tuned through three-photon resonance with the 4d[1/2]$_1$ state. Top trace was produced with counter-propagating beams. Bottom trace resulted from co-propagating beams. Beams were spacially and temprally overlapped in 102 Torr of pure Kr.

though at a slightly shifted resonant frequency. This is indeed the case, as shown in the excitation profiles of Fig. 2 in the upper and lower traces. In the experiment, two dye lasers are pumped with one Nd:YAG laser in 10 ns pulses. Excitation probabilities are monitored by measuring the wave-length dependence of the ionization rate with a charge collection electrode within the sample. The lower trace is the result produced by excitation with co-propagating beams and the upper by excitation with counter-propagation of the two laser sources (laser two is tuned through three-photon resonance with the 4d[1/2]$_1$ at $2\omega_1 + \omega_2$ with laser one held fixed at 249.6nm). Thus the wave-mixing- interference mechanism behaves as expected (concentration 50 Torr). The Kr state is *not* resonantly excited by coherent excitation with co-directional photons because of an internally generated destructive interference created between two paths. If the beams are counter-propagated, excitation occurs normally with a small shift that is below the limit of resolution in the present instance.

It is obvious from the results in Fig. 2 and from many other earlier studies of the same type, that coherent excitation can be turned on and off by the interference mechanism. In the usual discussion of coherent control by interfering paths, it is assumed that two degenerate excited states coexist in the target system and that the interference will affect the amplitudes unequally. However, unlike many molecules [5,6], simple noble gases do not have excited states that are superpositions of states that may be connected to different final products. In order to make a demonstration of IGIP control in a "chemical" sense, we introduce a "reactant" and create another explicit excitation route whereby the same Kr state can be excited along a new pathway. The alternative (degenerate) path involves a laser-driven reaction between Xe and Kr through another well studied process usually designated as light induced collisional energy transfer (LICET). This mechanism is depicted in Fig. 1(a) and (b).

The LICET process can be viewed from two perspectives. For the present case where two atomic systems are the reactants, Xe is first excited by two photons from laser 1 to produce Xe* (in the $6p[1/2]_0$ state in our example). The second laser is tuned to the energy difference between the Xe* level and the $4d[1/2]_1$ state of Kr. Then, during a collision of Xe* with Kr, the energy of one photon from laser 2 and the excitation energy of Xe* are absorbed by Kr to produce a ground state Xe and and excited Kr; that is, in the last step $Xe^* + h\omega_2/2\pi + Kr \rightarrow Xe + Kr^*$. The nonradiative return of Xe to the ground state is indicated by the curved arrow in Fig. 1(a). We note that an intermediate state in Kr occurs at an energy close to the $6p[1/2]_0$ level of Xe. We show this $5s[3/2]_2^{\circ}$ level in Kr with a thin line since, although parity forbidden in two-photons, it facilitates the transfer process during collision [19]. Alternatively, the colliding atoms can be viewed as a system AB in a molecular picture. A first excitation produces A*B (that correlates with A*+B at infinite separation). A second absorption carries the system to a higher state AB* (which correlates with A+B* at infinite separation). Again, we do not repeat any of the analysis of this process, but note that it is a much studied phenomenon [7]. We invoke this phenomenon to create an obviously different path in producing Kr excitation, thereby creating a coherently controlled "reaction" [7].

Specifically, we add Xe to a Kr sample and fire two co-directional dye laser beams simultaneously into the two-component system. Laser 1 is tuned to resonantly excite the $6p[1/2]_0$ state of Xe (energy 80,119.47 cm^{-1}) by two-photon absorption (Photons at 249.63 nm, 100 µJ per pulse from a frequency-doubled dye laser). Laser 2, which is spatially and temporally overlapped co-directionally with laser 1, is tuned through three-photon resonance with the Kr $4d[1/2]_1$ level as before. We know from the previous results [Fig. 2(lower)] that at concentrations above a few Torr, the wave-mixing-interference process prevents direct excitation of pure Kr. But if two photons first excite Xe, and then if a photon from the second laser along with the excitation energy of Xe* is absorbed by Kr (through LICET), the condition for destructive interference will not be met for a transition along this path, though the energy is degenerate with direct $2\omega_1 + \omega_2$ pumping of Kr. We can depict the process as, $Xe + Kr + 2h\nu_1 \rightarrow Xe^* + Kr$, and $Xe^* + Kr + h\nu_2 \rightarrow Xe + Kr^*$. Note that Kr excitation occurs only if the Xe is excited first and then transfers its excitation energy to the Kr along this alternate path [Fig. 1(a)]. The IGIP process prevents direct Kr excitation along the coherent three-photon path. Thus, the Xe acts in the manner of a catalyst. It starts and ends in its ground state, but the reaction will not go without it.

A series of experiments verifies these predictions. We have already shown in the lower trace in Fig. 2 that pure Kr shows no resonant excitation at the $4d[1/2]_1$ level with unidirectional pumping. This result is repeated in the bottom trace of Fig. 3, which shows the excitation profile in 90 Torr of pure Kr with co-propagation of the two laser beams. Laser 1 was set at 249.63 nm while laser 2 was scanned. No resonant excitation appears at 589.40nm. Under the same experimental conditions the ionization signal in 100 Torr of pure Xe is shown in the middle trace in Fig. 3. Laser 1 is set at the same 249.63 nm wavelength which corresponds to two-photon resonance with the $6p[1/2]_0$ state and laser 2 is scanned over the same wavelength region. This scan shows resonant excitation of the 14p state of Xe at 589.22nm, which is close to, but easily distinguished from, our chosen Kr state. Next we introduce a mixed sample of Xe and Kr. Laser 1 is again set to two-photon resonance in Xe and co-directional laser 2 is scanned across the Kr three-photon resonance. Now we expect to see Kr excitation occur through the LICET path. Indeed this is the case as shown in the top trace in Fig. 3. Now resonant excitation of Kr $4d[1/2]_1$ occurs with co-propagating pump beams, contrary to Fig. 2 where the transition was produced only when counterpropagating beams were employed. Here excitation is mediated along the alternative LICET path. Now the LICET process produces Kr excitation at 589.45 nm in addition to the Xe 16p line at 589.22 nm.

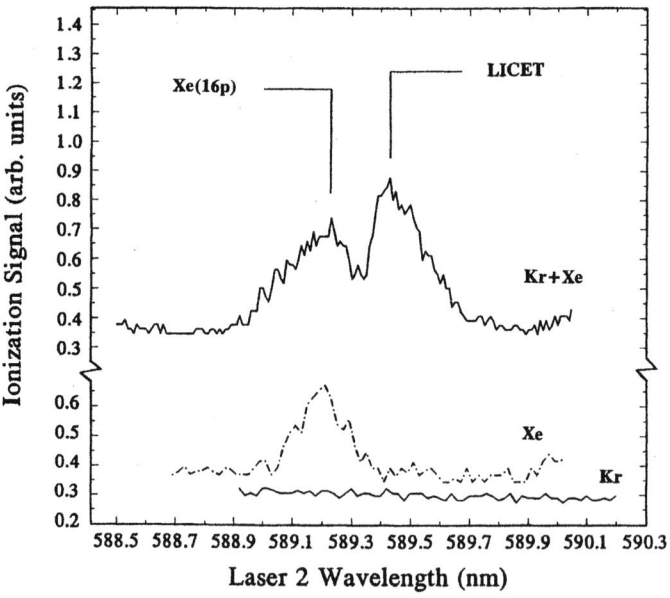

FIGURE 3. Results from experiments in 10 Torr of pure Kr (bottom trace); 100 Torr of pure Xe (middle trace); and 8 Torr of Xe mixed with 93 Torr of Kr (top trace). The Kr peak occurs only when two-photon excitation in Xe is employed with LICET to excite the Kr $4d[1/2]_1$ state.

Finally, since the two laser beams are co-propagating, the Kr resonance profile should disappear if laser 1 is detuned from the Xe resonance. This is true because the

LICET pathway will then be eliminated and the IGIP mechanism prevents direct three-photon excitation of the Kr level, independent of whether Xe is present or not. This expected effect was completely confirmed in the experiments.

CONCLUSION

In conclusion, we wish to stress that the interfering pathways approach to coherent control of a reaction can be implemented, with considerable flexibility, by making use of internally generated wave-mixing fields. In this context, we demonstrated how an alternate pathway in an interfering loop can be created by the previously studied method of changing laser beam geometries and by a new method involving introduction of an alternative path through a collision partner. This also represents the first time a primitive type of reaction between two chemically separate "reactants" have been carried out under some type of coherent control. We note that the concept of providing a more general "reactant" as control partner is an obvious extension of the method. Also note that a target species could be present in a "resonant" nonlinear medium in which the laser fields produce a wave-mixing field with a chosen interfering character lying in the continuum of the admixed target molecule. In principle the IGIP methodology can also be implemented in a unimolecular process where the alternative path can be provided by an intermediate state in a molecular system identical to the mode suggested in the original IP scheme [5,6].

REFERENCES

1. Miller, J. C., Compton, R. N., Payne, and Garrett, W. R. *Phys. Rev. Lett.* **45**, 114 (1980), Payne, M.G. and Garrett, W.R. *Phys. Rev. A* **26**, 356 (1982).
2. Garrett, W. R., Hart, R. C., Wray, J.C., Datskou, I., and Payne, M.G., *Phys. Rev. Lett.* **64**, 1717 (1990).
3. Garrett, W. R., Hart, R. C., Miller, J. C., Payne, M. G., and Wray, J. E., *Optics Comm.* **86**, 205 (1991).
4. Deng, L., Zhang, J. Y., Payne, M. G., and Garrett, W. R., *Phys. Rev. Lett.* **15**, 2035 (1994)
5. Garrett, W. R., and Zhu, Y., *J. Chem. Phys.* **106**, 2045 (1996).
6. Garrett, W. R., Zhu, Y., Deng, L., and Payne, M.G., *Optics Comm.* **128**, 66 (1996).
7. Brumer, P., and Shapiro, M., *Chem. Phys. Lett.* **126**, 541 (1986); J. Chem. Phys. **84**, 4103 (1986). Shapiro, M., Hepburn, J.W., and Brumer, P. *Chem. Phys. Lett.* **149**, 451 (1988).
8. The generic process referred to here as LICET was first described by L.I. Gudzenko and S.I. Yakovlenko, *Zh. Eksp. Teo. Fiz.* **62**, 1686 (1972) [Sov. Phys. - *JETP* **35**, 877 (1972)]; S.E. Harris and D.B. Lidow, *Phys. Rev. Lett.* **33**, 644 (1974).

Nonlinear Circular Dichroism in Chiral Molecules

S.M. Mahurin[1], R. Sullivan[1], R.N. Compton[1,2]

[1]Department of Physics, University of Tennessee, Knoxville, TN 37996
[2]Department of Chemistry and Physics, University of Tennessee, Knoxville, TN 37996

Abstract. Optical activity is manifested through circular dichroism (CD) and optical rotatory dispersion (ORD) which is the differential absorption of right and left circularly polarized light. Single photon CD has been studied for many years and is of practical benefit as a characterization method. Herein, non-linear circular dichroism in resonantly enhanced multiphoton ionization is studied for (S)-(+)- and (R)-(-)-2-butylamine. Measurements of mass resolved resonant and non-resonant multiphoton ionization of 2- butylamine entrained in a nozzle jet expansion into a linear time-of-flight mass spectrometer using right and left circularly polarized light from a high resolution OPO laser will be presented. The difference in ionization rate for right and left circularly polarized light for each enantiomer represents a measurement of a non-linear circular dichroism. In addition, circular-to-linear ratios were measured for resonantly enhanced multiphoton ionization of 2-butylamine through a 3s Rydberg state.

INTRODUCTION

Circular dichroism (CD), which is the differential absorption of right and left circularly polarized light, is a well-established tool for characterizing chiral molecules. Single photon CD effects are generally on the order of $\Delta\varepsilon/\varepsilon \sim 10^{-3}$ and are typically larger at shorter wavelengths. An unusually large circular dichroism of approximately 10% was recently reported[1] in the one-photon ionization of tyrosine nanocrystals. The chiral tyrosine also crystallizes into a chiral space group which presumably explains the large circular dichroism achieved in this experiment. Non-linear CD studies of chiral molecules have not been reported, however, two studies of non-linear optical rotatory dispersion (ORD) have appeared.[2,3] Circular dichroism effects have been predicted for angular distributions of photoelectrons ejected from optically active[4,5] and oriented[6] molecules. Circular dichroism in angular distributions of photoelectrons (CDAD) can be used to extract information about the alignment or orientation of laser excited states.[7] Appling et al.[8] used CDAD to probe the alignment of NO molecules excited to the A $^2\Sigma^+$ state by linearly polarized light and Cuellar et al.[9] presented the first CDAD measurements for an atom. The later study

CP584, *Resonance Ionization Spectroscopy 2000: 10th Int'l. Symp.*, edited by J. E. Parks and J. P. Young
© 2001 American Institute of Physics 0-7354-0024-5/01/$18.00

demonstrated m_j mixing of the intermediate $7p_{3/2}$ state of cesium during the laser pulse duration. Circular dichroism in the multiphoton ionization of chiral molecules in the gas phase with no alignment has yet to be reported.

EXPERIMENTAL

Multiphoton ionization mass spectra of the (S)-(+)- and (R)-(-)-2-butylamine enantiomers were obtained through supersonic expansion of the sample in an argon carrier gas into a linear time-of-flight mass spectrometer with a backing pressure of approximately 100 Torr. Accurate wavelength calibration was obtained using a Burleigh WA4500 pulsed wavemeter. The laser system used in these experiments consisted of a Nd:YAG pumped Continuum OPO Mirage laser which had the advantages of high resolution ($\sim.02$ cm^{-1}) and tunability. The laser was typically operated with 9 mJ/pulse focused with a 7.5 cm lens to produce a power density of approximately 10^{11} W/cm^2. The signal detected by two Galileo microchannel plates mounted in a chevron configuration was routed through a preamp, a boxcar averager and into a Hewlett-Packard Infinium digital oscilloscope. Samples of the (R)-(-) and (S)-(+) 2-butylamine were purchased from Aldrich (99% purity) and used without further purification.

RESULTS AND DISCUSSION

A typical time-of-flight mass spectrum of the 2-butylamine (taken at 440.62 nm) is shown in Figure 1. Tuning the laser around this wavelength (430 nm to 465 nm) produced no resonantly enhanced features and a mass spectrum similar

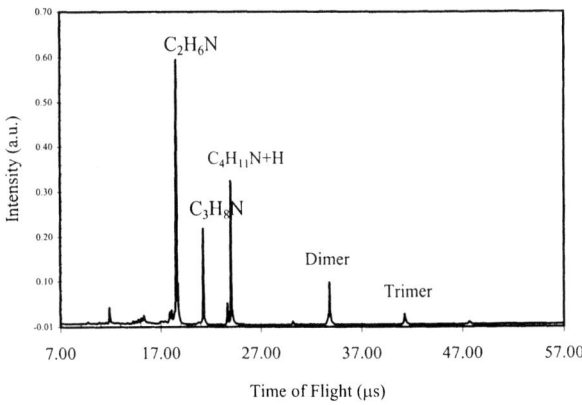

Figure 1. Nonresonant multiphoton ionization mass spectrum of 2-butylamine at a wavelength of 440.62 nm.

to that shown in Figure 1. Previous observation[10] of a (2+2) resonantly enhanced multiphoton ionization via an n→3s Rydberg transition at approximately 474 nm prompted closer inspection of this region in order to locate a resonance transition at which to perform the circular dichroism measurements. Figure 2 shows the resonance enhanced features. The two larger peaks (473.4 nm and 465.5 nm) are separated by 680 cm^{-1} which agrees with the spacing of approximately 700 cm^{-1} in the vibrational progression previously reported. The other peaks shown in this

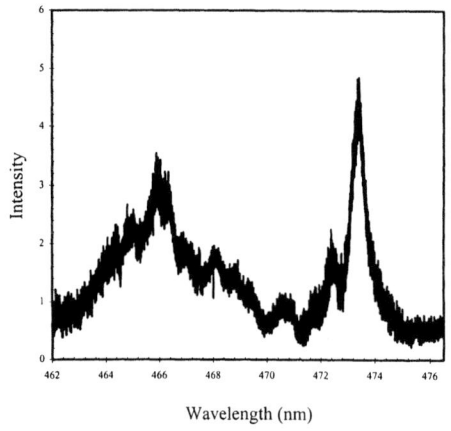

Figure 2. Wavelength dependence of racemic 2-butylamine showing the (2+2) REMPI through the 3s Rydberg state.

spectrum are attributed to other ro-vibrational frequencies.

Right-circularly polarized light (RCPL) and left-circularly polarized light (LCPL) were obtained using a single Fresnel rhomb and double Fresnel rhomb combination. Rotating the linearly polarized beam that exited the double Fresnel rhomb into the single Fresnel rhomb alternately produced RCPL and LCPL. The degree of circular polarization was assessed by studying the (3+2) REMPI of xenon through the 6s intermediate state. This transition is allowed for linearly polarized light, but forbidden for circularly polarized light. Disappearance of the ionization signal with circularly polarized light indicated the high degree of circular polarization. With the present experimental setup, the ionization signal did not completely disappear due to slight depolarization of the laser beam as it passed through the entrance window of the mass spectrometer. The circularly polarized beam was found to be >95% pure.

The rates of ionization for right circular, left circular, and linear polarizations were obtained by averaging one hundred laser shots at each polarization. The process was then repeated ten times to complete one experiment. Figure 3 shows the polarization data for the (R)-(-) and the (S)-(+)-2-butylamine taken at 473.35 nm. For (R)-(-)-2-butylamine, the ionization rates for the three different laser polarizations were measured to be $S_{RCPL} = 1.15\pm.04$, $S_{LCPL} = 1.12\pm.02$, and $S_{Lin}=1.00\pm.03$, where S represents the magnitude of the m/z=44 peak. The peak heights have been normalized such that the magnitude of the ionization for linearly polarized light is equal to one. From the statistical error of the experiment, the magnitude of the ionization for right- and left-circularly polarized

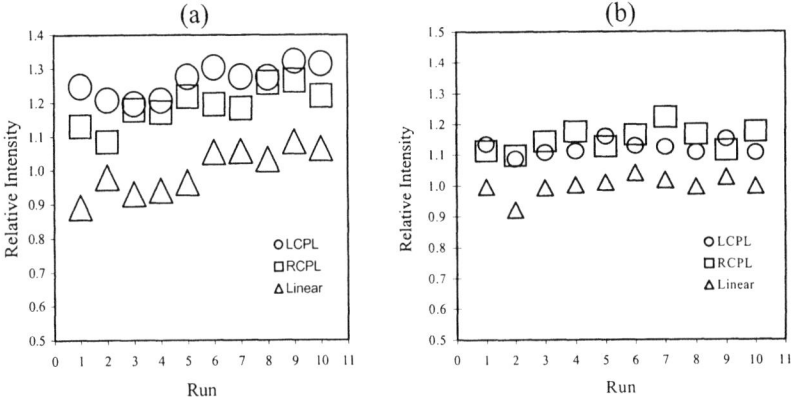

Figure 3. Polarization dependence of the rate of ionization for (a) (S)-(+)-2-butylamine and (b) (R)-(-)-2-butylamine. The relative intensity of each signal level is normalized to one for the signal level of linearly polarized light and the size of the data points reflect the size of the error bars.

light are identical, i.e., there is no detectable circular dichroism in the multiphoton ionization of 2-butylamine since the error bars overlap. However, it is instructive to note that the magnitude of the ionization for right-circularly polarized light is consistently higher than the ionization signal for left-circularly polarized light.

The magnitude of the ionization of (S)-(+)-2-butylamine for the different polarizations were measured to be $S_{RCPL} = 1.19\pm.06$, $S_{LCPL} = 1.26\pm.05$, and $S_{Lin} = 1.00\pm.07$. Once again, the ionization signal for right- and left-circularly polarized light were the same within the experimental error. It is interesting that left-circularly polarized light now consistently yields a higher ionization signal than right-circularly polarized light, which is exactly opposite to the case of (R)-(-)-2-butylamine. The ionization intensities for the two enantiomers cannot be directly compared since the intensities were obtained under different conditions, i.e., the amplifier settings and pressure were different. However, the intensities for right, left, and linear polarizations for a particular enantiomer can be compared since the experimental conditions were identical. It should be pointed out that since high precision is required to perform CD measurements, continuous wave sources are generally used as the light source. Inherent shot-to-shot variations in laser intensity with pulsed laser sources produced enough statistical error in the measurements to mask any circular dichroism effects.

Figure 4 shows the circular dichroism spectrum obtained for (R)-(-)-2-butylamine using an Aviv Model 202 circular dichroism spectrometer over the wavelength range from 220 nm to 350 nm. The CD spectrum for this molecule is quite simple with only one peak centered at approximately 226 nm. A possible mechanism for circular dichroism in the multiphoton ionization of a chiral molecule is that the resonance state might become differentially populated if that

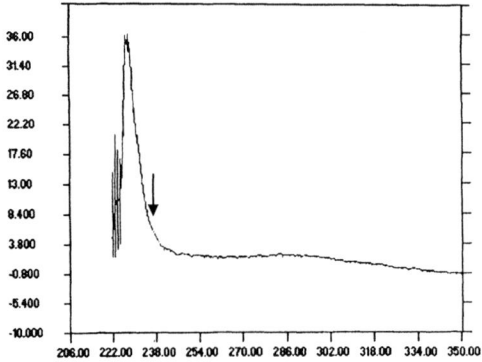

Figure 4. One-photon circular dichroism spectrum of (R)-(-)-2-butylamine. The arrow denotes the position of the two-photon excited 3s level.

state exhibits a single-photon circular dichroism. In this experiment, light of 473.35 nm (which corresponds to 236.7 nm at the one-photon level) was used to populate the 3s Rydberg level. From the CD spectrum, this particular level lies on the edge of the peak. Thus if the two-photon CD is of the same order of magnitude as the one-photon CD, only a small MPI-CD would be expected. However, it should be pointed out that it is not clear what effect each photon will have, i.e., it is likely that there will be a differential absorption of each photon leading up to and including ionization. Each step could thus serve to enhance or reduce the total circular dichroism. Clearly, more work, both theoretical and experimental, is required to determine the magnitude of these effects. The data presented in this work is indicative of a MPI-CD but not totally conclusive.

Circular-to-linear ratios for the two enantiomers were measured to be $S_C/S_{Lin}=1.23\pm.06$. Circular-to-linear ratios are typically used to determine the symmetry of the resonance state in a multiphoton experiment and are not confined to chiral molecules. McClain[11] has shown that, in two-photon absorption, transitions which preserve the symmetry of the ground state yield circular-to-linear ratios less than 3/2 (typically much less). Transitions that lower the symmetry of the ground state, i.e., the electronic state has less than the full molecular symmetry, produce circular-to-linear ratios approximately equal to 3/2.

The ratio obtained for 2-butylamine indicates that the transition preserves the symmetry of the ground state, which is not unusual for a Rydberg state.

REFERENCES

1. J. Paul, K. Siegman, *Chem. Phys. Lett.*, **304**, 23 (1999).
2. A. Gedanken, M. Tamir, *Rev. Sci. Inst.*, **58**, 950 (1987).
3. R. Cameron, G.C. Tabisz, *Mol. Phys.*, **90**, 159 (1997).
4. B. Ritchie, *Phys. Rev. A*, **13**, 1411 (1976).
5. B. Ritchie, *Phys. Rev. A*, **14**, 359 (1976).
6. B. Ritchie, *Phys. Rev. A*, **12**, 567 (1975).
7. R.L. Dubs, S.N. Dixit, V.M. McCoy, *J. Chem. Phys.*, **85**, 565 (1986): ibid. **85**, 6267 (1986).
8. J.R. Appling, M.G. White, T.M. Orlando, S.L. Anderson, *J. Chem. Phys.*, **85**, 6803 (1986).
9. L.E. Cuellar, C.S. Feigerle, H.S. Carman, R.N. Compton, *Phys. Rev. A*, **43**, 6437 (1991).
10. G. Siuzdak, J.J. BelBruno, *J. Phys. Chem.*, **94**, 4559 (1990).
11. W.M. McClain, *J. Chem. Phys.*, **55**, 2789 (1971).

Probing Single Ion Luminescence In Rare-Earth Doped Nanocrystals

Michael D. Barnes,[1] Adosh Mehta,[2] and Thomas Thundat,[2] and
Ramesh Bhargava[3]

[1] Chemical and Analytical Sciences Division, Oak Ridge National Laboratory
Oak Ridge, Tennessee 37831-6142

[2] Life Sciences Division, Oak Ridge National Laboratory
Oak Ridge, Tennessee 37831-6142

[3] Nanocrystals Technology, Briarcliff Manor, NY 10510

ABSTRACT. Recently there has been a great deal of attention focused on rare-earth doped nanocrystals (DNCs) as a new class of luminescent nanomaterials with novel and tunable optical properties. Such species have properties that make them attractive candidates for biological tags such as narrow spectral width and very high photochemical stability. However, the transitions that give rise to visible luminescence of rare-earth ions are nominally forbidden making luminescence from single ions very difficult to detect. In our experiments, single europium and terbium ions in isolated yttrium oxide nanocrystals (2 –15 nm diam.) were probed using time-resolved fluorescence microscopy techniques. In contrast with luminescence from larger crystals containing several ions, small particles believed to contain single ions show fascinating on-off behavior on a variable time scale as well as multiple discrete luminescence intensity levels. The latter behavior suggests interesting application in optical data storage.

INTRODUCTION

The past 15 years has seen amazing progress in analytical applications of ultrasensitive optical probes of luminescent species in condensed phase. Once barely considered as a possibility, single-molecule fluorescence detection and imaging using commercially available microscopy instrumentation is now fairly commonplace. Most condensed phase single-molecule fluorescence detection schemes trace their roots to Tomas Hirschfeld's landmark paper[1] in 1976 using multi-chromophore fluorescent proteins. However, about the same time Hurst, Nayfeh, and Young showed that single atoms in the gas phase could be detected using a resonance ionization format.[2] The latter approach has the important advantage of high specificity since the laser

CP584, *Resonance Ionization Spectroscopy 2000: 10th Int'l. Symp.*, edited by J. E. Parks and J. P. Young
© 2001 American Institute of Physics 0-7354-0024-5/01/$18.00

wavelength can be tuned to match a particular atomic transition of interest where background species do not absorb.

A primary disadvantage, however, is the complex experimental setup and gas-phase format. For many applications of single molecule/single-ion probes, one would like to have a convenient condensed phase format where the target species may be immobilized to a certain extent. Another key difference between single-molecule fluorescence probes and the resonance ionization approach is that, in fluorescence, one can detect several hundred to several thousand fluorescence photons from the molecule of interest, while an RIS approach involves the much more difficult problem of detecting a single ion. The focus of this paper briefly summarizes application of ultrasensitive time-resolved fluorescence imaging techniques to probe individual rare-earth ions trapped in metal-oxide nanocrystals.[3]

Recently, there has been a growing interest in rare-earth ion doped nanoparticles for multicolor phosphor applications. These species represent an interesting new class of nanoscale materials where the optical properties are defined by the dopant ion and are not strongly sensitive to the choice of host material. Of particular interest are quantum confinement (QC) effects and enhanced luminescence efficiency where the particle dimensions are comparable to Bohr radii of the atomic wavefunctions. Hybridization of electronic structure has been postulated as a mechanism for extraordinary enhancement (5 – 6 orders of magnitude) in luminescence efficiency of dopant ions in a nanoparticle environment. [4,5]

In these experiments, dry $Eu^{3+}:Y_2O_3$ nanoparticles on a quartz slide were continuously illuminated with the 514.5 nm line from an Argon ion laser and probed by time-resolved fluorescence imaging techniques similar to those used for single molecule studies in free solution,[6] polymer gels,[7] and polymer microspheres. [8] We used a Nikon TE300 inverted microscope in an epi-illumination configuration with a 0.85 N.A. 60x (dry) collection objective to simultaneously image the luminescence from several particles within a 15-μm diameter field of view. The characteristic red europium luminescence (near 620 nm) derives from the $^5D_0 \rightarrow {}^7F_2$ transition involving f electrons.[9,10] Eu^{3+} has the interesting feature of having two different excited levels accessible by 514 nm excitation . We used a 635-nm bandpass filter (55-nm bandwidth) to integrate all of the $^5D_0 \rightarrow {}^7F_2$ luminescence.

Figure 1 shows line scans taken from fluorescence images from three different $Eu^{3+}:Y_2O_3$ nanoparticles (200 ms integration) within the field of view. Luminescence from Eu^{3+} doped nanocrystals appeared as diffraction-limited spots with intensities ranging from 200 – 1000 counts above background per frame (200 ms exposure). As shown in Figure 1, on-off blinking was easily observed on a variable time scale as well as significant variation in duty-factor ("on"-time as a percentage of total measurement time) from particle to particle. Larger crystals or nanoparticle clusters (between 50 and 500 nm) showed significantly larger signals that were continuous and scaled linearly with pump power. In general, we observed an inverse correlation between luminescence intensity and dark-state persistence time: Some particles flashed brightly for only a few frames, while other particles showed decreased intensity with longer persistence times.

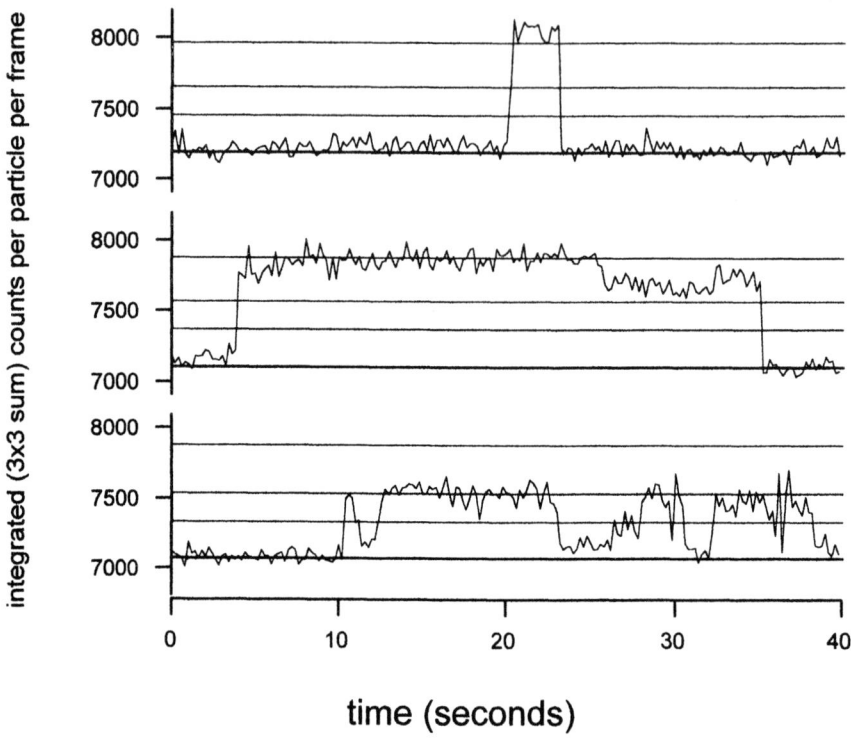

FIGURE 1. Line scans from three different Eu3+ doped nanocrystals showing on-off blinking behavoir. Each point is an integrated luminescence intensity taken from a 3x3 bin about the centroid of the diffraction-limited fluorescence spot. The different dashed lines indicate the dark and three bright-state luminescence intensity levels.

The mechanism for blinking behavior in the luminescence of Eu^{3+}: Y_2O_3 nanoparticles is clearly different than that of semiconductor quantum dots or fluorescent molecules. For semiconductor species, the mechanism derives from exciton formation followed by carrier trapping at defect sites in the crystal. In our system, no excitons are formed from (single-photon) absorption of 514.5 nm radiation because of the high bandgap of the yttrium oxide host material (6.2 eV); for larger particles, no evidence of two-photon absorption was observed. We believe that the blinking and multiple bright state effects derive from local symmetry fluctuations at the site of the dopant ion. For a Eu^{3+} ion, the E1 transition moment associated with the $^5D_0 \rightarrow {}^7F_2$ transition is zero in octahedral symmetry, while the transition moment may be restored in sites of reduced symmetry. We also observe "random-access" behavior in switching between different bright states; any one particular bright state can be accessed from any other. This would seem to be further proof of single-ion luminescence: If the multiple bright states were due to having say 3 ions within the particle, one would expect that **3** ↔ **0** transitions would take place stepwise through **1** and **2** intermediates. We observe that the most common transitions involving level **3** are direct transitions between **3** and **0**.

Since a significant fraction of the input photon energy must be dissipated in phonon modes of the crystal, it is conceivable that fluctuations between different stable or metastable symmetry configurations sites with correspondingly different transition moments could be induced by excess thermal energy provided by the pump laser. To test the idea of 'thermal activation' induced by the pump laser, we examined the luminescence of a nanoparticle-doped polymer thin film. In this case, we observed a complete disappearance of particle blinking as well as a significant reduction in overall luminescence efficiency. The thermal coupling of the particle with polymers in the thin film apparently results in enhanced thermal energy dissipation, and essentially 'freezes' the particle in different configurations that result in an overall decrease in luminescence intensity.

Fluorescence imaging of individual Eu^{3+} ions in Y_2O_3 nanoparticles have revealed striking on-off blinking and multiple-bright state luminescence that appears to be related to thermal activation of the particle by the pump laser. Such behavior cannot be observed in ensemble or multi-chromophore measurements, and is clearly different from the binary on-off behavior observed for single fluorescent molecules and quantum dots. We propose that the discrete multi-level luminescence - and oscillatory switching between levels - from these particles derives from local symmetry fluctuations that modulate the electric dipole transition moment. These single-particle luminescence-imaging experiments may provide further new insights into quantum confinement effects in doped nanocrystals that might be obscured under ensemble averaging. Ultimately, if access to different levels can be controlled, these species may have important applications in optical logic and information storage, and nanoscale photonics.

ACKNOWLEDGMENTS

This research was sponsored by the Division of Chemical Sciences, Office of Basic Energy Sciences, U.S. Department of Energy, under contract DE-AC05-00OR22725 with Oak Ridge National Laboratory, managed and operated by UT-Battelle, LLC. Adosh Mehta acknowledges support of the ORNL Postdoctoral Research Program administered through Oak Ridge Institute of Science and Engineering (ORISE).

REFERENCES

1. T. Hirschfeld, Appl Optics **15**, 2965 (1976)
2. G. S. Hurst, M. H. Nayfeh, and J. P. Young, Appl. Phys. Lett. **30**, 229 (1977).
3. M. D. Barnes, et al. J. Phys Chem. B **104**, 6099 (2000).
4. R. N. Bhargava, et al. Phys. Rev. Lett **72**, 416 (1994)
5. R. N. Bhargava, et al. Phys Status Solidi B **210**, 621-629 (1998).
6. X. Xu, and E. S. Yeung, Science **275**, 1106 (1997).
7. R. M. Dickson, D. J. Norris, Y-L. Tzeng, and W. E. Moerner Science **274**, 966 (1996).
8. M. D. Barnes, et al. Cytometry **36**, 169 (1999).
9. See, for example, G. H. Diecke *Spectra and Energy Levels of Rare Earth Ions in Crystals*, Interscience New York (1968).
10. F. Parsapour, D. F. Kelley, and R. S. Williams, J. Phys. Chem. B **102**, 7971 (1998).

SESSION VII

Resonant Laser Mass Spectrometry for Environmental and Industrial Chemical Trace Analysis

Ulrich Boesl[a], Jörg Rink[a], Volker Distelrath[a], Peter Püffel[b]

[a] Technische Universität München, Institut für Physikalische Chemie, 85747 Garching, Germany
[b] BMW-AG, Abteilung EA 82, 80788 München, Germany

Abstract. A promising new method for pollutant trace analysis is resonant laser mass spectrometry. It combines selectivity, sensitivity, and speed of measurement. In this paper, two examples of application are presented: exhaust analysis of combustion engines and analysis of polycylcic aromatic compounds in soil samples. The sensitivity of small, mobile instruments is discussed as well as alternative laser-based techniques in the case formation of cations by nanosecond lasers is improbable.

INTRODUCTION

Resonant laser mass spectrometry is particularly well adapted to fast on-line analysis of trace compounds in complex gas mixtures. Time resolution in the range of 10 ms is possible using modern commercial lasers. These are ideal preconditions for the trace analysis of pollutants in the exhaust of combustion engines in highly dynamic working conditions, such as starting , acceleration and deceleration processes, switching gears or single failures of ignition. Often, the electronic motor controlling system cannot follow these fast processes so that unwanted operating conditions are reached for short time intervals resulting in the emission of large amounts of pollutants. Particular attention should be paid to the effect of these short term disturbances on the operation of catalytic converters in motorised vehicles. Even traces of the motor oil can be detected in the exhaust by resonant laser mass spectrometry. This can be used for the study of processes responsible for the consumption of motor oil and is discussed here. Fast analysis not only allows the study of dynamic processes but also the investigation of large amounts of samples in a reasonable span of time. A typical case is the screening of industrial areas for pollution e.g. by polycylcic aromatics. By combination of laser desorption with resonance ionisation such a fast screening of soil samples is achievable.

SENSITIVITY

There exist highly sophisticated conventional techniques for the analysis of organic trace compounds with exceptional sensitivity and selectivity. However, they are time consuming (due to extensive sample preparation and clean-up) and laboratory-bound.

CP584, *Resonance Ionization Spectroscopy 2000: 10th Int'l. Symp.*, edited by J. E. Parks and J. P. Young
© 2001 American Institute of Physics 0-7354-0024-5/01/$18.00

Resonant laser ionisation offers the possibility of very fast measurement on the time-scale of minutes, seconds or even milliseconds (e.g. with modern 500Hz excimer lasers) and of reduced size. Both are preconditions for mobile analytic instruments, which allow on-line and on-site measurements. It's this feature, which may help to make resonance ionisation mass spectrometry a successful commercial technique. In the following, an estimation of the sensitivity is given for the case of pulsed lasers with reduced size (and therefore reduced power) and of a small vacuum system. Let us assume the use of a nanosecond-laser with 100 μJ UV-pulse energy and a time-of-flight mass spectrometer with a 150 l/s turbomolecular pump.

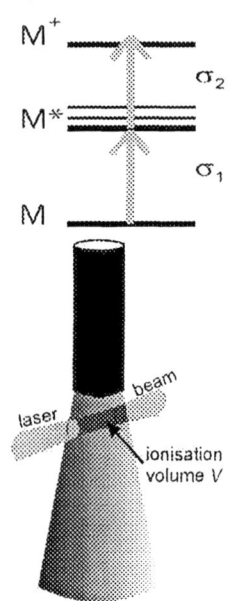

FIGURE 1. Excitation scheme and definition of ionisation volume V.

(A) If the first absorption step is not saturated ($\sigma_1 I \tau \ll 1$) the equation $[M^+] = [M] \, 1/2 \, \sigma_1 \sigma_2 I^2 \tau^2$ is valid. Assuming typical molecular absorption cross sections $\sigma_1 = 10^{-18}$ cm^2 and $\sigma_2 = 10^{-17}$ cm^2, laser pulse length $\tau = 10$ ns and laser intensity $I = 10^7$ W/cm^2 (0.1 mm^2 focus, 100 μJ & 10 ns) results in a ionisation efficiency $[M+]/[M] = 10^{-1}$ within the laser focus.

(B) Assuming furthermore, that at least 10 ions per laser pulse can be detected, the number of 100 neutral molecules within the laser focus should be detectable.

(C) If ionisation takes place in a molecular beam with 1mm diameter, the ionisation volume V is 0.1 mm^3. A pressure of p $= 10^{-3}$ mbar (3 10^{10} molecules / mm^3) within the gas beam then gives a total number of 3 10^9 molecules in volume V.

(D) Comparison of minimum detectable trace molecules (B) and total number of molecules within the ionisation volume (C) gives a sensitivity of $\approx$ 30 ppb. This number can easily be improved by a factor of 10^2 or 10^3 using higher laser pulse energies, higher gas beam densities or by averaging over several 100 laser pulses.

TRACE ANALYSIS: EXHAUST OF COMBUSTION ENGINES

Resonant laser mass spectrometry has been applied to time-resolved analysis of trace pollutants in the exhaust emissions of motor cars by our group [1,2,3]. Thus, situations of fast changes of motor conditions, dynamic effects in catalytic converters, or the emission of pollutants at cold start of combustion engines has been studied for single molecular species for the first time. These species are NO and acetaldehyde (synchronously measured products of complete and uncompleted combustion), aromatic compounds (e.g. BTXE = benzene, toluene, xylene, ethylbenzene etc.) or polycylcic aromatic compounds. The study of components of motor oil [4] detected in the exhaust gas will be described in the following as an illustrative example for the exciting new possibilities resonant laser mass spectrometry supplies to analytical research and developing engineers. In Figure 2, mass spectra of polycyclic aromatic compounds PAH (up to mass 240) measured in the exhaust of combustion engines (Otto motor) are represented. To find out if the origin of these PAH is gasoline or motor oil, studies at hydrogen fuelled engines have been performed (see the two mass spectra at top of Figure 2). A mass pattern typical for PAH and methylated PAH has been found which prove (i) that unburned components of motor oil actually are present in the exhaust and (ii) that resonance ionisation and time-of-flight mass spectrometry are sensitive enough to detect these traces on a fast time scale.

In addition, the two mass spectra in the upper part of Figure 2 show that different motor conditions lead to different PAH mass patterns. By conventional analytical means the total concentration of PAH has been determined to amount to 1ppm corresponding to 10 ppb on-line sensitivity for a single PAH. The two mass spectra in Figure 2/b are due to exhaust from a gasoline-fuelled engine. In this case, mass peaks at smaller masses (typically between 70 and 120 mass units) show up which are so strong that the ion detector saturates and larger masses are not detectable anymore. A mass gate has been used in order to study the motor oil components (larger PAH). This mass gate supplies a deflection voltage which is switched on when ions of mass 70 to 120 pass by thus preventing them from reaching the ion detector. Now larger masses also appear in the mass spectrum with a similar pattern as in Figure 2/a. Even more interesting is that a tracer molecule (e.g. pyrene) is clearly found in the exhaust gas again. At the bottom of Figure 2/b a spectrum due to 1% of pyrene added to the motor oil is displayed.

The tracer enables one to study dynamic and stationary processes of motor oil consumption. It was found that (i) the main source of oil consumption is due to loss in the combustion chamber, (ii) about 90% of evaporated oil is not burnt when reaching the exhaust, (iii) the tracer represents the unburned part and thus nearly the total amount of oil consumption. In summary, the tracer is an ideal indicator to study dynamic processes but also to measure motor oil consumption in a quantitative way.

Two applications of this method to determine motor oil consumption quantitatively are displayed in Figure 3. The first concerns fast

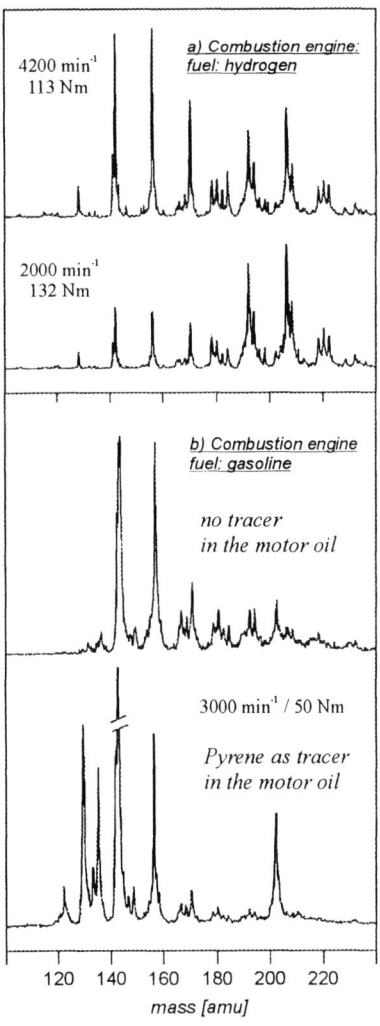

FIGURE 2. Mass spectra of PAH from motor oil emitted in the exhaust of hydrogen and gasoline fuelled engines. In the mass spectrum at the bottom a tracer (1% of pyrene) has been added to the motor oil.

dynamic effects of oil consumption (Fig. 3a), the other represents measurements at different stationary motor conditions, i.e. a speed-load plot of oil consumption (Fig. 3b). Figure 3a shows four signals as a function of time: (a) the speed between 1000 and 3000 rev./min, (b) the load between 20 and 200 Nm, (c) the tracer signal, and (d) the relative oxygen concentration. It can be clearly seen that high motor oil consumption occurs when both speed and load have rapidly decreased from a high level to a minimum value correlated to a strong decrease of the combustion process. These short moments of reduction of thrust force are characterised by sharp oxygen peaks. The sharp increase of motor oil consumption at these moments can be explained by a sudden pressure reduction in the cylinder inducing enhanced evaporation of oil components from the walls within the cylinder.

The results in Figure 3b differ from those of Figure 3a in several aspects. The motor now operates under stationary conditions, which are characterised by a pair of values of speed and load and which are kept constant for some minutes. The motor oil consumption is measured for each of these stationary

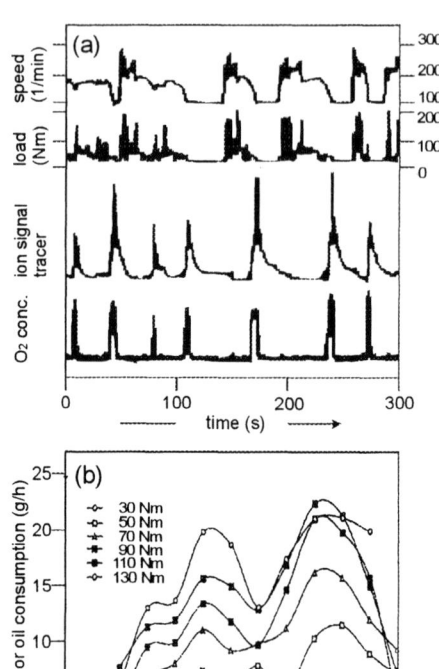

FIGURE 3: Oil consumption measured by recording the tracer signal. (a) The effect of fast dynamic changes of speed and load. (b) Oil consumption plot at stationary motor conditions.

(The decrease at high speed is artificial due to enhanced dissociation of the tracer.)

conditions and could now be quantified (given in g/h). For different speed and constant load a single curve of oil consumption is obtained. Varying the load resulted in groups of oil consumption curves. Instead of a smooth, steady increase of oil consumption with speed (which one might expect), these curves show a more or less periodic modulation of oil consumption. One explanation is that the piston rings not only move with the piston but are subject to secondary forces resulting in rotation and twisting. This influences the function of the piston ring as a scraper of the oil and finally leads to a varying thickness of the oil film at the inner walls of the cylinder. Thick films then give rise to an increased amount of evaporated oil and thus an increased motor oil consumption. One should notice, that the measurement of the whole oil consumption / speed / load plot took one day with resonant laser mass spectrometry, but takes a month or even longer with conventional techniques (i.e. weighing of oil before and after running the motor at one stationary motor condition).

In addition to exhaust gases from combustion engines emissions of other industrial combustion processes have been investigated by our research group in co-operation with the GSF-research centre for environment and health in Munich/Neuherberg. E.g., the PAH mass pattern of the rough gas of waste incinerator plants (measurement directly performed in the combustion chamber before gas cleaning) gives interesting information about the formation processes of small and large aromatic molecules [5] (e.g. formation of large PAH on the surface of chamber walls and dust particles). In a different on-site and on-line experiment it was shown that the selective detection of monochloro-benzene can be a valuable indicator for the so-called toxicity equivalent which is due to an integrated concentration of chlorinated dioxins where each of the very numerous isomers is weighted by its specific toxicity. While the former is a molecular component which is measurable by on-line resonant laser mass spectrometry, the on-line measurement of the latter is highly improbable due to the extremely low concentrations and exceptionally large number of isomers of chlorinated dioxins [6].

THE PAH-TRACE ANALYSIS OF SOLID SAMPLES: SOIL

Resonant laser mass spectrometry enables one to measure traces of pollutants or other chemical compounds not only in complex gas mixtures but also in complicate solid matrices (as soil is) on a fast time scale. This is achieved by combining resonance ionisation and laser desorption of neutral molecules. Figure 4 shows the experimental setup. The soil sample is prepared by forming little pellets with a diameter of 2 mm in a small press. They are fixed on the surface of a movable sample support by gluing tape. The sample support is a thin rod which can be introduced into the working mass spectrometer via a vacuum lock system. A pulsed laser desorbs neutral molecules (e.g. PAH) from the sample surface; they are carried away by a pulsed gas beam which is confined by a short cone shaped opening in a solid block of Teflon. This beam guide has several advantageous features: It results in a thermalisation of the internal motion of the desorbed molecules and of the direction and amount of their velocity. This results in a strongly reduced fluctuation of the ionisation yield (even allowing spectroscopy of desorbed molecules), as well as in an enhanced ionisation yield due to an efficient transport of desorbed molecules from the point

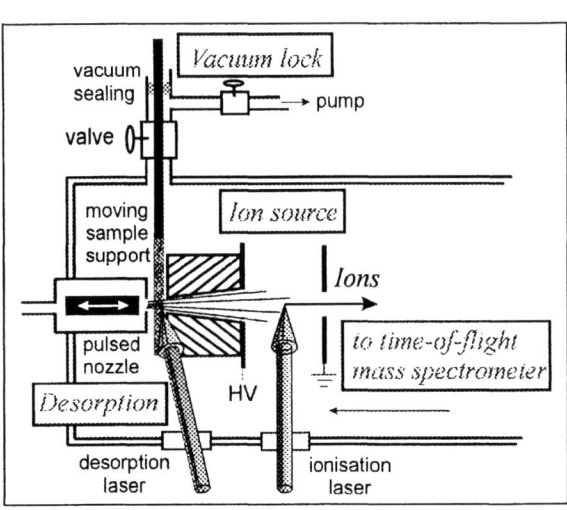

FIGURE 4: Setup of resonance ionisation combined with laser desorption of neutrals and time-of-flight mass separation.

of desorption to that of ionisation. The molecular ions formed selectively by resonance ionisation then are analysed and detected in a time-of-flight mass spectrum. The whole procedure (from pressing pellets to obtaining a mass spectrum) takes about 5 minutes. As an example, a soil sample from the bottom of an American river with a certificate of the EPA (environmental protection agency) has been investigated.

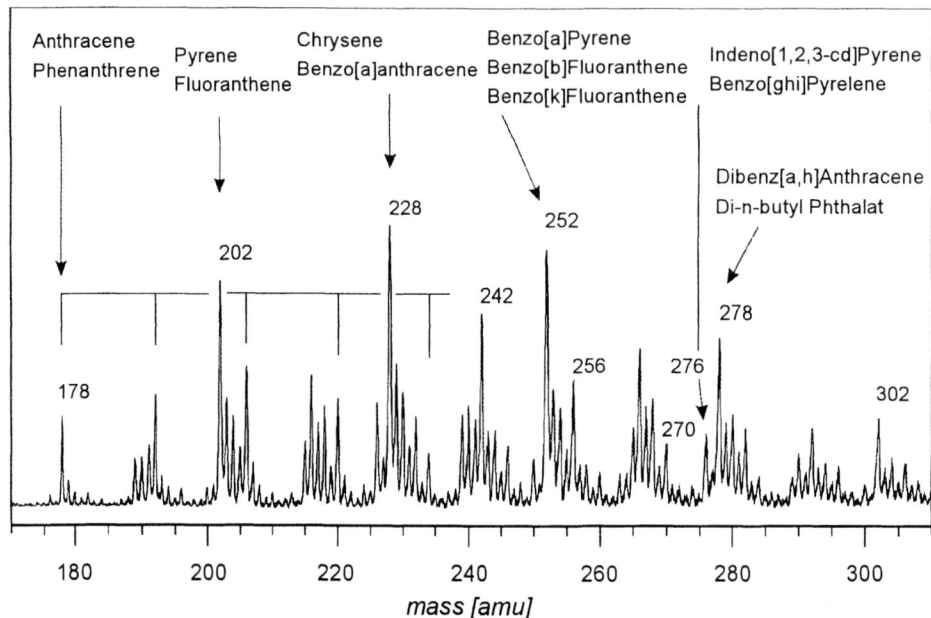

FIGURE 5: Mass spectrum of a certificate soil sample. The signal intensities correspond to typical concentrations of few ppm. From the signal to noise ratio a sensitivity of significantly better than 100 ppb for this fast quasi-on-line (< 5 min. for sample pellet and mass spectrum) analysis is deduced.

A special advantage of the confined gas beam is the strong reduction of fluctuations of molecular internal level population and of velocities. Laser desorption now does not disturb spectroscopy severely. In Fig. 6, a conventional UV-gas phase spectrum (measured in a homebuilt special absorption cell) is compared with the REMPI-spectrum, the agreement is obvious. The increase of the ion signal at wavelength shorter than 270 nm is caused by more and more favourable Franck-Condon factors. This feature of our laser desorption -laser post-ionisation assembly is useful for isomer selective

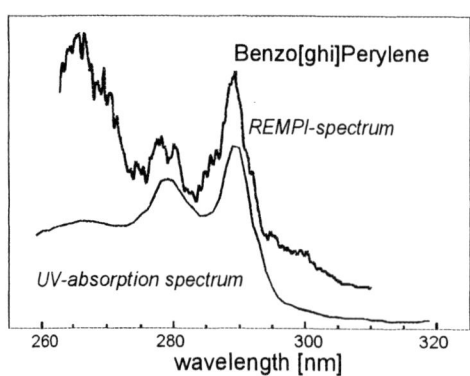

FIGURE 6: UV-gas phase absorption spectrum and REMPI-spectrum of laser desorbed molecules.

detection. The spectral resolution is still high enough. In Figure 7, mass spectra of a mixture of Benzo[b]Fluoranthene and Benzo[ghi]Perylene are diplayed for the laser wavelengths 290nm and 265 nm. A strong change of the relative signal intensities can be observed. Similar behaviour is found for the isomers of Benzo[b]Fluoranthene: Benzo[a]Pyrene and Benzo[k]Fluoranthene. Ratios for three different wavelengths of these three isomers (mass 252) are given in the table in Fi. 7.

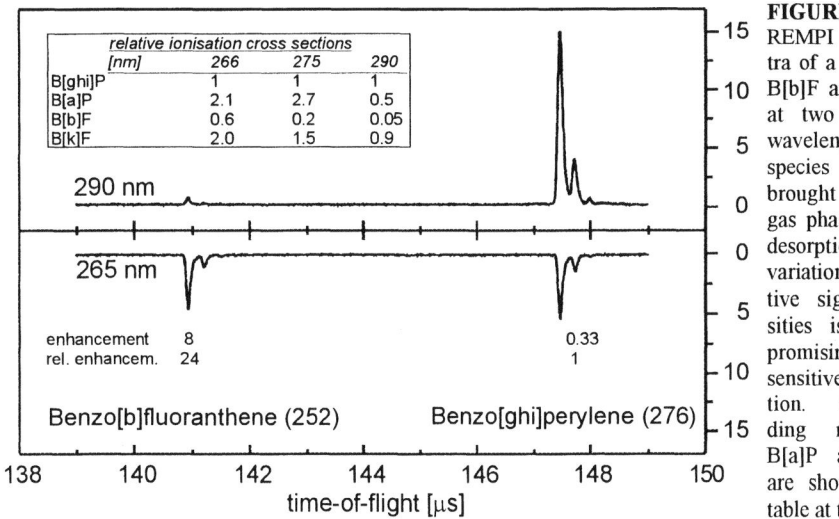

relative ionisation cross sections [nm]	266	275	290
B[ghi]P	1	1	1
B[a]P	2.1	2.7	0.5
B[b]F	0.6	0.2	0.05
B[k]F	2.0	1.5	0.9

FIGURE 7: REMPI mass spectra of a mixture of B[b]F and B[ghi]P at two ionisation wavelengths. Both species have been brought into the gas phase by laser desorption. Strong variation of relative signal intensities is observed promising isomer sensitive ionisation. Corresponding ratios for B[a]P and B[k]F are shown in the table at the top.

PROBLEMS WITH REMPI AND POSSIBLE SOLUTIONS

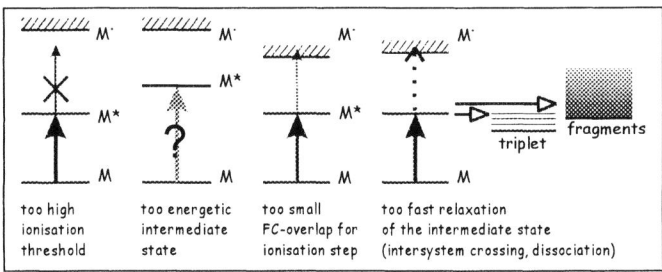

FIGURE 8: Four examples of molecular situations which may cause problems for resonance ionisation.

Some problems with resonance ionisation of molecules are caused by unfavourable energies of intermediate states or ionisation threshold for which no appropriate laser wave lengths are available. One countermeasure is to use multi-photon absorption of higher order (e.g. 2+1, 2+2, or even 3+1) for ionisation at the expense of sensitivity. One severe problem, however, is often induced by fast relaxation processes in intermediate states. In particular fast dissociation processes may even totally prevent the appearance of molecular ions in the resonance ionisation mass spectrum. This is the reason why femtosecond lasers are going to be used to ionise such problematic molecular species, as demonstrated successfully in this volume [7,8]. Another possibility is electron attachment and detection of negatively charged ions (similarly to chemical ionisation sources). Our approach, is negative ion formation at low pressures (i.e. in a supersonic molecular beam) and with low energetic electrons which are formed by laser induced electron emission from metal surfaces. A laser photon energy is used which only is slightly larger than the work function of this metal giving rise to electrons in the some tenths to one eV range [9]. For nitroaromatic as well as polychlorinated aromatic molecules mass spectra without fragmentation products have been obtained (see Figure 9). Of course, electron attachment is not a selective ionisation technique (apart from the fact, that some molecules have negative electron affinity). However, the analysis of photodetached electrons allows photoelectron spectroscopy. This detachment and therefore the corresponding photoelectron spectra can be performed in a mass selective way, and for every single ionisation cycle. In summary, the anion mass spectra deliver a kind of overview information while the photoelectron spectra deliver highly specific (even isomer selective) information of selected mass peaks. The technique combines laser induced electron attachment, time-of-flight mass spectrometry, and time-of-flight photoelectron spectroscopy [10].

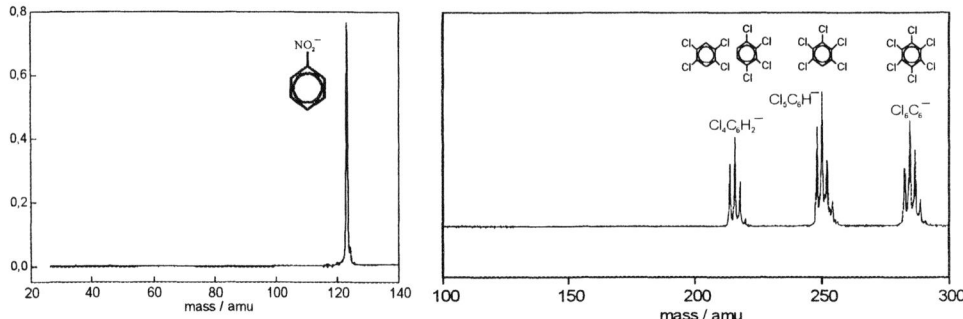

FIGURE 9: Anion mass spectrum of nitrobenzene (on the left), and of tetra-, penta- and hexachlorobenzene (on the right). The tetrachlorine benzene signal is due to two different isomers.

REFERENCES
1 Boesl, U., Nagel, H., Weickhardt, C., Frey, R., Schlag, E.W., *in Encyclopedia of Environmental Analysis and Remediation*, edited by R.A.Meyers R.A., John Wiley & Sons, 1998, pp. 5000-5022.
2 Boesl, U., Heger, H.J., Zimmermann, R., Püffel, P.K., Nagel, H. *Encyclopedia of Chemical Analysis* edited by R.A. Meyers, John Wiley & Sons, Chichester, 2000, pp. 2087 - 2118.
3 Boesl, U., *J. Mass Spectrom.* **35**, 289 - 304 (2000).
4 Püffel, P.K., Thiel, W., Frey, R., Boesl, U., *SAE Technical Paper Series* **982438**, 1 - 7 (1998).
5 Heger, H.J., Zimmermann, R., Dorfner, R., Beckmann, M., Griebel, H., Kettrup, A., Boesl, U., *Analytical Chemistry* **71**, 46-57 (1999).
6 Zimmermann, R., Heger, H.J., Blumenstock, H., Dorfner, R., Schramm, K.W., Boesl, U., Kettrup, A., *Rapid-Comm. Mass-Spectrom.* **13**, 307-314 (1999).
7 Grun, C., Schmidt, S., Toennies, C., Weickharedt, C., Grotemeyer, J., Session VII this Volume
8 Hankin, S.M., Ledingham K.W.D., and others Session I, this Volume
9 Boesl, U., Bäßmann, C., Drechsler, G., Distelrath, V., *Eur. Mass Spectrom.* **5**, 455 - 470 (1999)
10 Boesl, U., Bäßmann, C., Schlag, E.W. in *Advanced Series in Physical Chemistry Vol 10: Photoionization & Photodetachment* ed. by Cheuk-Yiu Ng World Scientific, Singapore,2000, pp. 809 - 853

Environment-dependent desorption of benzene molecules

Chad A. Meserole*, Erno Vandeweert¶, Yusheng Dou*, Zbigniew
Postawa§, and Nick Winograd*

*Department of Chemistry, The Pennsylvania State University, University Park, PA 16802
¶Laboratorium voor Vaste-Stoffysica en Magnetisme, Celestijnenlaan 200 D, B-3001 Leuven, Belgium
§Institute of Physics, Jagellonian University, ul. Reymonta 4, PL 30-059 Krakow 16 Poland

Abstract. We have investigated the dependence of the 8 keV Ar^+ ion beam-induced desorption of neutral benzene (C_6H_6) molecules on the local environment. The probed system was comprised of carefully constructed combinations of C_6H_6 and sec-Butyl Alcohol (sBA) deposited on Ag(111). A pulsed 8 keV Ar^+ ion beam at 45° incidence was used to stimulate the desorption of adsorbed molecules. Using state-selective resonant ionization spectroscopy and time-of-flight (TOF) mass spectrometry, C_6H_6 molecules in the ground state and in the first level of a vibrationally excited state were probed. TOF distributions were measured versus the exposure of the sBA. The results show that the ejection of the C_6H_6 depends strongly on both the internal energy and thickness of the formed layer. In general, as more sBA is added to the probed system, the TOF distributions tend to broaden and shift to higher times, and the relative C_6H_6 yields diminish rapidly. The vibrationally excited C_6H_6 is more sensitive to the addition of sBA, and an interpretation of the combined data suggests that the C_6H_6 desorbed in an excited state originates from the organic/vacuum interface.

INTRODUCTION

When a surface is irradiated with energetic primary ions, molecules and atoms in a variety of charge and internal energy states are liberated from the surface [1]. Such a process is the basis for surface analytical techniques, such as secondary ion mass spectrometry [1,2]. However, the vast majority of the desorbed species are in a neutral charge state [2]. Because we use mass spectrometry to study the Ar^+ ion stimulated-desorption, an efficient means of ionizing the desorbed neutral species must be employed to gain as much information about the desorption process as possible per primary ion. The utilization of laser postionization, especially when resonance-enhanced ionization schemes are used, provides an efficient means of generating ions [1]. The combination of time-of-flight mass spectrometry and laser postionization is a powerful method to study the interaction of energetic primary ions with surfaces and the desorption events that follow the impact of the primary ion.

The phenomena of the desorption of intact molecules from metals surfaces has been studied in the past using benzene (C_6H_6) adsorbed onto a Ag(111) surface that is subjected to 8 keV Ar^+ ion bombardment [3]. Benzene was selected because it has

CP584, *Resonance Ionization Spectroscopy 2000: 10th Int'l. Symp.*, edited by J. E. Parks and J. P. Young
© 2001 American Institute of Physics 0-7354-0024-5/01/$18.00

many attractive attributes that are conducive to such studies of molecular desorption [3]. Noticeable changes in the energy distributions were observed for both ground state benzene molecules (C_6H_6) and vibrationally excited benzene molecules (C_6H_6*) as the benzene exposure was increased [3]. In order to probe how the local chemical environment affects the desorption characteristics of the benzene molecules, additional experiments have been performed. These experiments consist of systematically making the system more organic in nature by depositing a sec-butyl alcohol (sBA) layer of increasing thickness on top of a monolayer of benzene on the Ag(111) surface. By varying the sBA layer thickness we can make the benzene layer closer or farther away from the organic/vacuum interface and effectively change the system from a thin organic film on a metal surface to one that is much more organic in nature. Moreover, an additional layer of benzene can be placed on top of the sBA layer so that a fresh sample of benzene is once again near the surface/vacuum interface. This final step probes the behavior of analyte molecules that are on an organic matrix and near the surface/vacuum interface.

EXPERIMENTAL SETUP

The experimental setup and procedure are described in detail elsewhere [4,5]. Several freeze-pump-thaw cycles are applied to both the benzene and the sBA before dosing either gas to remove dissolved contaminants. Before dosing the first monolayer of benzene, which is a 7L exposure (1L = 1x10^{-6} Torr sec), the single crystal of silver is sputter cleaned, annealed, and then cooled to 120 K and held at 120 K for the duration of the experiments. The exposures of benzene and of sBA were controlled by monitoring the chamber pressure and the dosing time.

The desorption process is initiated by bombarding the sample with an 8 keV, 200 ns Ar^+ ion pulse. The Ar^+ ion beam from which the pulses are derived is directed at the sample at 45° incidence and focused to a spot size of 3 mm. The primary ion dose is kept sufficiently low (~10^{11} Ar^+ ions/cm^2) in an attempt to minimize accumulated surface damage. Desorbed particles are detected by multiphoton ionization in combination with time-of-flight mass spectrometry. The postionization laser beam is focused to a ribbon shape about 1 cm above the crystal surface. As the time delay between the primary ion impact and laser ionization event is systematically varied, the density of the particles in the laser ionization volume is recorded as function of time delay. Resonant two-photon one-color photoionization of the ejected benzene molecules is achieved by tuning the frequency-doubled output of Nd:YAG-pumped dye laser to drive the 6^1_0 transition at 259.01 nm originating from the zero level of the molecular ground state, and the 6^0_1 transition at 266.82 nm starting from the first quantum of the $v^"_6$ vibration [6].

RESULTS AND DISCUSSION

A generalized representation of the experimental sequence (A through C) used to make the system take on a more organic character is illustrated in Fig. 1. The first 7L exposure of benzene is applied to the surface to serve as a reference and so that the changes to the desorption characteristics of benzene as the benzene layer is systematically moved away from the surface/vacuum interface can be monitored. The layer of sBA creates an organic partition between

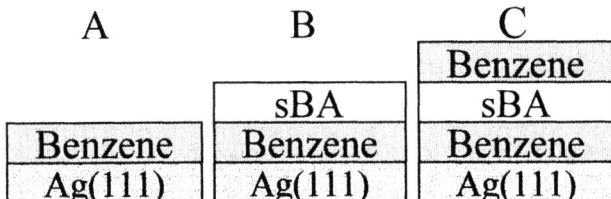

	A	B	C
			Benzene
		sBA	sBA
	Benzene	Benzene	Benzene
	Ag(111)	Ag(111)	Ag(111)

Figure 1. Illustration of the general experimental scheme. First, benzene (7L) is dosed (A) and spectra are obtained. Then, sec-butyl alcohol (sBA) is dosed (B) and spectra are obtained. Finally, an additional 7L of benzene is dosed (C) and spectra are taken.

the first monolayer of benzene and the vacuum. The layer of benzene on top of the sBA layer creates a system that is largely organic with the analyte molecules in very close proximity to the vacuum interface.

Time distributions of benzene molecules desorbed in both the ground state and in the vibrationally excited state as a function of sBA exposure from steps (A) and (B) are illustrated in Fig. 2. The plots are peak normalized so that the shifts toward higher times as more sBA is applied to the surface is more evident. As the sBA layer becomes thicker due to increased exposure, the signal of the vibrationally excited benzene molecules suffers significantly (as seen by the substantial reduction in signal to noise). However, the ground state signal is much less affected. These two findings suggest that as

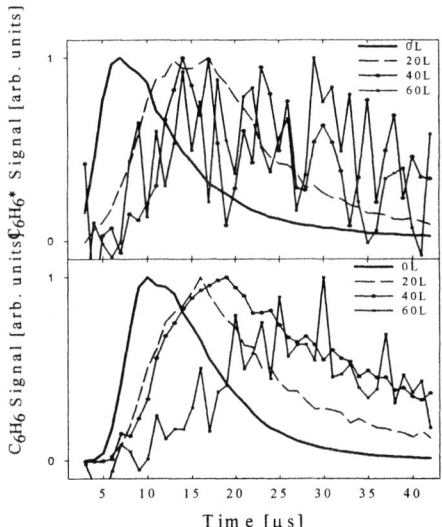

Figure 2. Peak normalized time distributions for vibrationally excited benzene (TOP) and ground state benzene (BOTTOM). The distributions are for a variety of sBA exposures from 0L to 60L. The vibrationally excited benzene signal degrades more rapidly (as a function of sBA exposure) than the signal from the ground state benzene.

the benzene is moved farther from the surface/vacuum interface, the probability that benzene molecules in internally excited states will be desorbed and detected becomes smaller. As the benzene is buried deeper in the organic matrix, the time distributions shift toward higher times, indicating that the presence of the organic matrix depletes the benzene molecules of translational energy as well.

Time distributions for the benzene layer applied to the top of the sBA layer are illustrated in Fig. 3 (step C). A significant finding from these plots is that the signal from the excited state benzene molecules is resurrected (TOP panel). In other words, as long as there are analyte molecules near the surface/vacuum interface, then some portion of the molecules can be desorbed and detected as internally excited molecules even as the system becomes more organic in nature. Another striking feature is that the time distributions of vibrationally excited molecules tend to shift to higher times and broaden more readily than do the distributions of the ground state molecules. Such an observation implies that the vibrationally excited molecules are desorbed with less translational energy than the ground state molecules as the system becomes one that resembles an organic matrix, as opposed to one resembling a thin organic film on a metal surface.

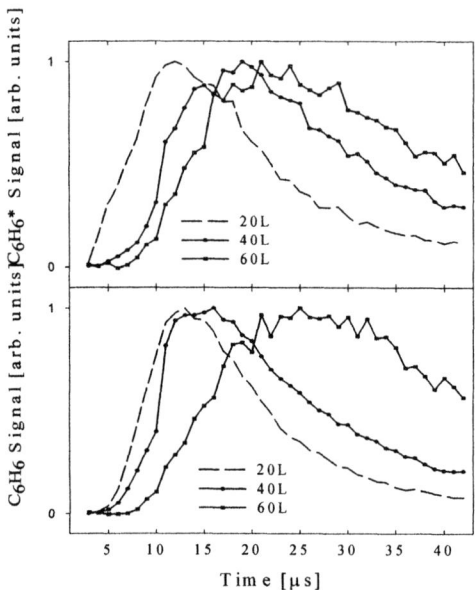

Figure 3. Peak normalized time distributions for vibrationally excited benzene (TOP) and ground state benzene (BOTTOM) from benzene deposited on top of the sBA layer. The sBA layers are from exposures ranging from 0L to 60L.

CONCLUSIONS

As the layer of analyte molecules becomes buried beneath an organic matrix, the time distributions for both excited state and ground state benzene molecules shift to higher times and broaden. Such changes indicate that the overlayer of the sBA tends to deplete the benzene molecules of translational energy. Because the $C_6H_6^*$ signal is readily diminished by the presence of the sBA overlayer, it is reasonable to suggest that molecules buried deep in the organic matrix will not be desorbed and detected in an internally excited state. However, if analyte molecules are present near the surface/vacuum interface, then molecules in the ground state and internally excited states may be desorbed. Molecules from the upper-most layer near the vacuum interface tend to be desorbed with less translational energy as the matrix progresses to one that is more organic.

ACKNOWLEDGEMENTS

The authors gratefully acknowledge the financial support of the National Institutes of Health, of the National Science Foundation, and of the Office of Naval Research, as well as the Polish Committee for Scientific Research Fund, Instytut Fizyki UJ, and Maria Sklodowska-Curie Fund MEN/NSF-97-304. E.V. is a postdoctoral fellow of the Fund for Scientific Research-Flanders.

REFERENCES

1. Winograd, N., *Anal. Chem.* **65**, 622A-629A (1993).
2. Benninghoven, A., Hagenhoff, B., and Niehuis, E., *Anal. Chem.* **65**, 630A-640A (1993).
3. Vandeweert, E., Meserole, C.A., Sostarecz, A., Dou, Y., Winograd, N., and Postawa, Z., *Nucl. Instrum. Methods Phys. Res. B* **164-165**, 820-826 (2000)
4. Kobrin, P.H., Schick, G.A., Baxter, J.P., and Winograd, N., *Rev. Sci. Instr.* **57**, 1354-1362 (1986).
5. Chatterjee, R., Riederer, D.E., Postawa, Z., and Winograd, N., *J. Phys. Chem. B* **102**, 4176-4182 (1998).
6. Callomon, J.H., Dunn, T.M., and Mills, I.M., *Phil. Trans. Roy. Soc. London* **259A**, 499-532 (1966).

Multi Photon Ionization Mass Spectrometry of Carbamate Pesticides, Herbicides and Fungicides

Carsten Grun, Marcelle König, Jürgen Grotemeyer

Institut für Physikalische Chemie, Christian-Albrechts Universität zu Kiel,
Olshausenstraße 40, D-24098 Kiel, Germany

Abstract. Pesticides and herbicides are useful for a wide range of applications today. The determination of these substances either in the pure form or in complex matrices is of high analytical interest. Especially since these substances can by found in every day products. The combination of multi photon ionization (MUPI) and time of flight laser mass spectrometry may be a powerful tool for achieving fast well interpretable mass spectra for analytical purposes. In this paper we will discuss the mass spectra of several pesticides and herbicides accessed by MUPI-time-of-flight mass spectrometry. The influence of the laser pulse duration on the mass spectra are discussed

Introduction

The application of pesticides and herbicides in agriculture is a common way protecting monocultures today . Although the application of these substances is generally not harmful to mammals and humans if applied in a correct way, the detection and quantification is of high interest in analytical chemistry. Usually these substances can be divided by several ways: the application method in the field, its toxic potential or the chemical substance class. The standard analytical procedures for the detection of these compounds are usually gas chromatography (GC,GC/MS)[1] or high performance liquid chromatography (HPLC)[2]. In this short contribution we are presenting some results on the measurement of these substances with multi photon ionization followed by time-of-flight mass spectrometry [3]. The main point to be addressed is the influence of the laser pulse duration as well as the possibility to detect these sample molecules.

CP584, *Resonance Ionization Spectroscopy 2000: 10th Int'l. Symp.*, edited by J. E. Parks and J. P. Young

Experimental

The experimental setup used for the investigations presented here is described in detail elsewhere[4]. Only a brief discussion of the used components is given here. Mass spectrometric investigations were performed with a Bruker TOF 1 reflectron mass spectrometer (Bruker Daltonic GmbH, Bremen, Germany). Neutral laser desorption with a CO_2-laser (Alltech, Lübeck, Germany) at 10.6 µm is used to vaporize target molecules with an insufficient vapor pressure at room temperature. As a result of the mounting of the probe tip in front of the nozzle of a supersonic beam expansion the neutral molecules are transferred with low internal energy into the ionization chamber through a skimmer. Here ionization is performed either by a nanosecond tunable MOPO laser system (MOPO 750, Spectra physics, Darmstadt) or by an excimer pumped femtosecond dye laser system (LLG 500, Laserlabor Göttingen). The first laser system produces UV pulses with energies up to 1.2 mJ distributed over a pulse duration of about 5 ns, the latter delivered 100 - 500 fs pulses with energies of approximately 40 µJ. Both laser beams can be focused into the ionization region simultaneously for a direct comparison of the ionization process with the different pulse length. The wavelength is tuned between 255 nm to 268 nm at both ns-ionization and fs-ionization.

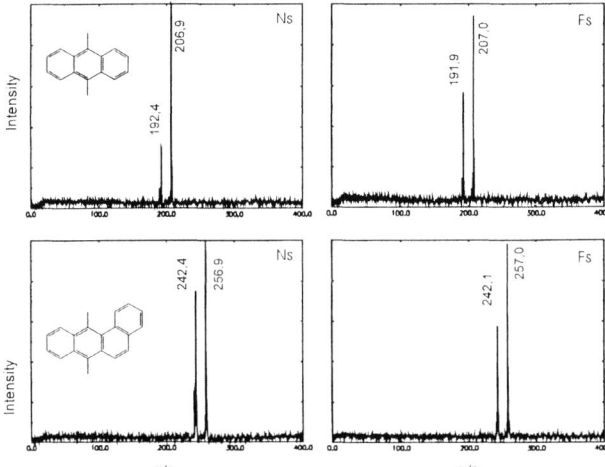

Chemicals were purchased commercially in analytical quality (Riedel de Haen) and used without further purification. For laser desorption the pure substances were mixed with polyethylene and pressed to pellets.

Figure 1 Multi photon ionization mass spectra of 9,10 Dimethylanthracene (top) and 7,12 Dimethyl-benz(a)anthracene (left spectra ns-activation, right spectra fs-activation)

Results and Discussion

Investigations of pesticides, herbicides and further compound of this family showing a nearly similar behavior for the fragmentation pattern in the mass spectra. This is valid both for applying ns-laser pluses as well as fs-laser pulses respectively. In general the fragmentation pattern follows standard rules of organic mass spectrometry [5], thus allowing an easy interpretation. It should be noted that in contrast to

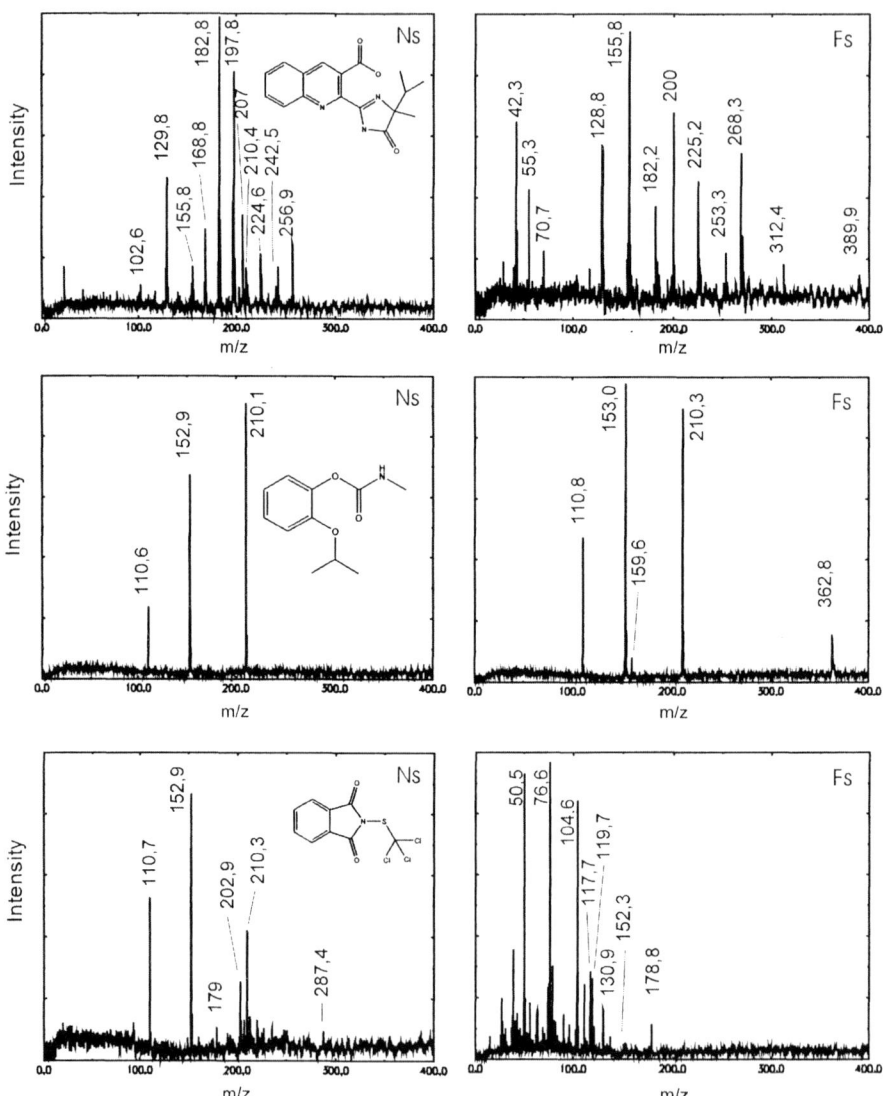

Figure 2 Multi photon ionization mass spectra of Imazaquin (top), Propoxur and Folpet (bottom). (left side : ns-activation, right side : fs-activation)

electron ionization multi photon processes can be steered in such a way that pre-dominantly the molecular ion can be observed. But this is dependents strongly on the photo physical processes in the excited electronic state. It has been shown that either intersystem crossing or neutral fragmentation of the excited electronic state is the general behavior of a large number of molecules, thus preventing ionization of the molecules through nanosecond (ns) laser pulses. In these cases multi photon ioniza-tion with femtosecond (fs) pulse duration (500 - 100 fs) yield usually in the intense preparation of the molecular ion. Unfortunately for analytical purposes this behavior is been found for aromatic molecules with heteroatom containing side chains, such as ketones, aldehydes and carboxylic acids, or molecules with heavy atoms and so on.

Pure aromatic systems like benzene or poly aromatic hydrocarbons (PAH) show with nanosecond and femtosecond activation clear and easy interpretable mass spectra. In figure 1 the mass spectra of 9,10-Dimethylanthracene and 7,12-Dimethyl-benz(a)anthracene are shown.

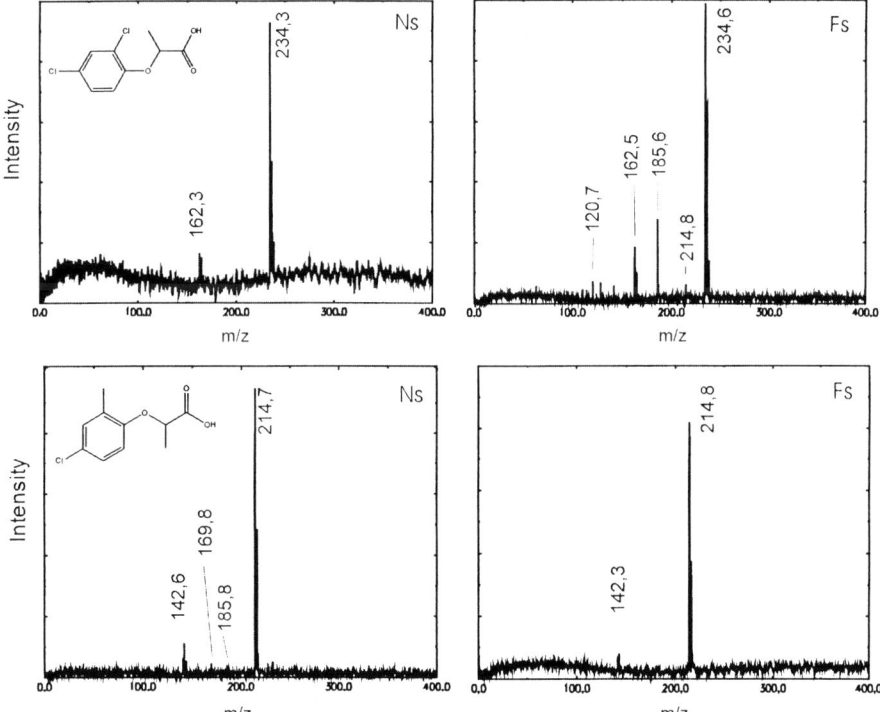

Figure 3 Multi photon ionization mass spectra of Dichlorprop (top) and Mecoprop (bottom).

The mass spectra are almost the same by either ns- or fs-activation. The molecular ion is for both samples at both pulse durations the base peak. In both samples the fragment ion is due to the loss of a methyl group. This behavior reflects the fact, that these aromatic compounds have long-living intermediate electronic states thus yielding no difference for the different laser pulse durations.

In figure 2 the mass spectra of Imazaquine, Propoxur and Folpet are shown. These compounds belong to the group of carbamates and phenolic pesticides. As seen from the mass spectra the formation of the molecular ion as well as the fragment ions is different with respect to the laser pulse duration.

The Imazaquine mass spectra display of a wide range of different fragmentation reactions. The main difference between ns- and fs-activation is the observation of the molecular ion. Only with the fs-laser pulse the Imazaquine display the molecular ion at mass 312. The high mass signal at mass 389 is due to a fragmentation from the dimer.

The mass spectra of Propoxur display the molecular ion with both laser pulses. There is also no difference in the fragmentation pattern. In the femtosecond ionization a signal is been found at m/z 362 corresponding to the dimer ion losing the side chain (209 + 153). It should be noted that this class of compounds are very similar to the polyaromatic hydrocarbons. Here the active group is a phenolic moiety. Compounds with this group display intense molecular ions since they have long living electronic states.

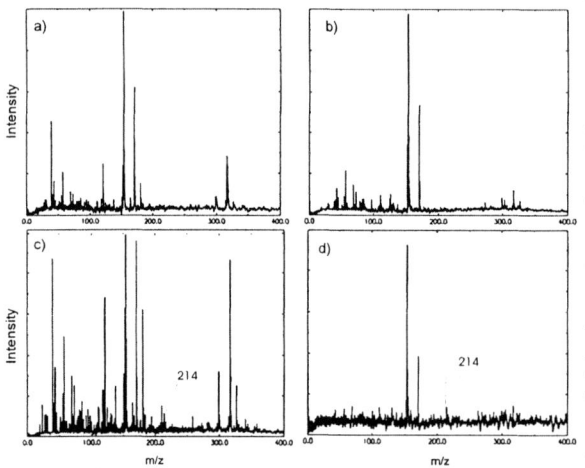

The mass spectra of Folpet display in case of ns-laser ionization no molecular ion while the femtosecond laser pulses yield in the formation of a molecular ion with low intensity. The fragmentation pattern is typical for molecular systems stemming from Phtalic acid derivates.

Figure 4 a) ns-mass spectrum of a pure maple leaf; b) fs-mass spectrum of the same sample; c) ns-mass spectrum of a maple leaf spiked with Mecoprop; d) fs-mass spectrum of a maple leaf spiked with Mecoprop

In figure 3 the mass spectra of Dichlorprop and Mecoprop are displayed. The mass spectra of Dichlorprop yield in the molecule ion as base peak accompanied by a fragmentation pattern of the side chain. The main difference is the better signal noise in the fs mass spectra than in the ns spectra. In the Mecoprop mass spectra the molecule ion is the base peak.

To demonstrate the usefulness of multi photon ionization in the quantification of the samples the figure 4. Here the pure extract from a maple leaf is investigated by different laser pulse durations. The spectra are qualitatively similar. But if an extract from a leaf treated with Mecoprop is investigated the signal at 214 (molecular ion) is found with high intensity. This demonstrate that the application of a femtosecond laser pulse is superior in the detection of substances in complex mixtures.

Conclusions

The multi photon ionization mass spectra of some pesticides and herbicides have been investigated by using laser pulses with 10 ns and 500 fs duration. Depending on the nature of the excited electronic state of the chromophoric group the resulting mass spectra show molecular ions with different intensities. In general femtosecond laser pulses are superior in the detection of different molecules in complex mixtures.

Acknowledgment

This work is supported in part by the Deutsche Forschungsgemeinschaft and the Christian Albrechts Universität zu Kiel. Support from the Fonds der Chemischen Industrie is gratefully acknowledged.

References

1.   G. Schomburg „Gaschromatographie", Wiley-VCH, Weinheim, 1987.
2.   J. Lewandowski, S. Leitschuh, V. Koß „Schadstoffe im Boden" Springer Verlag, Berlin, 1997
3.   C. Grun, M. König, J. Grotemeyer; in preparation.
4.   U. Boesl, K. Walter, J. Grotemeyer, E.W. Schlag; Anal. Instrum. **16**, 151 (1987).
5.   F.W. McLafferty, F. Turecek „Interpretation of Mass Spectra", University Science Books, Mill Valley, CA, 1996.

Laser Analysis and Restoration of Nineteenth Century Daguerreotypes

Valerie V. Golovlev[a], John C. Miller[a#], Grant Romer[b], Paul Messier[c]

[a]Life Sciences Division, Oak Ridge National Laboratory, P. O. Box 2008, Oak Ridge, TN 37831-6125
[b]International Museum of Photography and Film, George Eastman House, 900 East Ave., Rochester, NY 14607-2298
[c]Boston Art Conservation, 60 Oak Square Ave., Boston, MA 02135
Author for Correspondence: John C. Miller (423) 574-6239 – Tel.; (423) 576-4407 – Fax
e-mail: millerjc@ornl.gov

Abstract. Laser ablation mass spectrometry and surface-enhanced Raman spectroscopy have been used to characterize both modern and 150 year old daguerreotypes. Such investigations are a necessary prelude to attempts to clean them of tarnish by laser ablation. A demonstration of the successful cleaning of a damaged daguerreotype is presented.

INTRODUCTION

Laser desorption/ablation describes a process whereby an intense, pulsed laser beam interacts with a solid surface leading to the ejection of atoms, molecules, ions, clusters and macroscopic particles (1,2). Electromagnetic energy is converted to electronic, thermal, chemical and mechanical energy. The process is extremely complex since different mechanisms operate at different values of laser wavelength, pulse duration and intensity. In general, the term laser desorption is used when particle emission occurs with no visible microscopic alteration of the surface composition or structure. In contrast, laser ablation is a much more energetic phenomenon accompanied by high temperatures, gas dynamic effects, plasma formation and surface cratering. In practice, these two processes are not completely distinct, but reflect a continuum of physical effects indexed to the laser power density and various material properties.

Shortly after the birth of the laser in 1960, a novel application was demonstrated by A. L. Shawlow, one of the pioneers of the early history of laser invention and development (3). He showed that typewriter print could be selectively removed from paper by laser ablation. This "laser eraser" was based on the selective absorption of the laser energy by the black ink rather than the less-absorbing paper substrate.

CP584, *Resonance Ionization Spectroscopy 2000: 10th Int'l. Symp.*, edited by J. E. Parks and J. P. Young
© 2001 American Institute of Physics 0-7354-0024-5/01/$18.00

Although a somewhat whimsical demonstration at the time, the same principles have been applied to the use of laser ablation cleaning of a variety of substances. Of relevance to the present experiments, laser ablation cleaning has been applied to the restoration of various objects of artistic or historic value (4,5). Although most well-developed for the cleaning of stone or marble statues or building facades, laser ablation cleaning of such disparate art media as paintings, frescoes, parchment, leather, wood and stained glass has been demonstrated.

Because of the fragile and unique nature of such national treasures, as well as their cultural and monetary value, extensive characterization of the object is prudent in order to guide the cleaning and restoration process. Furthermore, laser ablation efficiency is strongly dependent on the composition and the optical and thermal properties of the material under study. Finally, in situ monitoring of the extent of the photoablation can be an important adjunct to effective restoration.

In the present study, we have employed laser ablation mass spectroscopy in order to characterize the nature of pristine and degraded daguerreotypes in preparation for attempts at restoration by desorption/ablation of the surface contaminants. We have also demonstrated the use of daguerreotypes as a substrate for surface-enhanced Raman spectroscopy (SERS). Finally we have demonstrated the feasibility of laser ablation cleaning of daguerreotypes. A successful example of such cleaning will be shown and discussed.

Daguerreotypes were the first form of photographs and were popular between 1840 and 1860, after which they were superseded by more modern techniques. The daguerreotype image is composed of silver/mercury microcrystals of varying size and density on a silver-coated copper substrate. Nineteenth century daguerreotypes, over the intervening 140 years, have suffered degradation and oxidation, which has greatly reduced their historic and artistic value. More details of the present study may be found in a recent paper (6). Related studies of image manipulation and optical properties of daguerreotypes (7), as well as the results of the laser cleaning (8) will be published elsewhere. Very recently, an independent study of laser cleaning of daguerreotypes has appeared in the conservation literature (9).

CHARACTERIZATION OF DAGUERREOTYPES

The physical aspects of daguerreotypes have been described in a series of papers by Barger and colleagues (10). In addition to the characterization tools used by those authors, we have employed laser ionization mass spectrometry and surface-enhanced Raman spectroscopy.

Examples of scanning electron microscope images of 150-year old and modern daguerreotypes may be found in references 7 and 10. Basically the image is formed by silver particles whose size and number density determine the amount of scattered light and hence the "darkness" of that portion of the image. Briefly, the image particles range from 0.1 to 50.0 microns in diameter and the density (particles/mm^2) varies from 20,000 to 1000 to 100 as one examines a highlight, midtone or shadow region of the daguerreotype (10).

Laser ablation mass spectroscopy was used for the first time in art conservation applications to study the chemical composition of the surface of daguerreotypes. In addition, as the laser-induced plasma at the surface is the basis for the cleaning process described later, laser ablation mass spectrometry adds insight into the laser-surface interaction process. The laser used was a Quanta Ray Nd:YAG laser operated in either the fundamental (1064 nm), second harmonic (532 nm), or third harmonic (355 nm) mode. The 5-10 ns laser beam was focussed by a 25 cm lens to about a $0.1 mm^2$ spot with adjustable fluences of 1-50 mJ/cm^2. Peak powers were consequently about 10^5 to 10^7 W/cm^2. The mass resolution ($m/\Delta m$ = ~100), although low, is sufficient for the separation of silver isotopes and the identification of the mostly atomic or small molecular ions. Representative spectra were taken from different spots on tarnished nineteenth century daguerreotypes (6). The Na^+, Ca^+, Hg^+, and Ag_n^+ ions were all present on mass spectra of pristine daguerreotypes and the presence of the Ag_nS^+ peaks on degraded areas indicated the presence of tarnish. The dark tarnish on silver items (including daguerreotypes) has long been attributed to the slow reaction of H_2S and other sulfur compounds present in low concentrations in the air. Mass spectra recorded after laser ablation cleaning show the silver sulfide peaks to be greatly reduced or eliminated. More details of these studies may be found in reference 6.

Surface enhanced Raman spectroscopy (SERS) is a variant on conventional Raman spectroscopy where interaction of the analyte molecules with surface irregularities or particles leads to a signal enhancement of from 10^6 to 10^8. The nanostructured surface of daguerreotypes suggests their use as a novel substrate for SERS. A SERS spectrum of the dye "fast cresyl violet" was observed after evaporating a dilute solution of the dye on an actual nineteenth century daguerreotype (8). In this case, the 514.5 nm line (~50 mW) of a cw argon ion laser (Lexel) was used and the spectrum was recorded on a Dilor XY triple monochromator. Peaks observed at 490 nm (weak), 590 nm (strong), and 675 nm (weak) matched in wavelength and intensity pattern those from a conventional substrate but the overall intensity was lower. The possibility of producing tailored nanostructured arrays of silver particles by daguerreian photography is intriguing. Alternately, the SERS effect could conceivably be used to enhance the images on faded daguerreotypes. The nonlinear nature of the Raman process might allow for "amplification" of a faded image.

LASER RESTORATION OF DAGUERREOTYPES

In order to optimize the experimental conditions for the efficient laser cleaning of daguerreotypes, it was first necessary to determine the best choices of laser power, wavelength and pulse length. Beam quality and spot size (focussed or not focussed) are also important for the best and most efficient cleaning. We have compared laser cleaning results using the fundamental (1064 nm), second harmonic (532 nm), and third harmonic (355 nm) of Nd:YAG lasers. Both nanosecond (Quanta Ray, 10ns) and picosecond (Continuum, 25ps) versions of the laser were tried.

To achieve uniform cleaning, the daguerreotype was moved in a zigzag trajectory in front of the laser beam with displacement increments of 0.2 mm between two laser shots. A computer controlled X-Y translation stage was used to move the sample. When possible, an unfocused laser beam was used in order to avoid focussing "hot spots" from a poor quality beam that could lead to streaking.

Although, within the accuracy of our measurements, the ablation threshold was the same (at the given laser wavelength) for 5-nsec and 25-psec laser pulses, we found the overall cleaning performance to be quite different. So far, the best cleaning results have been obtained using the second harmonic (532 nm) of the 25-psec Nd:YAG laser. Nanosecond pulses usually caused more difficulty with peeling of the silver layer and produced less uniform cleaning when the daguerreotype was exposed to large numbers of laser shots. We also observed a distinct difference in the cleaning depending on the wavelength for the case of the 25-ps laser. For the same energy fluence of 50 mJ/cm^2 and same laser exposure, the second harmonics (532 nm) produced a much better cleaning compared with the fundamental wavelength (1064 nm).

Figure 1 shows the results of laser ablation cleaning of a tarnished daguerreotype. The image on the left shows the 150-year old photograph in its original state. The image of the man is covered by a brownish tarnish and is surrounded by a blue/purple "tarnish ring" which indicates a thicker silver sulfide layer. The right portion of figure 1 shows the results of cleaning using the 30 ps, 532 nm laser in the central portion and the 1064 nm beam for the left edge. The right portion was left in its original condition for comparison.

FIGURE 1. Example of a nineteenth century daguerreotype before (left) and after (right) partial laser cleaning.

Clearly, as seen in the central portion of the daguerreotype, the laser technique can be effective in removing the tarnish layer without damaging the underlying delicate image. In some portions of the daguerreotype, however, laser cleaning did not reveal the underlying image and it is assumed that the thicker tarnish had already destroyed the image particles beyond recovery. Finally, the cleaning attempt with the YAG fundamental was not very effective at restoration. On this sample (and another shown in reference 8, the infrared light was clearly less suitable.

In conclusion, modern laser-based techniques can be very useful in applications to the understanding and restoration of objects of artistic or historical value. Laser spectroscopic techniques can characterize the surface of such objects and represent a powerful adjunct to the conservators more usual toolkit. Also, daguerreotypes can now be added to the inventory of such objects that are amenable to laser-ablation cleaning. The "proof of principle" given here and by others is clearly only the first step to developing a useful tool for the restorer but the successful application to daguerreotypes represent the more delicate and demanding example to date.

REFERENCES

1. Miller, J. C., Ed., *Laser Ablation: Principles and Applications,* Springer Series in Material Science, Vol. 28 (Springer-Verlag, Berlin, 1994).
2. Miller, J. C. and Haglund, Jr., R. F., Eds., *Laser Ablation and Desorption*, Experimental Methods in The Physical Sciences, Vol. 30 (Academic Press, New York, 1998).
3. Shawlow, A. L., *Science*, **149**,13 (1965).
4. Asmus, J. F., *IEEE Circuits and Devices Magazine*, **2**, 6 (1986).
5. Fotakis, C., *Opt. Photon. News*, **6**, 30 (1995).
6. Hogan, D. L., Golovlev, V. V., Gresalfi, M. J., Chaney, J. A., Feigerle, C. S., Miller, J. C., Romer, G., and Messier, P., *Appl. Spectrosc.* **53**, 1161 (1999).
7. Hogan, D. L., Golovlev, V. V., and Miller, J. C., Submitted for publication (2000).
8. Golovlev, V. V., Gresalfi, M. J., Miller, J. C., Romer, G., and Messier, P., J. *Cultural Heritage*, **1**, S xxx (2000).
9. Turovets, I., Maggen, M., and Lewis, A., *Studies In Conservation* **43**, 89 (1998).
10. Barger, M. S., Messier, R., White, W. B., *Photogr. Sci. Eng.* **26,** 285-291 (1982); **27**, 141-146 (1983); **28,** 172-174 (1984).

Indirect Absorption and Direct Fluorescence Detections in the Field of a Resonantly Excited Surface Plasmon

V. A. Lioubimov, Al. A. Kolomenskii, P. D. Gershon[*],
and H. A. Schuessler

Department of Physics, Texas A&M University, College Station,
TX 77843-4242
[*]Department of Biochemistry and Biophysics, Institute of Biosciences
and Technology, Texas A&M University, Houston, TX 77030-3303

Abstract. The identification of interacting biomolecules in surface plasmon resonance (SPR) measurements can be achieved by attaching small characteristic covalent fluorophores. To implement this technique we studied the SPR response in the vicinity of an absorption band of dye molecules in ethanol and aqueous solutions. It is shown that the indirect determination of the absorption is possible from changes it induces in the SPR characteristics. Another technique studied employs the direct monitoring of the fluorescence signal in the surface plasmon field. The fluorescence light is then collected in the flow cell near the surface of a gold film with an optical fiber and transmitted to the registering photomultiplier. A significant enhancement of the fluorescent signal due to the excitation in the localized surface plasmon field was observed.

INTRODUCTION

The resonance excitation of surface plasmons by a laser beam is highly sensitive to the dielectric properties of the media involved. In addition, the electromagnetic field is localized in the vicinity of a thin metal film [1] which strongly enhances the observed physical effects as compared to measuring without surface plasmon resonance (SPR). In the traditional SPR sensing technique [2] the variations of the refractive index of the sample medium induce changes in the angle of the resonance excitation, thus providing integrated information on the mass of the material present within the adsorption layer. This information is obtained by observing the SPR curve, namely by monitoring the minimum of the angular distribution of the reflected light. If a ligate contains several molecular species that can potentially interact with the immobilized ligand molecules, as for instance in a competition assay, information regarding the molecular identity of the interacting partners cannot be retrieved. We used two novel modes in SPR measurements, which have the potential to overcome the above limitation: one of them allows the identification of the components by registering the characteristic absorption bands of the fluorophores [3,4] which can be attached to biomolecules and detected with SPR by varying the excitation optical wavelength.

CP584, *Resonance Ionization Spectroscopy 2000: 10th Int'l. Symp.*, edited by J. E. Parks and J. P. Young
© 2001 American Institute of Physics 0-7354-0024-5/01/$18.00

The other technique employs the characteristic fluorescence signal of suitably chosen fluorophores excited in the surface plasmon field. This signal can be monitored with a high sensitivity, so that the detection of singular molecular interactions becomes possible [5].

THEORY

For a three layer system consisting of a glass prism (dielectric permittivity ε_0), a metal (gold) film with complex dielectric permittivity $\varepsilon_1 = \varepsilon_{1r} + i\varepsilon_{1i}$ and a thickness d_1, and a sample medium (dielectric permittivity ε_2, which is also assumed to be complex $\varepsilon_2 = \varepsilon_{2r} + i\varepsilon_{2i}$) the reflectivity can be presented in Lorentzian form:

$$R = 1 - \frac{4(\Gamma_{i1} + \Gamma_{i2})\Gamma_{rad}}{(k - k_p)^2 + (\Gamma_{i1} + \Gamma_{i2} + \Gamma_{rad})^2}, \tag{1}$$

where k is the modulus of the component of the light wave vector parallel to the surface, k_p is the surface plasmon wave number, the ratios (Γ_{i1}/k_p) and (Γ_{i2}/k_p) present dimensionless loss factors for the propagation of the surface plasmons and describe the internal losses in the metal and the sample medium; Γ_{rad} describes the radiative loss due to the transmission of light through the metal film. Equation (1) was derived in the following approximation: $\varepsilon_{1r} \gg \varepsilon_{1i}$ and $\varepsilon_{2r} \gg \varepsilon_{2i}$.

The change in reflectivity ΔR with the variation $\Delta\alpha$ of the absorption coefficient of the medium can be presented in the form:

$$\Delta R = R - R_0 = (1 - R_0)\Delta\Gamma_{i2} \frac{(\delta - 1) - R_0(\delta + 1)}{2\Gamma_i}, \tag{2}$$

where $\Delta\Gamma_{i2} = (\Delta\alpha\lambda_0 / 2\pi\varepsilon_{2r}^{3/2})k_p n_p^2$ and $\delta = \Gamma_i / \Gamma_{rad}$, $\Gamma_i = \Gamma_{i1} + \Gamma_{i2}$, $n_p = [\varepsilon_{1r}\varepsilon_{2r} / (\varepsilon_{1r} + \varepsilon_{2r})]^{1/2}$, λ_0 is the optical wavelength in vacuum.

Using Eq. (2) the change of the absorption coefficient is derived from the change in value of the SPR minimum as:

$$\Delta\alpha(\lambda) = F(\lambda)\Delta R_{\min}(\lambda), \text{ where } F(\lambda) = \frac{\pi\varepsilon_{2r}^{3/2}(\delta + 1)^3\Gamma_i}{2k_p n_p^2 \lambda_0 \delta(\delta - 1)}. \tag{3}$$

The changes in reflectivity due to variations of the absorption have maxima at two thicknesses of the metal film: $d_{1,2} = d_{opt} + (1/2k_p)(\varepsilon_{2r}/\varepsilon_{1r})^{1/2}\ln(2 \pm \sqrt{3})$, where d_{opt} is the optimal thickness for measurements of the changes due to variations of the refractive index of the sample medium. For a gold film and $\lambda_0 = 632.8$ nm an estimate provides $d_{opt} = 42$ nm, $d_1 = 60$ nm and $d_2 = 26$ nm.

The amplitude of the electromagnetic field near the surface of a metal film can be strongly enhanced by the resonance excitation of the surface plasmon. If fluorophores exhibit an absorption band for the optical wavelength used and are located within the penetration depth of the surface plasmon field, then the intensity of the fluorescence light will increase correspondingly. In the same approximation as above the enhancement factor [1] is

$$T = \frac{2|\varepsilon_{1r}|^2 [|\varepsilon_{1r}|(\varepsilon_0 - \varepsilon_{2r}) - \varepsilon_0]^{1/2}}{\varepsilon_{2r}\varepsilon_{1i}(1 + |\varepsilon_{1r}|)}. \tag{4}$$

For a gold film with a thickness of 47 nm at $\lambda_0 = 632.8$ nm an estimate based on Eq. (4) gives $T \sim 10$. For the optimal thickness of the gold film the value $T \sim 30$ can be reached.

EXPERIMENTAL SETUP

To perform measurements with *absorbing* samples (ethanol solutions of Rhodamine 700 dye) over a range of optical wavelenths covering the absorption band we used a white light source in conjunction with a monochromator (model H20 UV, American ISA Inc.). An optical system formed a narrow p-polarized convergent light beam incident on a hemicylindrical prism, which coupled the excitation light with the surface plasmon. Passing through a prism the light formed a spot of ~1.5 mm diameter at the surface of a gold film of (47±0.15) nm thickness. The angular distribution of the reflected light intensity was detected with a photodiode array. The wavelength of the excitation light passing through the monochromator was changed in small increments (~0.5 nm) using a stepping motor.

For *fluorescence* measurements we used ethanol and aqueous solutions of Nile Blue dye and Cy5 dye with the excitation at the fixed wavelength $\lambda = 632.8$ nm of a He-Ne laser with circular polarization. The laser beam was modulated and the fluorescence light extracted by a narrow band interference filter was registered by a photomultiplier. For an increase of the signal/noise ratio phase detection with lock-in amplifier was used. The output signal was recorded by a computer.

RESULTS

Figure 1 shows the dependence of the SPR angular shift on the wavelength for pure ethanol and for ethanol with dissolved dye. A characteristic feature of the data of Fig. 1 is that the relative angular shift in the position of the SPR minimum for the dye solution with respect to pure ethanol changes the sign at the wavelength close to the absorption band maximum. This angular shift is related to the dispersion induced

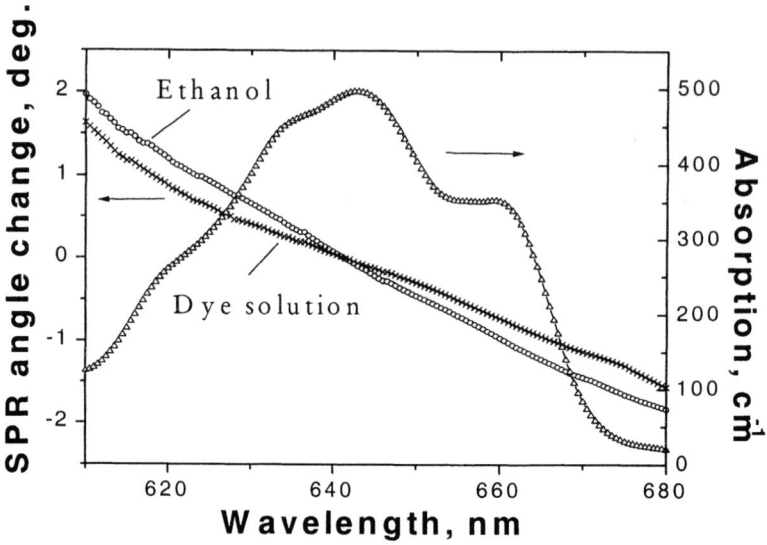

FIGURE 1. Changes in SPR angle (left scale) measured at different wavelength in the vicinity of the absorption band for the Rhodamine 700/ethanol solution and for pure ethanol. For comparison also the absorption coefficient is plotted (right scale).

by the absorption band. The absorption band with the central wavelength ~640 nm, which was determined from the SPR measurements using Eq. (3) is also shown in Fig. 1.

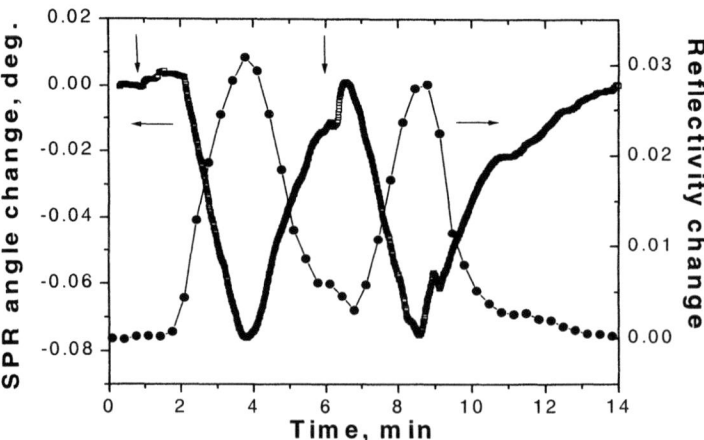

FIGURE 2. Variation in SPR angle and reflectivity at the SPR minimum during two injections of a Rhodamine 700/ethanol solution into a continuous flow of ethanol.

In order to observe changes of the SPR parameters at a concentration continuously varying in some interval, the dye solution was injected in the flow of pure ethanol. Then the concentration of the dye in the measuring cell gradually changed as result of mixing and diffusion processes. Fig. 2 shows the changes in the SPR dynamic response when two injections of 0.2 ml aliquots of the dye solution ~0.5 min duration each were made in a streaming flow of ethanol through the cell, the vertical arrows in Fig. 2 indicate the moments at which these injections were initiated.

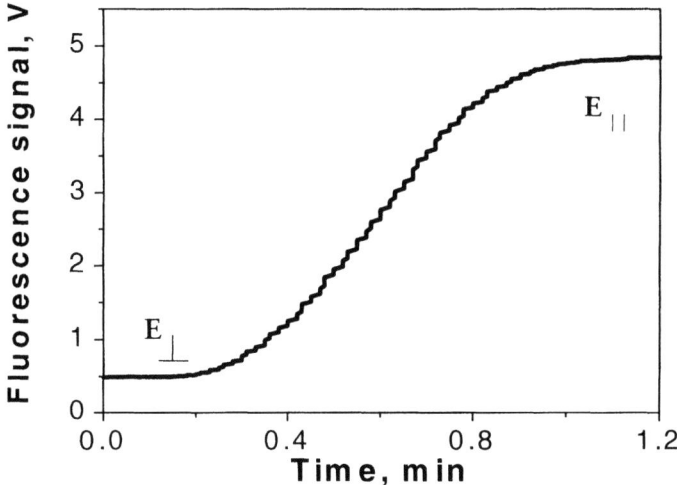

FIGURE 3. Increase of the fluorescence signal during a gradual rotation of the orientation of the electric field from normal to parallel to the incidence plane. In the latter case surface plasmons were resonantly excited. The laser wavelength was chosen to be close to the center of the absorption band to efficiently induce fluorescence.

Fig. 3 shows the observed change in the fluorescence signal when the polarization of the laser beam was changed. Initially the beam had e-polarization at which no resonance interaction of the laser beam with surface plasmons was possible. Then the orientation of the polarization was changed by 90° allowing resonance excitation of the surface plasmon. This change of the polarization was accompanied by an increase of the fluorescence signal by a factor of ~10 in agreement with the above estimate, thus demonstrating the enhancement of the signal in the strongly localized surface plasmon field.

ACKNOWLEDGMENTS

This work was supported in part by the Texas Advanced Technology Program under Grant No.010366-084.

217

REFERENCES

1. Raether, H. *Surface plasmons on smooth and rough surfaces and on gratings*, Springer Tracts in Modern Physics, v. 111, Heidelberg,: Springer, 1988, pp. 6-7.
2. Earp R. L., and Dessy, R. E., "Surface plasmon resonance ", in *Commercial Biosensors: Applications to Clinical, Bioprocess, and Enviromental Samples*, edited by G. Ramsay , New York: Wiley, 1998, pp. 99-164.
3. Pockrand, I. and Swallen, J. D. , *J. Opt. Soc. Am.* **68**, 1147-1151 (1978).
4. Kolomenskii, A.A., Gershon, P.D., and Schuessler, H.A., *Applied Optics* **39**, 3314-3320 (2000).
5. Yokota, H. , Saito, K. , and Yanagida, T. , *Phys. Rev. Lett.* **80**, 4606-4609 (1998).

RIMS Measurements for the Determination of the First Ionization Potential of the Actinides Actinium up to Einsteinium

Achim Waldek[a], Nicole Erdmann[a], Carsten Grüning[a], Gerhard Huber[b],
Peter Kunz[b], Jens Volker Kratz[a], Jens Lassen[b], Gerd Passler[b],
Norbert Trautmann[a]

[a]Institut für Kernchemie and [b]Institut für Physik, Universität Mainz, D-55099 Mainz, Germany

Abstract. The ionization potential is a fundamental property of an element. In order to determine the first ionization potential, a method based on the measurement of photoionization thresholds in the presence of a well-defined electrical field is used. By one or two step resonant laser excitation, the investigated atoms are promoted to a highly excited state. Ionization is obtained by scanning the wavelength of an additional tunable laser across the threshold W_{th}, which is detected by a sudden increase of the ion signal. Extrapolation of W_{th} to zero field strength yields directly the first ionization potential. Using this method the first ionization potentials of Ac, Am, Cm, Bk, Cf, and Es have been determined for the first time. Furthermore, the ionization potentials of Th, U, Np, and Pu were remeasured.

INTRODUCTION

Resonance Ionization Mass Spectrometry (RIMS) is a very sensitive and selective detection method for elements. Due to the sensitivity of RIMS, this method allows to investigate elements which are very rare and/or highly radioactive, and it is also used in ultra-trace analysis of environmental samples.

The ionization potential of an element is one of its fundamental properties. Its accurate knowledge is important for the classification of the element and its position within the periodic system of elements as well as for drawing conclusions on its electronic structure, which is directly connected to the atomic spectra [1]. The value of the first ionization potential is also dependent on relativistic effects which occur in the case of heavy elements like the actinides, due to the relativistic velocities of the inner shell electrons [2]. Furthermore, the precise determination of the first ionization potential of the actinides allows to test the predictions of multi configuration Dirac-Fock calculations [3].

Two approaches are found in the literature dealing with the calculation of the first ionization potential for the actinides. One is based on the extrapolation of spectral properties [4,5], the other on calculating the s-electron binding energies by the semi-empirical Slater-Condon method and ab initio Hartree-Fock calculations and making use of analogies between lanthanides and actinides [6].

CP584, *Resonance Ionization Spectroscopy 2000: 10th Int'l. Symp.*, edited by J. E. Parks and J. P. Young
© 2001 American Institute of Physics 0-7354-0024-5/01/$18.00

In this paper, we present the experimental determination of the first ionization potential of the actinides from Ac up to Es using laser spectroscopy for measuring the ionization threshold within an external electrical field. The sensitivity of this method allows the determination of the ionization potential of an element with samples of less than 10^{12} atoms.

EXPERIMENTAL

Experimental setup

An atomic beam of the element under investigation is produced by heating a sandwich filament. This consists of a tantalum backing foil, onto which the actinide is electrochemically deposited in form of its hydroxide which is subsequently covered with a thin metal layer. By heating such a filament, the hydroxide is converted to the oxide which is reduced to the metallic state by diffusing through the metallic layer. For the elements uranium up to einsteinium, a titanium layer is very well suited [7]. For actinium, the reduction by titanium was not efficient enough; therefore, in this case zirconium was used.

For the optical excitation and ionization, two laser systems have been used. Most of the measurements were performed with a high repetition rate dye laser system [8], consisting of three dye lasers which are pumped by two copper vapor lasers at a repetition rate of 6.6 kHz. More recently, a full solid-state laser system came into use, consisting of three Ti:Sa lasers pumped by a pulsed Nd:YAG laser at an equally high repetition rate. The measurements on actinium were done by a combination of one Ti:Sa laser for resonant excitation and one dye laser for the ionization step.

The wavelengths are measured by an ATOS 'Lambdameter' with an accuracy of $\Delta\lambda/\lambda = 10^{-6}$. Using prisms and dielectric mirrors and/or quartz fibers, the laser beams are focused into the apparatus, where they cross the atomic beam perpendicularly. The photo-ions are accelerated in a homogenous electric field, which also serves for lowering the ionization threshold. After passing a time-of-flight mass spectrometer (TOF), the ions are detected by a multichannel plate detector. The TOF is described in [9].

Photoionization in an electrical field

The measurements of the first ionization potential are based on the determination of photoionization thresholds in the presence of an external electrical field F. According to the classical saddle point model [10], the excitation energy $W(r)$ relative to the electronic ground state of an atom with one highly excited electron is – in a one-dimensional approximation-given by:

$$W(r) = IP - \frac{Z_{eff} \cdot e^2}{4\pi\varepsilon_0 r} - eFr \qquad (1)$$

where e is the electric charge of the electron, Z_{eff} the effective charge number of the core, r the distance of the excited electron from the nucleus, ε_0 the permittivity of the vacuum, and *IP* the first ionization potential. The ionization threshold W_{th}, which is the maximum value of $W(r)$, depends on the electrical field strength as follows:

$$W_{th}(F) = IP - 2 \cdot \sqrt{\frac{Z_{eff} \cdot e^3}{4\pi\varepsilon_0}} \cdot \sqrt{F} = IP - const \cdot \sqrt{F} \qquad (2)$$

Equation (2) shows a linear dependence of W_{th} on the square root of the electrical field strength F.

In the case of actinium a highly excited state is resonantly populated by a one-step excitation. For most other actinide elements two resonant steps were used instead. The ionization laser is scanned across the ionization threshold W_{th} in the presence of a well-defined electrical field F. The threshold W_{th} is indictated by a sudden increase of the ion count rate. Extrapolation of W_{th} to zero field strength yields directly the first ionization potential.

MEASUREMENTS AND RESULTS

Using the procedure described above, the first ionization potential (IP) of all actinide elements from actinium up to einsteinium (except for protactinium) has been experimentally determined; for the elements heavier than plutonium, no experimental values had been published before. The same holds for the IP of actinium, which is presented in this paper for the first time at all. Figure 1 shows the ionization thresholds of actinium at three different electrical field strengths. For the first time, measurements have been performed down to field strengths as low as 1.6 V/cm. Therefore, it is necessary to correct for field perturbations, which may occur from background suppression voltages, steering electrodes etc. Taking an offset of 12 V/cm into account, due to the penetration of electric fields into the interaction region, the thresholds perfectly fit a straight line (Fig. 2). The IP of actinium has been determined to be 43398(3) cm^{-1} = 5.3807(3) eV . A plot of the ionization thresholds for six actinide elements measured at Mainz using RIMS is shown in Fig 2.

Table 1 displays all results obtained at Mainz for the first ionization potentials of the actinide elements [11-15] compared to experimental values published in the literature [16-19]. The uncertainties for the experimental values in this work are statistical errors given as two standard deviations [2σ] derived from a least-squares fit including weighted errors for each data point.

In order to compare the experimental results with theoretical predictions based on extrapolated spectroscopic data [4,5] or Hartree-Fock calculations [6], the first ionization potentials have been normalized to the ionization process $5f^N 7s^2 \rightarrow 5f^N$ 7s. Figure 3 shows the normalized data plotted versus the number of 5f electrons from the experiment as well as from theoretical predictions. As pointed out in [6], the actinide ionization potentials follow the trend for binding energies of the s-electrons by forming two straight lines. For the heavier actinides of $Z > 94$ the ionization potentials follow such a straight line. However, for the lighter actinides the

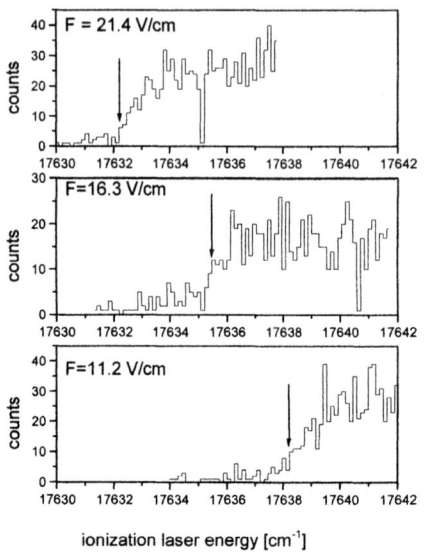

ionization laser energy [cm⁻¹]

FIGURE 1. Ionization thresholds of actinium at three different electrical field strengths. The thresholds are indicated by arrows.

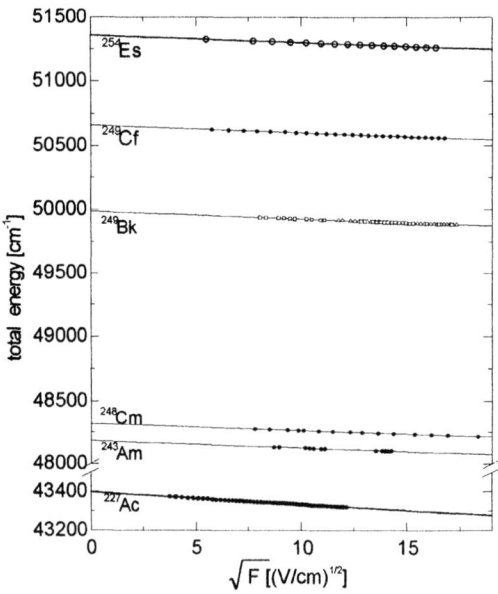

FIGURE 2. The measured ionization thresholds of six actinide elements plotted versus the square root of the electrical field strength.

TABLE 1. First ionization potentials of 10 actinide elements determined at Mainz.

Actinide element	IP$_{exp.}$ Mainz [cm^{-1}]	IP$_{exp.}$ Mainz [eV]	IP$_{lit.}$ [cm^{-1}][a]
Actinium	43398(3)	5.3807(3)	-
Thorium	50867(2)	6.3067(2)	50890(20)
Protactinium	-	-	-
Uranium	49957(1)	6.1939(1)	49958.4(5)
Neptunium	50535(2)	6.2655(2)	50536(4)
Plutonium	48601(2)	6.0258(2)	48604(1)
Americium	48180(3)	5.9736(3)	-
Curium	48324(2)	5.9914(2)	-
Berkelium	49989(2)	6.1979(2)	-
Californium	50663(2)	6.2814(2)	-
Einsteinium	51358(2)	6.3676(2)	-

Tabulated are the experimental data obtained at Mainz [11-15] and experimental data previously published [16-19], IP$_{lit.}$ [cm^{-1}][a].

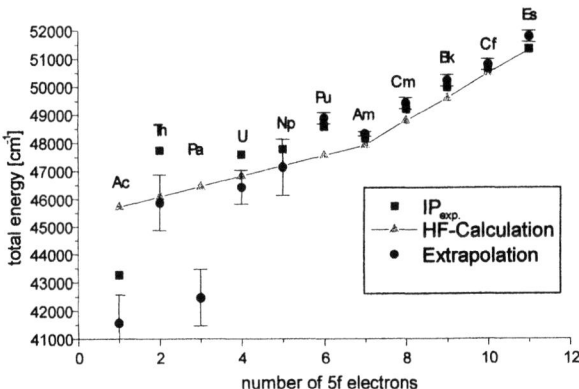

FIGURE 3. Comparison of the experimentally determined ionization potentials and two theoretical predictions normalized to the ionization process $5f^N 7s^2 \rightarrow 5f^N 7s$.

experimental values for the first ionization potential show strong deviations from a linear dependence of the number of 5f-electrons which is ascribed to configuration interactions. The experimental ionization potentials are higher for the lighter and lower for the heavier actinides than the extrapolated data.

OUTLOOK

The experimental determination of the first ionization potential of protactinium has not yet been performed, due to insufficient reduction of protactinium oxide during evaporation from the sandwich filament. Even by the use of thorium instead of titanium as covering layer, no significant reduction to the metallic state was achieved. Further investigations are in progress.

Measuring the isotopic dependence of the ionization potential could serve as a test of theoretical calculations. The differences for two isotopes of the same element can be calculated very precisely. For elements, where isotopes with quite a large difference in neutron number are accessible for the experiment, e.g. ^{232}U / ^{238}U or ^{236}Pu / ^{244}Pu, this effect might be observable, as it is expected to be in the order of 0.1 cm^{-1}/amu [20].

The elements beyond einsteinium are very rare and short-lived. The lack of spectroscopic data make the determination of an ionization threshold even more difficult. However, the determination of the first ionization potential of fermium (Z=100) will be attempted in the nearest future, using less than 10^{12} atoms of ^{255}Fm ($T_{1/2}$ = 21.8 h)

ACKNOWLEDGEMENTS

This work was funded by the Deutsche Forschungsgemeinschaft. The authors are indebted for the use of ^{249}Bk and ^{254}Es to the Office of Basic Energy Sciences, U.S. Department of Energy, through the transplutonium element production facilities at the Oak Ridge National Laboratory, managed by Lockheed Martin Energy Research Corporation.

REFERENCES

1. Pyykkö, P. and Descleaux J.P, *Acc.Chem. Res* **12**, 276 (1979).
2. Pyykkö, P., *Chem Rev.* **88**, 563 (1988).
3. Fricke, B., Johnson E and Rivera, G.M., *Radiochim. Acta* **62**, 17 (1993).
4. Sugar, J., *J. Chem. Phys.* **59**, 788 (1973).
5. Sugar, J., *J. Chem. Phys.* **60**, 4103 (1974).
6. Rajnak, K. and Shore B.W., *J. Opt. Soc. Am.* **68**, 360 (1978).
7. Eichler, B. et al., *Radiochim. Acta* **79**, 221 (1997).
8. Ruster, W. et al., *Nucl. Instr. Meth. Phys. Research A* **281**, 547 (1989).
9. Urban, F.-J. et al.in: *Resonance Ionization Spectroscopy,* Inst. Phys. Conf. Ser. 128, 1992 p. 233.
10. Bransden, B.H. and Joachain, C.J., *Physics of Atoms and Molecules,* London, Longman 1984.
11. Deisenberger, R. et al., *Angew. Chemie Int. Ed. Engl.* **34**, 814 (1995).
12. Riegel, J. et al., *Appl. Phys. Phys. Part B* **56**, 275 (1993).
13. Köhler S. et al., *Angew. Chemie Int. Ed. Engl.* **35**, 2856 (1996).
14. Köhler S. et al., *Spectrochimica Acta Part B* **52**, 717 (1997).
15. Erdmann, N. et al., *J. Alloys Comp.* ***271-273***, 837 (1998).
16. Johnson, S.G. et al., *Spectrochim. Acta Part B* **47**, 633 (1992).
17. Solarz, R.W. et al., *Phys. Rev. Part A* ***14***, 1129 (1976).
18. Worden, E.F. and Conway, J.G., *J. Opt. Soc. Am.* **69**, 733 (1979).
19. Worden, E.F. et al., *J. Opt. Soc. Am. Part B* ***10***, 1998 (1993).
20. Fricke, B., private communication.

A Nuclear Transition in the Optical Range: Search for the 3.5 eV Gamma Emission from Thorium-229m

J. P. Young,* R. W. Shaw,* and E. Tkalya[‡]

* Oak Ridge National Laboratory, Oak Ridge TN 37831, USA
[‡] Institute of Nuclear Physics, Moscow State University, Ru-119899, Moscow, Russia

Abstract. The reported photon emission from relaxation of the 3.5 eV excited nuclear state of Th-229 (1,2) has been proven to be incorrect (3-6). The status of current experiments to determine the energy of this nuclear isomeric state is described, and plans for a more direct excitation of the state are discussed.

INTRODUCTION

An anomalously low-energy, excited level for the ^{229}Th nucleus was reported some years ago by Helmer and Reich (7). Analysis of gamma transition energies to the ground and first excited states in ^{229}Th yielded the range of 3.5 +-1.0 eV for the energy of this level (7). Four research teams have published results of optical measurements of ^{229}Th in the last three years. The low energy isomeric level of ^{229}Th is populated approximately 2% of the time in the alpha decay of ^{233}U. A structured photon emission for a sample of ^{233}U in the near UV optical range was observed by two groups in 1997 and 1998 (1,2). The ^{229}Th basis for these optical spectra was questioned by Young et al. (3) in 1998, and Utter et al. (4), Shaw et al. (5) and Young et al. (6) in 1999. Alternative explanations for the spectral emission from ^{233}U samples were proven in the later work: namely, the near-UV emission is from nitrogen in the air (5), luminescence of uranyl ions (6), and fluorescence of the silica tube (6), all excited by the respective sample alpha activity.

The anomalously low-energy isomeric level in ^{229}Th has generated considerable interest because such a level should have some unusual properties, such as for example, an optical range gamma ray and an accompanying electron bridge decay route (8). The electron bridge raises the possibility of a highly effective nuclear

CP584, *Resonance Ionization Spectroscopy 2000: 10ᵗʰ Int'l. Symp.*, edited by J. E. Parks and J. P. Young excitation by optical photons (9); this could lead to a change in the alpha decay rate of [229]Th via population of the 3.5 eV level under low intensity laser radiation (10).

The decay channels of the 3.5 eV level and the half-life should depend strongly on the chemical environment (8). In an isolated Th atom, the 3.5 eV level should decay predominantly via the electron bridge channel (8). This is a third order process, which is essentially unobserved in other nuclear transitions. Within a metal sample, nonradiative decay via conduction electrons should dominate (11). This second order one-photon process is the inverse of the process of inelastic electron scattering by nuclei and can be viewed as the analog of internal conversion for these conduction electrons. In a wide energy-gap dielectric, such as [229]ThO$_2$, the first order process of nuclear gamma emission should be possible (12,13).

The 3.5 eV level in [229]Th is a "bridge" between low energy nuclear physics and many other fields of physics, such as optics, laser physics, solid state physics, physics of surfaces, and nuclear isomer interactions.

EXPERIMENTAL

Earlier experimental work concerned with locating the energy position of Th-229m emission was reported at the last RIS conference (4). Further emission studies of various compounds of isotopically pure [233]U (< 0.1 ppm [232]U) have been carried out. Weak emissions were recorded over the 250 to 1000 nm wavelength range. The samples have included [233]U$_3$O$_8$, [233]UO$_3$, [233]UO$_4$ 2H$_2$O and a [238]UO$_2$(NO$_3$)$_2$ solution spiked with [225]Ac. The last solution was prepared to have an alpha activity equivalent to that for a 10 ppm [232]U impurity. The samples were contained in one of three types of SiO$_2$ tubes. Commercial grade SiO$_2$ from an unidentified vendor was used for the first tests. Later Suprasil II or S-300 silica (Technical Glass Products, Mentor, OH) was used. The first two samples were sealed in the sample tubes under vacuum after flushing with He. The uranyl peroxide and nitrate solution samples were sealed in air with a teflon ball fixed into place with Torr Seal epoxy (Varian Vacuum Products, Lexington, MA). Details of the emission spectroscopic instrument have been described previously (4,5,6). The sample image was focused on the entrance slit of a Spex polychromator (Model HR 460); a 150 groove/mm or in a few cases a 1500 groove/mm diffraction grating, blazed at 500 nm, was the dispersing element in the spectrometer. The dispersed image was focused onto a 1024 by 256 pixel, liquid nitrogen-cooled, UV coated, CCD detector (Model EEV CCD15-11). All of these items were available from Instruments SA, Edison, MA. Data were collected via Spectramax (version 1.1D), a DOS program also available from Instruments SA. The program was modified to allow multiple averaged exposures or a single exposure for periods of time up to 4 days.

RESULTS AND DISCUSSION

We have carried out emission spectral studies of several ^{233}U oxides and have seen no reproducible emission that could be attributable to a gamma ray emission. The oxides have included $UO_4\cdot2H_2O$, a compound thought to be the sample (1) that was used in the original study and which gave a structured emission in the ultraviolet region. This structured emission was later identified as N_2 and N_2^+ emission in air, excited by alpha radiation from the sample (5,6). A broad visible emission in the 520 nm region was also observed for this sample (1,2); that was later proven to be UO_2^{++} luminescence (6).

The various uranium oxide samples we have studied and the emission results observed are summarized in Table 1. As can be seen from the table, emission from the SiO_2 container was seen in all the samples. Of note is that an atomic oxygen line (O I) was observed for the uranyl peroxide sample, $UO_4\cdot2H_2O$. This sample was sealed in air, but so was a sample of UO_3. Atomic oxygen was not observed in the latter sample. It would appear that O I is generated in the UO_4 sample, being a transient species coming from the radiation-induced decomposition of the $O_2^=$ moiety. Our earlier observation of N_2 emission (5), originating in the air above the sample, was seen above either $UO_4\cdot2H_2O$ or UO_3 when they were sealed in air, but was not observed when imaging only the solid in these cells.

TABLE 1. Summary of ^{233}U Oxide Emission Experiments[a]

Compound	Color	Sample atmosphere	Type of SiO$_2$ container	Comments
$UO_4 2H_2O$	orange	air	Commercial SiO2	SiO$_2$ emission O l emission
UO_3	yellow	air	Suprasil II	SiO$_2$ emission
UO_3	yellow	vacuum	Suprasil II	SiO$_2$ emission
U_3O_8[b]	black	vacuum	Commercial SiO$_2$	SiO$_2$ emission 314 nm emission
U_3O_8[b]	black	vacuum	Suprasil II	SiO$_2$ emission
U_3O_8[b]	black	vacuum	S-300	SiO$_2$ emission

[a] All spectral studies involved averaged 1 hour integrations for more than 18 hour total.
[b] The identical U_3O_8 powder sample was used in all three studies.

An apparently artifactual emission was seen at 314 nm in one sample of $^{233}U_3O_8$ sealed in an evacuated commercial SiO_2 container. The emission was sharp, but not slit-limited when observed using a 0.2 mm entrance slit in the polychromator. As noted in the table, when that same sample powder was observed in a sealed evacuated Suprasil II or S-300 (4 ppm OH concentration) sample container, the emission was not observed. Based on chemical analysis of that U_3O_8 sample, it contained, perhaps, unusual amounts of Al, (20 wt %); Pt, (2 wt %); and Nd, (5%). The amount of Al and

Pt is unusually large and was thought to be present because of the sampling handling. None of these three elements, however, would appear to be the basis for the single emission peak seen, however. Keep in mind, further, that U_3O_8 is visually black and would therefore be expected to absorb UV photons.

In all experiments with long-term, multi-hour, observations, SiO_2 emissions were seen. These emissions were most probably excited by alpha decay from ^{233}U, 1.6 x 10^5 year half-life. A typical SiO_2 emission spectrum is shown in Figure 1. This spectrum corresponds to a sample of $^{233}UO_4 \cdot 2H_2O$ sealed in air in a commercial silica tube. Note the broad emission peaks at 415, 550 and 650 nm. The energy position of these three bands correspond to the three bands reported for the cathodoluminescence of silica (14). The relative intensities of these bands do not match those seen in that study. They are, however, consistent with the view that they arise from silica (15); the relative intensities would be expected to vary with composition of silica and the type of excitation source. Note that the most intense SiO_2 peak is seen near 550 nm for this sample and silica type.

The emission spectrum of Suprasil II SiO_2 is shown in Figure 2. Note that the emission in the 420 nm region is almost nonexistent; the peak at 550 nm is weak; and the most intense peak is at 620 nm. The sample in this case was $^{233}U_3O_8$ sealed in a vacuum. A similar emission spectrum was recorded for $^{233}UO_3$ in an evacuated Supersil II tube.

As pointed out in Table I, an unidentified emission peak was observed at 314 nm in one sample of $^{233}U_3O_8$ sealed in an evacuated commercial SiO2 tube. In this sample of U_3O_8, an emission was also seen in the 620 nm region, but a more intense broad emission was seen that extended from 350 to 600 nm. Whether the composition of this tube was the same as that used in the study shown in Figure 1 is unclear. As noted in the table of results, however, when this same sample was later contained in an evacuated Supersil II tube, the emission peak at 314 nm was not observed. Further the emission spectrum of SiO_2 seen was essentially identical to that seen in Figure 1. It seems to be established that the emission of Suprasil II is reproducible with ^{233}U sample excitation.

The same sample of $^{233}U_3O_8$ was next transferred to and sealed in an evacuated S-300 SiO_2 tube. As noted earlier, S-300 contains a very low concentration of OH radicals (4 ppm). In our loading process, the SiO_2 was sealed by heating in a gas flame. We had hoped that the emission at 620 nm, thought to be related to OH radicals, would be minimal. This was not the case and the assumption can be made that sealing such SiO_2 with a gas flame introduces OH into the SiO_2. The peak at 314 nm was not seen. It was only observed in a commercial SiO_2 tube and must be considered to be the result of that material.

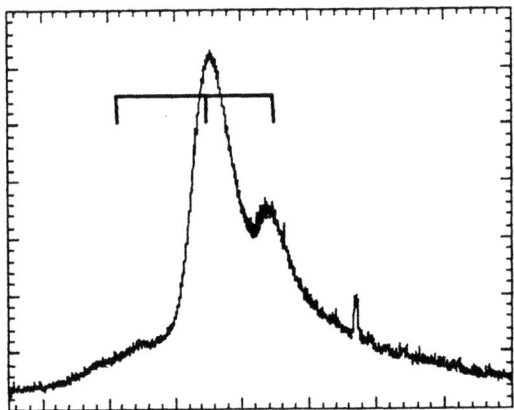

FIGURE 1. Emission spectrum of $^{233}UO_4 \cdot 2H_2O$ sealed in a commercial grade SiO_2 tube. The date was obtained by averaging one hour observations totaling more than 18 hours. The SiO_2 emission band positinos (14) are indicated. The sharp emission at 777 nm is due to O 1

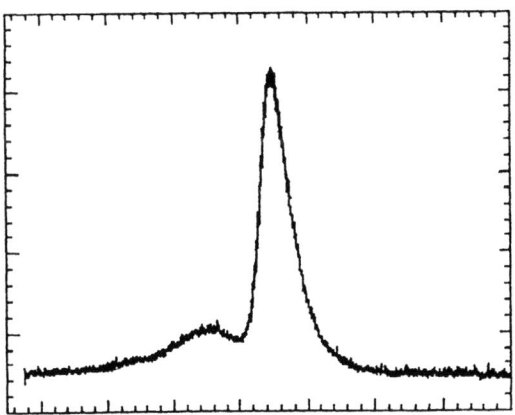

FIGURE 2. Emission spectrum of $^{233}U_3O_8$ in an evacuated Suprasil II SiO_2 tube. The data was obtained by averaging one hour observations totaling more than 50 hours.

CONCLUSIONS AND FUTURE WORK

No UV-visible emission has yet been observed that can be attributed to a gamma emission from ^{229m}Th when that isomeric state is generated from ^{233}U alpha decay. We have, however, observed other emissions from various ^{233}U compounds, and most of these can be ascribed to SiO_2 emissions excited by alpha decay of the sample contained therein.

A different approach is needed to investigate the presence of the isomeric state of ^{229}Th. Proposals for a better experiment have been suggested in the literature by Tkalya (12,13) and in an presentation by Young and Shaw (16). The approaches are similar in several respects. Both choose to study ^{229}ThO$_2$ as the sample material, probably as a dried rather than as an ignited oxide. ThO$_2$ is transparent in the region below 6 eV (17). Both approaches would use a synchrotron source to excite the isomeric state from ground state ^{299}Th. Both would look for the gamma emission from the isomeric state as the goal of the study. A theoretical basis for this approach has been presented by Tkalya (12). He proposed to excite the intermediate nuclear levels $5/2^+$ (29.19 KeV) and $7/2^+$ (71.78 KeV) by monochromatic synchrotron radiation. From the 5/2+ level, then, the isomeric state could be populated and observed. In the Young and Shaw approach (16), the ^{229}ThO$_2$ contained in an evacuated Suprasil II container would be excited by broad band synchrotron energy from 4.5 to 2.5 eV, or broader if necessary. Observation of the gamma emission would be observed after an irradiation time or between excitation cycles. Once the emission has been identified, an attempt to excite the isomeric level with a carefully tuned UV laser could be made. Much useful information about isomeric nucleii and methods of excitation can be realized if laser coupling can be accomplished from the isomeric state that nature has provided.

ACKNOWLEDGEMENTS

Research sponsored by Laboratory Directed Research and Development Program. U-233 and Ac-225 provided by U.S. Department of Energy Isotope Production and Distribution Program. Partial funding also provided by the Department of Energy, Office of Science, Laboratory Technology Research Program. Oak Ridge National Laboratory is managed by UT-Battelle, LLC under Contract DE-AC05-00OR22725 with the U.S. Department of Energy. E.V. Tkalya's effort supported partly by the Russian Foundation for Basic Research, Grant No. 98-02-16070a and Grant No. 00-15-96651 in Support of the Leading Scientific Schools.

REFERENCES

1. Irwin, G. M. and Kim, K. H., *Phys. Rev. Lett.* **79**, 990 (1997).
2. Richardson, D. S.; Benton, D. M.; Evans, D. E.; Griffith, J. A. R.; and Tungate, G., *Phys. Rev. Lett.* **80**, 3206 (1998).
3. Young, J. P.; Shaw, R. W.; and Webb, O. F., "Spectral Studies Related to the 3.5 eV Isomeric State of Th-229" AIP Conference Proceedings 454, New York, 1998, pp 235-240.
4. Utter, S. B.; Beiersdorfer, P.; Barnes, A.; Lougheed, R. W.; Crespo Lopez-Urrutia, J. R.; Becker, J. A.; and Weiss, M. S., *Phys. Rev. Lett.* **82** 505 (1999).
5. Shaw, R. W.; Young, J. P.; Cooper, S. P.; and Webb, O. F., *Phys. Rev. Lett.* **82** 1109 (1999).
6. Young, J. P.; Shaw, R. W.; Webb, O. F., *Inorg. Chem.* **38** 5192 (1999).
7. Helmer, R. G. and Reich, C. W., *Phys. Rev. C* **49** 1845 (1994).
8. Strizhov, V. F. and Tkalya, E. V. [Zh. Eksp. Teor. Fiz. **99 697** (1991].
9. Tkalya, E. V.; Varlamov, V. O.; Lomonosov, V. V.; and Nikulin, S. A., *Phys. Scr.* **53** 296 (1996).
10. Dykhne, A. M.; Eremin, N. V.; and Tkalya, E. V., *JETP Lett.* **64** 345 (1996) [Pis'ma Zh. Eksp. Tdeor. Fiz. **64** 319 (1996)].
11. Tkalya, E. V., *JETP Lett.* **70** 371 (1999)[Pis'ma Zh. Eksp. Teor. Fiz. **70** 367 (1999)].
12 Tkalya, E. V. Ibid **71** 311 (2000)[ibid. **71** 449 (2000)].
13. Tkalya, E. V.; Zherikhin, A. N.; and Zhudov, V. I., *Phys. Rev. C* **61** 064308 (2000).
14. Morimoto, Y.; Weeks, R. A.; Barnes, A. V.; Tolk, N. H.; and Zuhr, R. A., *J. Non-Cryst. Solids* **203** 62 (1996).
15. Weeks, R. A., private communication.
16. Young, J. P. and Shaw, R. W., oral presentation, 219th Nat'l. Mtg., American Chemical Society, San Francisco, CA., March 26, (2000).
17. Sviridova, A. I. and N. V. Suikovskaya, N. V., *Opt. Spectrosk.* **22** 940 (1967)[*Opt. Spectrosc.* **22** 509 (1967)].

SESSION VIII

In-Situ Resonance-Enhanced Multiphoton Ionization (REMPI) Measurements Using an Optical Fiber Probe

S.M. Angel[1], R.C. Chinni[1], B.M. Cullum[2], S.K. Shealy[1], D.M. Gold[3], D. LeSage[3], S.B. Brown[3], and B.W. Colston[3]

1. *Department of Chemistry and Biochemistry, University of South Carolina, Columbia, SC 29208*
2. *Oak Ridge National Laboratory, Oak Ridge, TN 37831*
3. *Lawrence Livermore National Laboratory, Livermore, CA 94550*

Abstract. Two fiber-optic REMPI probes have been developed to determine the feasibility of using fiber-optics for *in-situ* determination of volatile organic compounds. In both designs, an optical fiber transmits a high-powered laser pulse to the sample, causing it to ionize, and the subsequent electrons are collected by a platinum electrode. One probe contains focusing optics while the other contains no focusing optics other than the fiber. Excitation using a 2+2 scheme was used for all measurements because visible excitation has optimal transmission of light through fiber-optics. The non-lensed probe was tested by measuring toluene and benzene and the response was compared to that using the lensed probe.

INTRODUCTION

Volatile organic compounds (VOCs) in the environment are of concern due to the hazards they pose to human health. Many have acute and long-term toxic effects and some, like benzene, are carcinogens.[1,2] VOCs have been accidentally and intentionally introduced into the environment and contaminate the surrounding soil and ground water supplies. The most common method used for trace detection of VOCs is gas chromatography (GC).[2,3] GC VOC analysis usually requires a great deal of time due to the preconcentration step that is used to extract the compound from water. Extraction methods include the use of activated charcoal, liquid/liquid extraction, and solid phase microextraction.[4,5]

We are investigating more rapid, in-situ techniques for measurement of VOCs in groundwater and soil. One technique that looks promising is REMPI, which can give quantitative and qualitative information about VOCs. REMPI is a spectroscopic technique that can be used to selectively detect and identify volatile organic compounds (VOCs) even in the presence of other species in a sample.[6] In REMPI, a tunable laser is used to selectively excite and ionize the contaminant molecule using one or more photons, and an electrode is used to collect the subsequent electrons that are produced.[7] The current is a measure of contaminant concentration and the excitation wavelength

CP584, *Resonance Ionization Spectroscopy 2000: 10th Int'l. Symp.*, edited by J. E. Parks and J. P. Young
© 2001 American Institute of Physics 0-7354-0024-5/01/$18.00

depends on the specific contaminant.. The REMPI technique has the potential to be useful along with other in-situ techniques such as on-line GC and on-line GC/MS. Typical REMPI measurements take approximately one minute, require no sample preparation, and are sensitive down to a few parts-per-billion.[8]

EXPERIMENTAL

The REMPI System: The REMPI system consists of an Nd:YAG laser operating at 266 or 532 nm (New Wave Research Model Minilase-10) or an Nd:YAG-pumped OPO laser operating in the range of 480-560 nm (Spectra Physics MOPO 730 pumped by Spectra Physics GCR 250-10), a high voltage power supply (SRS Inc. Model PS325), the cell/electrode electronics, a fast digital oscilloscope (Tektronix TDS 684B), and a readout computer (Dell OptiPlex GXI w/ Labview). The laser beam is launched into a fiber-optic where it is focused to cause ionization of the sample in the REMPI cell. The electrons are collected by the electrode biased at high positive potential and the current is measured as a voltage drop over a 250 kΩ resistor with a digital oscilloscope.

Lensed Fiber-optic REMPI Probe: The lensed fiber-optic probe (Figure 1(a)) consists of an optical fiber (3M model FG-550-UER) for to deliver the laser excitation, two lenses, and a platinum electrode biased at a high positive potential for electron collection.

Non-Lensed Fiber-Optic REMPI Probe: The non-lensed probe (Figure 1(b)) consists of an optical fiber (3M model FG-600-UER) encased in a stainless steel syringe, and is polished flat. In this probe, the stainless steel syringe serves as the electrode.

Preparation of Samples: Solutions of toluene and benzene were prepared by mixing the pure solvent with methanol (Aldrich). It should be noted that methanol has no resonance peaks in the region from 480-560 nm. Different amounts of each solution were injected into the cell, allowed to vaporize, and equilibrate for 20 minutes. The concentrations are reported on a mole/mole basis relative to the amount of air in the cell.

Analysis of Data: All of the data are shown without averaging or smoothing other than what was done by accumulating the signal. All of the REMPI excitation profiles are shown from 480-560 nm in 0.2 nm increments using an average of 20 shots at each wavelength. All of the calibration data are shown by accumulating either 1000 laser pulses using the lensed REMPI probe or 300 laser pulses using the non-lensed REMPI probe. For all measurements, the intensity of the REMPI signal is based on the peak height and all error bars displayed represent ±1 standard deviation of five replicate measurements. The data is plotted with Igor Pro Version 3.16 data analysis software from Wavemetrics, Inc. For calibration plots, the linear best fit was calculated and the detection limits are taken as three times the standard deviation of the lowest concentration divided by the slope of the linear calibration curve (e.g. a 3σ detection limit).

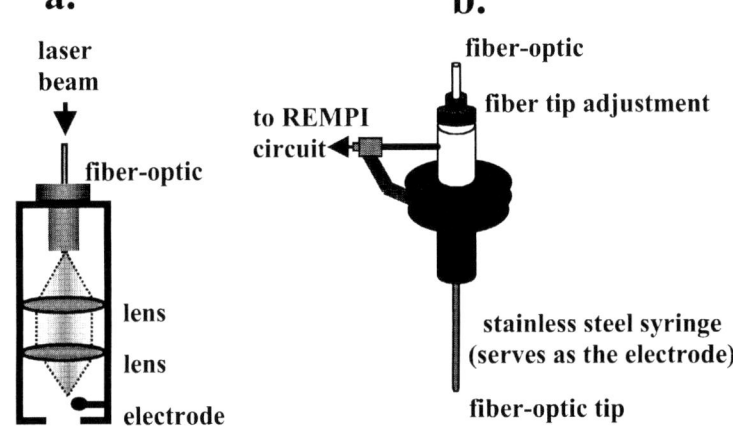

Figure 1. (a) Schematic of the lensed fiber-optic REMPI probe. The lenses are 1-in diameter f/1.5 plano convex and the platinum electrode is biased at +1000 V with respect to the aluminum housing. (b) Schematic of the non-lensed fiber-optic REMPI probe. The stainless steel syringe serving as the electrode is biased at +2000 V.

RESULTS AND DISCUSSION

Initial studies of toluene using the lensed fiber-optic REMPI probe show the feasibility of making *in-situ* REMPI measurements using fiber-optics. Figure 2 shows calibration curves using 1+1 (a) and 2+2 (b) REMPI excitation. The detection limits for toluene were 1.47±0.002 ppb using 1+1 excitation and 8.5±0.4 ppb using 2+2 excitation. These initial results also show the potential use of higher order excitation schemes without significantly affecting the sensitivity. This is important when using optical fibers due to the higher transmission of visible light in the optical fiber.

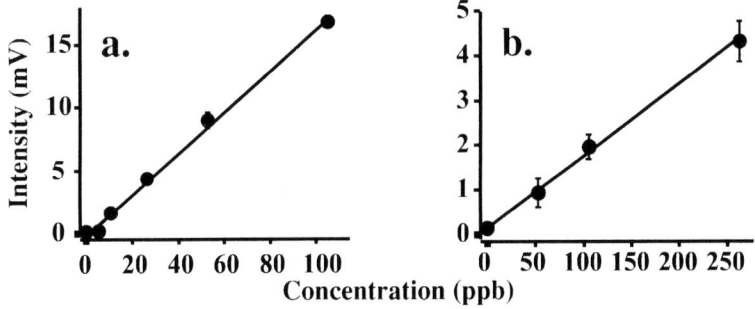

Figure 2. Calibration curve for toluene exciting at (a) 266 nm (1+1 excitation) and at (b) 532 nm (2+2 excitation) using the lensed fiber-optic REMPI probe

The results obtained using the lensed fiber-optic probe are very promising. However, the probe is relatively large. In an attempt to reduce the size of the fiber-optic REMPI probe, we are testing a non-lensed design. This probe design contains no focusing optics and is thus easier to align. In this design, the optical fiber is encased in a stainless steel syringe which serves as both protection to the fiber and as the electrode. The optical fiber that is used in the non-lensed probe is a 600 μm core diameter fiber. This produces a high enough power density at the tip of the fiber to cause REMPI 2+2 excitation of both toluene and benzene, but it is not necessarily the optimum size.

To test the non-lensed REMPI probe, toluene and benzene solutions were injected into the cell, allowed to vaporize, and measured with the probe. Figures 3 (a) and Figure 4 (a) show the calibration curves for benzene and toluene, respectively. Each concentration shown is the average of five replicate measurements. The detection limit and sensitivity for toluene were 43 ppb and 0.0028 mV/ppb, respectively. The detection limit and sensitivity for benzene were 20 ppb and 0.0030 mV/ppb, respectively. A comparison of the results from the lensed probe and the non-lensed probe using 2+2 REMPI excitation is shown in Table 1. It can be seen that the sensitivity of the non-lensed REMPI probe is approximately an order of magnitude less than the sensitivity of the lensed probe. Also, detection limits obtained with the non-lensed probe are higher than the lensed probe. However, these results are preliminary and the probe geometry and fiber size have not yet been optimized.

Excitation scans for toluene and benzene are shown in Figures 3(b) and 4(b), respectively. These scans were taken by introducing the vapor into the ionization cell, focusing the laser beam into the cell with a parabolic mirror, and scanning the laser. These scans show the unique excitation profiles and wavelength dependence for each of these compounds. This also shows that it is possible to use different excitation wavelengths to discriminate these two compounds.

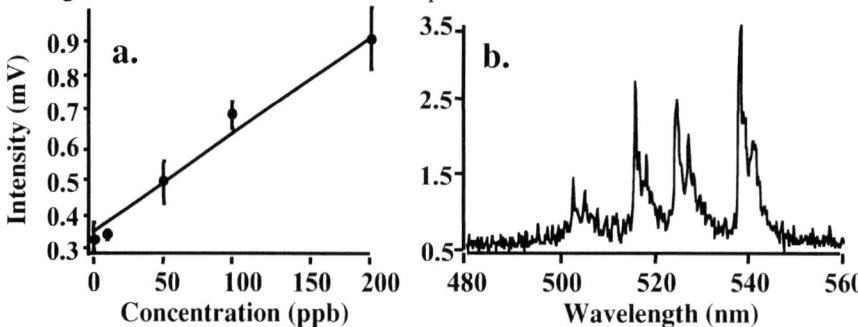

Figure 3 (a) Toluene calibration with the non-lensed REMPI probe with excitation at 512.0 nm. (b) REMPI 2+2 excitation scan for toluene.

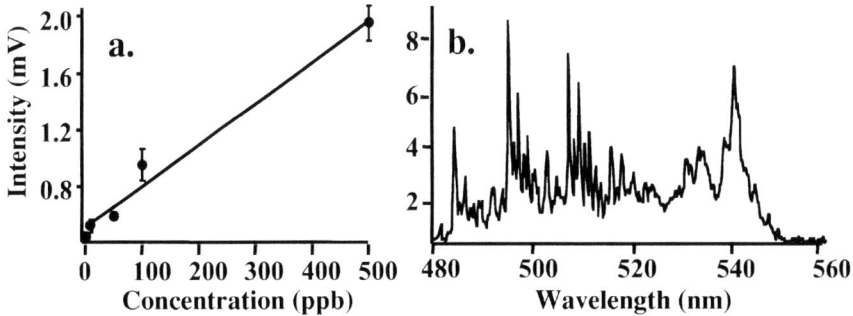

Figure 4. (a) Benzene calibration with the non-lensed REMPI probe with excitation at 508.8 nm. (b) REMPI 2+2 excitation scan for benzene.

TABLE 1. Comparison of the Results from Lensed REMPI Probe and the Non-Lensed REMPI probe using 2+2 excitation.

Probe	Detection Limits (ppb)	Sensitivity (mV/ppb)
Preliminary REMPI Probe[a]	8.5	0.014
New REMPI probe [a]	43	0.0028
New REMPI probe [b]	20	0.0030

[a]Toluene data.

[b]Benzene Data.

The linear dynamic range of the non-lensed REMPI probe was determined by measuring toluene and benzene at concentrations ranging from 10 ppb to 10 ppm, and making five replicate measurements at each concentration. The results show that the non-lensed REMPI probe is linear over approximately four orders of magnitude for both toluene and benzene (Figure 5).

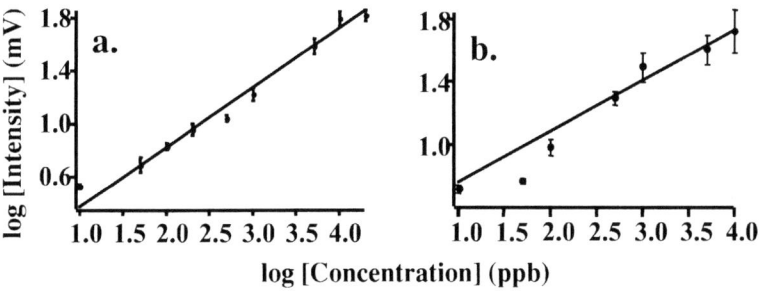

Figure 5. Graph showing the linear dynamic range for the non-lensed REMPI probe for detection of (a) toluene and (b) benzene.

CONCLUSIONS

In-situ REMPI measurements of benzene and toluene are demonstrated using two different fiber-optic probe designs, a lensed and a non-lensed probe. Detection limits suitable for many environmental analyses were obtained. The non-lensed probe has a large linear dynamic range. This probe also has the advantage of no focusing optics which allows for easier alignment. REMPI 2+2 excitation scans show unique excitation spectra for these two compounds.

FUTURE WORK

Future work will consist of optimizing the geometry of the non-lensed fiber-optic REMPI probe to improve the detection limits and sensitivity. Studies will also be performed using mixtures of VOCs to determine optimal excitation wavelengths and possible interferences. A long-term goal of this project is to incorporate a fiber-optic REMPI probe into a cone penetrometer for soil vapor and ground water analysis.

ACKNOWLEDGMENTS

We would like to thank the U.S. Department of Energy for support of this work under Grant Number DE-FG0799ER62881.

REFERENCES

1. *The Merck Index. An Encyclopedia of Chemicals, Drugs, and Biologicals.* 11th ed. Ed. S. Budavari. Merck and Co. Inc, Rahway, NJ. 1989.
2. Neff. J.M. *Chem. Engr. Prog.* **83**, 27-33, (1987).
3. Dean, J.R.; Tomlinson, W.R; Makovskaya, V.; Cumming, R.; Hetheridge, M.; Comber, M. *Anal. Chem.* **68**, 130-133, (1996).
4. Potter, D.W. and Pawliszyn, J. *Environ. Sci. Technol.* **28**, 298-305, (1994).
5. Zhang, Z. and Pawliszyn, J. *J. High Resolut. Chromatogr.* **16**, 689-692, (1993).
6. Lubman, D.L. *Mass Spectrom. Rev.* **7**, 535-554, (1988).
7. Cool, T.A. *Appl. Opt.* **23**, 1559--1572, (1984).
8. Cullum, B.M.; Shealy, S.K.; Angel, S.M. *Appl. Spec.*, **53**, 1646-1650, (1999).

REMPI-TOFMS for On-Line Monitoring and Controlling the Coffee Roasting Process

Ralph Dorfner[1], Thomas Ferge[1], Chahan Yeretzian[2], Ralf Zimmermann[1*],
Antonius Kettrup[1]

[1] GSF Forschungszentrum für Umwelt und Gesundheit GmbH, Institut für Ökologische Chemie,
Ingolstädter Landstraße. 1, D-85764 Neuherberg, Germany and Technische Universität München,
Lehrstuhl für Ökologische Chemie und Umweltanalytik, D-85748 Freising, Germany

[2] Nestlé Research Center, Nestec Ltd., Vers-chez-les-Blanc, P.O. Box 44,
CH-1000 Lausanne 26, Switzerland

* corresponding author

Abstract. REMPI@266nm-TOFMS is used for on-line analysis of the coffee roasting process. Volatile and flavor active compounds of coffee were ionized by REMPI@266nm and monitored on-line and in real-time by TOFMS during the coffee roasting process. The phenol and 4-vinylguaiacol time-intensity profiles, for example, show typical behavior for different roasting temperatures and provide an indicator to the achieved degree of roasting. The impact of the moisture level of the green coffee beans on the time shift of a typical (commercial) roasting time, correlates with REMPI-TOFMS measurements and literature data.

INTRODUCTION

The headspace-gas of food products is a rich and versatile source of information. As a matter of fact, all food products, throughout the whole food processing chain, emit volatile compounds. Furthermore, volatile compounds are - in spite of the fact that they represent only a very small fraction of a food product - related to important properties of foods like their flavor, their age or shelf-life, origin (geographical or genetic), history of treatment and processing, as well as quality. Finally, analyzing the headspace-gas can be done non-invasively. All of this makes on-line headspace analysis very suitable for process monitoring and control in industrial contexts. Real-time monitoring of volatiles can help to better understand and control these processes, and hence optimize product quality, improve process efficiency and reduce costs. Particularly noteworthy are food processing steps that involve flavor generation in enzymatic or thermal reactions, optimization of concentration or drying processes, or aroma extraction and recovery processes.

In order to be effective, time-resolved headspace analysis methods have to be fast, sensitive and selective. This can be implemented by REMPI-TOFMS@266nm, if one aims at analyzing mainly volatile aromatic and heterocyclic compounds. Using other

CP584, *Resonance Ionization Spectroscopy 2000: 10th Int'l. Symp.*, edited by J. E. Parks and J. P. Young
© 2001 American Institute of Physics 0-7354-0024-5/01/$18.00

wavelengths [1] or ionization techniques (SPI, LEI or PTR) [2-5], the detection of several other compound classes is possible.

EXPERIMENTAL

Laser mass spectrometry is a well suited method for on-line analysis since it is a two dimensional analytical method. It combines the optical selectivity of the ionization process as well as the mass selectivity of a time-of-flight mass spectrometer. Using different gas inlet systems the selectivity can be tuned from substance class selectivity up to single isomer selectivity [6-10]. For our investigations we use a compact and mobile device, which contains the laser unit, the TOF-MS and a computer for experiment control and data acquisition. Mass calibration and absolute quantification is realized by using an external gas standard consisting of a series of diffusion and permeation tubes. For further details see e.g. Heger et. al. [11].

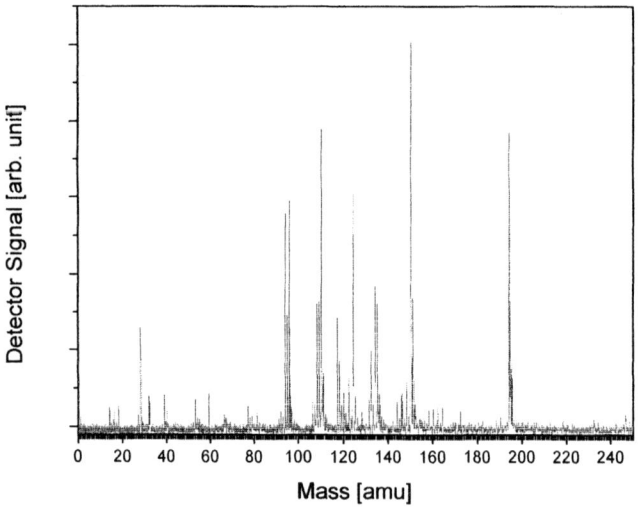

FIGURE 1. Typical mass spectrum measured with REMPI@266nm during a coffee roasting simulation experiment at 225°C. Several molecular compounds like that of phenol (94 amu), furfural (96 amu), dihydroxybenzene (110 amu), guaiacol (124 amu), 4-vinylguaiacol (150 amu) or caffeine (194 amu) are detected.

Roasting experiments in the laboratory were performed using a small lab-scale roaster (Probat, BRZ 2, electrically heated). With this setup, batches of 80 – 100 g green beans were roasted at 200°C, 225°C and 250°C. The roaster was preheated to its roasting temperature. Then the green beans – arabica coffee from Columbia – were filled in the drum and the heated probe of the sampling system was then pushed into the front opening of the roasting drum. During the roasting process the roasting temperature was not readjusted. The process was then monitored for times ranging between 450 sec and 900 sec, depending on the roasting temperature.

RESULTS AND DISCUSSION

Figure 1 shows a typical mass spectrum of volatiles released of arabica coffee being roasted at 225°C and measured by REMPI@266nm. Using this wavelength, typically, heterocyclic and aromatic compounds like phenol, furfural, dihydroxybenzene, guaiacol, vinyl-guaiacol and caffeine are detectable [12,13].

Comparing the time profiles of e.g. 4-vinylguaiacol or phenol (Figure 2) of several measurements (roasting temperature always 225°C), one recognizes that all time profiles show the same behavior. Only the starting point of the pyrolysis reaction varies due to some differences in the intial conditions, such as temperature variations of the roaster by a few degrees, differences in starting time of the data acquisition or others. Clearly, there is a good correlation between the different runs. This is a good indication that the analyzing method works reproducible, and therefore is very suitable for application in industrial processes.

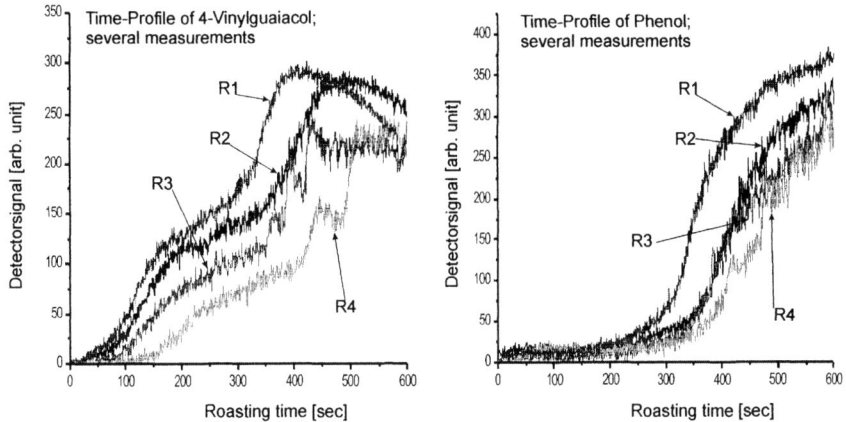

FIGURE 2. Time-Intensity-profiles of 4-vinylguaiacol, 150amu, (left) and phenol, 94 amu, (right) of several roasting experiments at 225°C, run number 1-4. For details see text.

As mentioned before, roasting experiments at different temperatures (200°C, 225°C, 250°C) were done. When the preheated roasting drum was filled with green beans, an initial decrease of the temperature was observed in every case. That is due to the additional thermal mass and the initial drying process of the beans before the pyrolysis reaction starts. Figure 3 shows the dependence of the time-intensity profiles for phenol and 4-vinylguaiacol on the selected roasting temperature. The profiles of phenol and 4-vinylguaiacol are again shown as an example. The shift of the signal increase to earlier times using higher roasting temperatures is clearly visible. The higher the roasting temperature the earlier the aroma forming reactions start. At first glance the nearly perfect correlation of the 225°C and 250°C roasting experiments is surprising. However, it can be explained with heat transfer conditions in the roasting drum: The observed temperature decrease after filling the drum is about 50°C in every roasting experiment, independent of the selected roasting temperature. The reactions occurring during pyrolysis are mostly exothermic in nature. A definite activation energy is needed to start the reactions. For roasting coffee these reactions proceed in a small temperature window of about 160°C – 250°C. The highest reaction efficiency is achievable within the bounds of 210°C – 240°C [14].

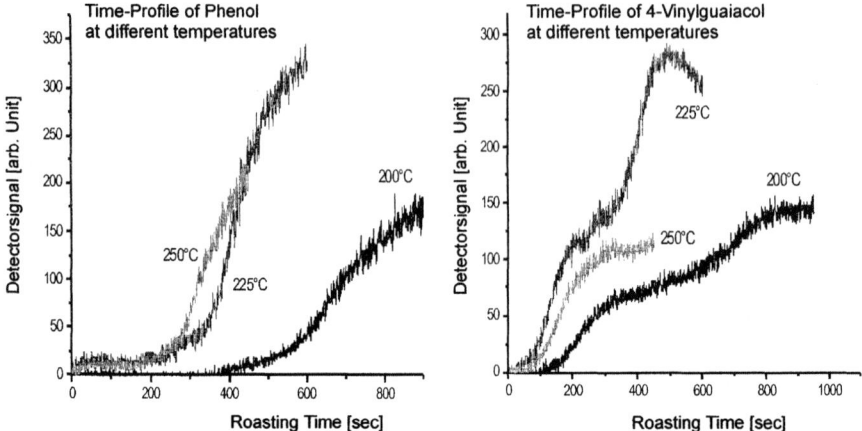

FIGURE 3. Time-intensity-profile of phenol (left) and 4-vinylguaiacol (right) at three different roasting temperatures (200°C [900sec monitoring time], 225°C [600sec monitoring time] and 250°C [450sec monitoring time]). Note the similar behavior of the three time-profile. In this case differences in the times of reaction-onset are significant.

Knowing this, figure 3 should be interpreted differently: For a 200°C roasting experiment, prior to addition of the green beans, the drum temperature is decreased to about 150°C after a few seconds when the beans are added. About 3 minutes later the temperature is increased again to about 180°C and the pyrolysis reactions start (formation and release of 4-vinylguaiacol is one of the first reactions). Since the temperature does not exceed the 200°C mark, the effect of the exothermicity is not as severe as at higher roasting temperatures - the pyrolysis reaction is much slower than at higher temperatures. That is why the shape of the 200°C 4-vinylguaiacol profile is flatter than the others and why at 200°C the production of phenol clearly only starts later.

The resemblance of the 225°C and the 250°C profile can be explained in a similar way: In both cases the temperatures decrease after starting the roasting process in both cases. Nevertheless, the temperature is always high enough to activate the pyrolysis reaction. Therefore the 4-vinylguaiacol signal starts to increase for both roasting temperatures already 1.5 min after the beginning. Since the temperatures are both within the reaction temperature window, the profiles are very similar. That is clearly visible in the nearly simultaneous increase of the phenol signals.

We also tried to establish a degree of roasting to a batch of roasted beans. For this we roasted green beans at several temperatures and varied the roasting time. Then we determined the percentage total mass loss as an indicator for the degree of roasting [15]. Figure 4, for example, shows the phenol profile for different roasting temperatures superimposed with the time window for the indicated commercial roasting degree. The corresponding commercial roasting time at 200°C found by Gianturco[16] is marked as well. The different time window for a 200°C roasting experiment, established by us and found by Gianturco, can be explained and corrected as follows: The determination of the roast degree, using the mass loss method, is normally based on the assumption of 12% humidity in green beans. However, the beans we used were not humidity sealed. So the beans are drying with increasing duration of storage. Supposing a lower level of humidity the best correlation will be found for a humidity level of about 7%, which seems to be acceptable for our beans. That means for comparison with the commercial roasting time of Gianturco (12% humidity) the roasting time of our roasting experiments have to be shifted to a shorter value. Then our results are in good correspondence with literature data. Taking the signal intensity as an indicator for the roast degree we are able to establish the roast degree for the 225°C and the 250°C experiment too. These roasting times are in good correlation with the roast degree determined by the mass loss method.

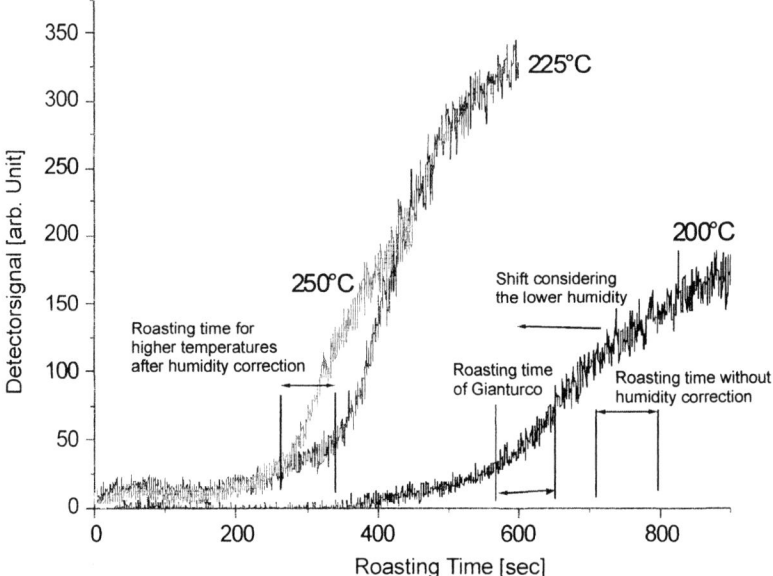

FIGURE 4. Correlation of the roast degree and the signal intensity using the time-profile of phenol.

CONCLUSION

REMPI-TOFMS is well suited for on-line, real-time anaylsis of volatiles in food production processes. Several flavor active and aroma relevant compounds of the coffee roasting process were monitored. Due to different time behavior of several compounds, information about important process conditions like roast level and influence of several start conditions like humidity level can be detected.

ACKNOWLEDGMENTS

R. Dorfner thanks the Max-Buchner-Forschungsstifung for the scholarship. K. Hafner, F. Mühlberger and Th. Hauler are thanked for support and motivating discussions.

REFERENCES

[1] R. Dorfner, R. Zimmermann, C. Yeretzian et al., "Laser Ionisation Mass Spectrometry (REMPI-TOFMS) for On-line Analysis of Volatiles in Food Science: Coffee-Roasting and Headspace Experiments," presented at the 19th International Symposium on Resonance Ionization Spectroscopy, Manchester, United Kingdom, 309-312 (1998).

[2] R. Dorfner, Th. Ferge, C. Yeretzian et al., "Laser-Mass-Spectrometry as an On-Line Analysis Technique for Monitoring of the Coffee Roasting Process," Analytical Chemistry, in preparation (2000).

[3] W. Lindinger, A. Hansel, and A. Jordan, "On-line monitoring of volatile organic compounds at pptv levels by means of Proton-Transfer-Reaction Mass Spectrometry (PTR-MS) Medical applications, food control and environmental research," International Journal of Mass Spectrometry and Ion Processes 173, 191-241 (1998).

[4] C. Yeretzian, A. Jordan, H. Brevard et al., "Time-Resolved Headspace Analysis by Proton-Transfer-Reaction Mass-Spectrometry," presented at the ACS Symposium 763 on Flavour Release, (2000).

[5] C. Yeretzian, A. Jordan, H. Brevard et al., "On-line Monitoring of Coffee Roasting by Proton-Transfer-Reaction Mass-Spectrometry," presented at the ACS Symposium 763 on Flavour Release, (2000).

[6] J. B. Anderson, R. P. Andres, and J. B. Fenn, "Supersonic nozzle beams," Advances in Chemical Physics X, 275-315 (1966).

[7] H. J. Heger, R. Zimmermann, E. R. Rohwer et al., "A novel interface for Optimized Coupling of Gas Chromatography and Supersonic Jet Spectroscopy," Journal of High Resolution Chromatography, submitted for publication (1999).

[8] R. Tembreull and D. M. Lubman, "Use of Two-Photon Ionization with Supersonic Beam Mass Spectrometry in the Discrimination of Cresol Isomers," Analytical Chemistry 56, 1962-1967 (1984).

[9] J.M. Hayes, "Analytical spectroscopy in supersonic expansions," Chemical Reviews 87, 745 (1987).

[10] R. Zimmermann, H. J. Heger, A. Kettrup et al., "Laser Induced Jet-REMPI-Mass Spectrometry as Isomer Selective Capillary Gas Chromatographic Detector: A Novel Approach for Rapid Environmental Analysis.," presented at the 18th International Symposium on Capillary Chromatography, Riva, Italy, 1996 (unpublished).

[11] H. J. Heger, R. Zimmermann, R. Dorfner et al., "On-line Emission Analysis of Polycyclic Aromatic Hydrocarbons down to pptv Concentration Levels in the Flue Gas of an Incineration Pilot Plant with a Mobile Resonance-Enhanced Multiphoton Ionization Time-of-Flight Mass Spectrometer," Analytical Chemistry 71, 46-57 (1999).

[12] R. Zimmermann, R. Dorfner, and A. Kettrup, "Direct analysis of products from plant material pyrolysis," Journal of Analytical and Applied Pyrolysis 49 (Elsevier), 257-266 (1999).

[13] R. Dorfner, C. Yeretzian, R. Zimmermann et al., "On-Line Analysis of Food Processing Gases by Resonance Laser Spectroscopy (REMPI-TOFMS): Coffee Roasting and Related Applications," presented at the 18th International Conference on Coffee Science (ASIC `99), Helsinki, Finnland, 136-142 (2000).

[14] J. J. Clarke and R. Macrae, *Technology* (Elsevier Applied Science Publisher, London, 1987).

[15] M. Sivetz and Desrosier N. W., *Coffee Technology* (AVI Publishing Co., Westport Conn., 1979).

[16] M. Gianturco, "Coffee Flavour," in *The Chemistry and Physiology of Flavours*, edited by H. W. Schultz, E. A. Day and L. M. Libbery (AVI Publishing Co., Westport Conn., 1967), 431-449.

POSTERS

Gadolinium Trace Determination in Biomedical Samples by Diode-Laser-Based Multi-Step Resonance Ionization Mass Spectrometry

Ch. Geppert[*], K. Blaum[*], S. Diel[*],
P. Müller[*], W.G. Schreiber[†] and K. Wendt[*]

[*]Institut für Physik, Johannes Gutenberg-Universität, D-55099 Mainz, Germany
[†] Universitätsklinik Mainz, Klinik für Radiologie, D-55131 Mainz, Germany

Abstract. Diode laser based multi-step resonance ionization mass spectrometry (RIMS), which has been developed primarily for ultra trace analysis of long lived radioactive isotopes has been adapted for the application to elements within the sequence of the rare earths. First investigations concern Gd isotopes. Here high suppression of isobars, as provided by RIMS, is mandatory. Using a three step resonant excitation scheme into an autoionizing state, which has been the subject of preparatory spectroscopic investigations, high efficiency of $>1 \times 10^{-6}$ and good isobaric selectivity $>10^7$ was realized. Additionally the linearity of the method has been demonstrated over six orders of magnitude. Avoiding contaminations from the Titanium-carrier foil resulted in a suppression of background of more than one order of magnitude and a correspondingly low detection limit of 4×10^9 atoms, equivalent to 1pg of Gd. The technique has been applied for trace determination of the Gd-content in animal tissue. Bio-medical micro samples were analyzed shortly after Gd-chelat, which is used as the primary contrast medium for magnetic resonance imaging (MRI) in biomedical investigations, has been injected. Correlated in-vivo magnetic resonance images have been taken. The RIMS measurements show high reproducibility as a well as good precision, and contribute to new insight into the distribution and kinetics of Gd within different healthy and cancerous tissues.

INTRODUCTION

Trace determination of elements in the region of the rare earths is hampered by their widely identical chemically behavior. Chemical separation processes are rather inselective and thus conventional trace determination techniques are limited by elemental as well as additional molecular isobaric interferences. A suitable alternative is found in selective laser ionization. Our diode laser based resonance ionization mass spectrometry (RIMS) has been developed primarily for selective ultra-trace determination of long lived radioactive isotopes, e.g. of ^{41}Calcium [1,2]. This system has been adapted for the selective determination of individual rare earth elements.

Because of various applications of interest, first investigations concerned Gadolinium isotopes. Precise determination of the minor isotope ratio ^{152}Gd/^{154}Gd in single grain meteorite inclusions is one field of research. This ratio is of interest for cosmo-chemical studies on stellar temperatures during the s-process [3]. A different application for Gd-analytics which shall be reported here are biomedical investigations. Contrast enhanced MRI is an effective and non-invasive medical

CP584, *Resonance Ionization Spectroscopy 2000: 10th Int'l. Symp.*, edited by J. E. Parks and J. P. Young
© 2001 American Institute of Physics 0-7354-0024-5/01/$18.00

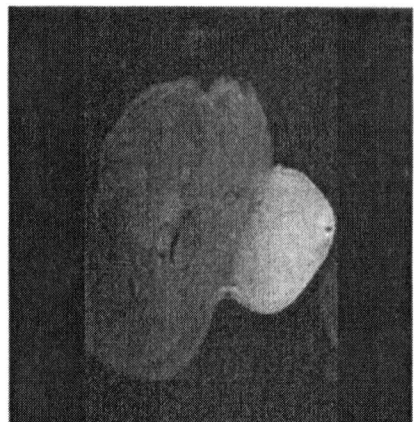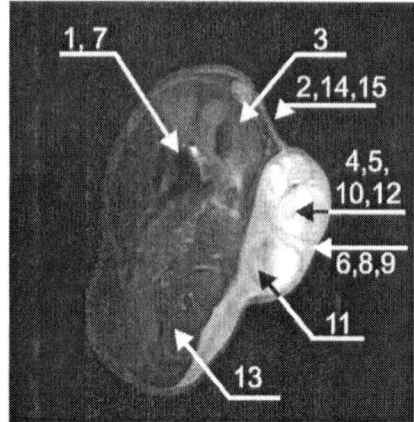

Figure 1. Cross section MRI images of a nude mouse, measured before (a) and immediately after (b) the injection of Gd-containing contrast medium. White areas on the right side of both pictures represent an tumor which developed after implantation of tumor cells. After application of Gd-DTPA for contrast enhancement, various organs and structures within the tumor can be clearly identified. Numbers in Fig. 1b indicate the different positions from where samples of 10 µg tissue have been dissected.

technique for detecting and characterizing tumors and for providing information about tissue microcirculation and function. A strong enhancement can be seen on the two in-vivo MRI images of a nude mouse with an experimental lung tumor acquired before (Fig. 1a) and immediately after (Fig. 1b) injection of Gd-DTPA as contrast medium. However, quantitative analysis of the amount of Gd-DTPA within a volume element (voxel) is difficult because many physical and biophysical factors influence the relation between Gd-DTPA concentration and MRI signal intensity. For a better knowledge of biophysical aspects it is mandatory to determine the precise overall Gd-content in different tissue samples after Gd-DTPA injection.

For all of these applications a high suppression of isobars is essential. Using triple step resonant laser excitation in combination with mass selection by a quadrupole mass filter we attain elemental and isotopic selectivities of both better than 10^7 and a high overall detection efficiency of up to 10^{-5}.

EXPERIMENTAL SETUP AND METHOD

The experimental apparatus used for analytical measurements is shown in Fig. 2a. The sample of interest is brought into a self collimating electrothermally heated graphite crucible (GC). Therein the sample is evaporated to form a well collimated atomic beam. Up to three continuous lasers are used for isotope selective triple step excitation of Gd atoms in the atomic beam. The optical excitation and ionization scheme, as given in Fig. 2b, uses compact external cavity diode-lasers for all three transitions applying laser wavelengths of 422.7 nm → 790.6 nm → 859.2 nm. Precise control and adjustment of all laser wavelengths is achieved via a computer controlled stabilization system constructed around a 300 MHz confocal Fabry-Perot interferometer (CFI). Prior to the analytical investigations, the efficient excitation

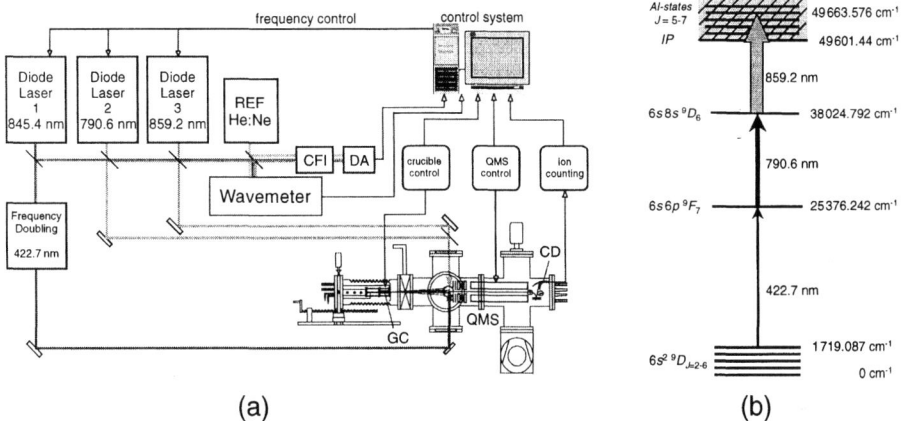

Figure 2. (a) Schematic diagram of the experimental setup used, including the diode laser system for triple-resonance ionization with the confocal Fabry-Perot interferometer (CFI) and detector assembly (DA) for laser frequency stabilization, as well as the vacuum chamber with graphite crucible (GC), quadrupole mass spectrometer (QMS) and channeltron detector (CD).

scheme and the choice of the autoionizing state has been determined and characterized in a number of preparatory spectroscopic investigations [4,5]. The resulting photoions are mass-selected in a commercial quadrupole mass spectrometer (QMS) and subsequently counted in an off-axis channeltron detector system.

All analytical measurements were performed in the following way: A typical quantity of only 10 mg of primary tissue sample is used permitting for a high lateral resolution. The different biological tissues were homogenized in an ultra centrifuge producing about 10 µl of liquid solution to which an aliquot of 10 µl of 3 M nitric acid was added. No chemical treatment or separation was performed. Samples were directly introduced into the graphite furnace, dried at atmospheric pressure at temperatures of about 150°C for 20 minutes and transferred into the vacuum system. Following a rapid temperature ramp, efficient atomization in the graphite crucible was carried out on a fixed temperature of 1900°C during the full measurement period of 400 seconds. During this period all three diode lasers were fixed on resonance of ^{158}Gd and the ion signal was integrated. The absolute gadolinium concentration in each sample was calculated by multiplying the integrated count rate with the known isotope ratio and the overall detection efficiency, which was determined by regular calibration measurements. Additional measurements of blank samples and background were carried out regularly during the measurement sequence of analytical samples.

SPECIFICATIONS OF THE DETECTION METHOD

In preparatory investigations several synthetic samples artificially enriched with different known amounts of Gd-DTPA have been studied. By measuring these samples in the way described above, a good linearity in the detection over six orders of magnitude was demonstrated. This linearity is illustrated by the linear fit to the data

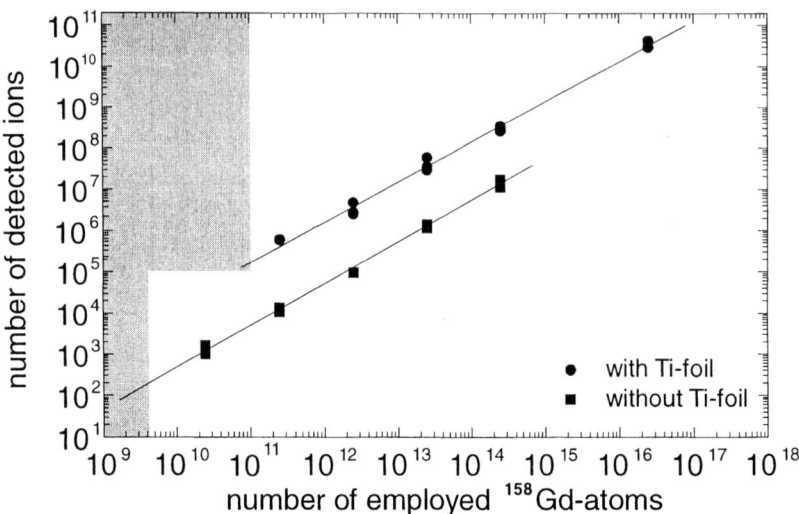

Figure 3. Calibration curve for Gd trace analysis by RIMS, given in double-logarithmical scale. Calibrated synthetical samples were used, error bars are much smaller than the size of the symbols. The right edges of the gray shaded areas represent the detection limits. In case of the use of Ti-foil as a carrier for the samples this level of detection is caused by background effects. If the samples are brought directly into the graphite crucible without any carrier the limitation is caused by the efficiency and counting statistics. Further explanation in the text.

points in Fig. 3, the slight variance of individual data points demonstrates the a very good reproducibility of the detection method of about 10% along the full dynamical range. The two different data-sets as shown result from two different way of sample introduction into the graphite furnace, which has been investigated. If a Titanium-foil is used for efficient atomization an efficiency of 1.5×10^{-6} was reached (dots in Fig. 3). But even though a foil of maximum purity, specified higher than 99,999%, has been selected, unspecific Gd background from this material resulted in a detection limit of about 10^{11} atoms Gd per sample, marked by the upper gray shaded area in the plot. Therefore we abandoned the use of this technique and brought the samples directly into the graphite crucible without any additional carrier material (squares in Fig. 3). As illustrated by the offset of the linear fit curves the efficiency of the atomization in that way decreased by about a factor of 30. However due to the absence of any unspecific background the detection limit was nevertheless decreased down to $4 \cdot 10^9$ atoms per sample, corresponding to about 1 pg of Gd atoms.

RESULTS AND DISCUSSION

For demonstration of the applicability of RIMS to real bio-medical investigations Gd ultra trace determination studies by RIMS have been performed on tissue samples. These were taken at the Mainz university clinics from a nude mouse with implanted tumor which acts as a case study. Shortly after injection of Gd-chelate in-vivo magnetic resonance images were acquired. After that the mice were killed and tissue samples (10 mg) were obtained from a variety of healthy and tumor tissues (Fig. 1b). Gd trace content in each sample was measured at least twice and in arbitrary order. Variations of more than two orders of magnitude, between a few 10^{12} to 10^{14} atoms of Gd per 10 mg initial sample material were recorded in the different tissues under study. All results lie more than 3 orders of magnitude above the detection limit of our technique. Correspondingly, the measurements show good statistical agreement within a relative precision of about 5%. Additionally, an overall uncertainty of ~20% has to be considered for absolute numbers, resulting from variations in the overall efficiency. A graphical representation of quantitative results for the different tissues is given in Fig. 4.

The measured tissue concentrations are in the expected range. As expected, the largest amount of Gd was found in the kidney sample (#3), where Gd-DTPA excretion from the body takes place. Muscle and liver tissue as well as liquid from necrotic areas (#13, #7 and #11) show rather similar and still high Gd concentration, being lowered by only about one order of magnitude. For samples taken from the the tumor periphery

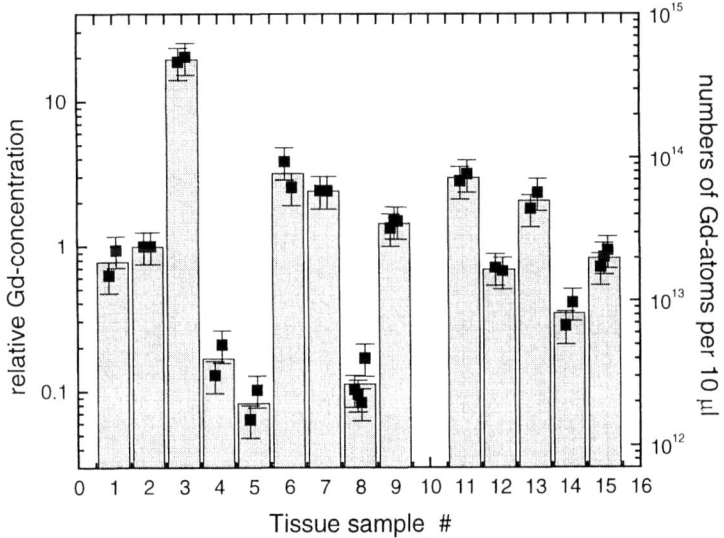

Figure 4. Experimental data points for individual measurements (dots with error bares) and averages (gray blocks) for Gd trace content in various bio-medical tissue samples from a naked mouse. Further discussion in the text.

(#6, #8 and #9), two out of three samples show a comparable high Gd content significantly above that in the inner region (#4, #5 and #12). This is related to an increased blood flow due to newly developed blood vessels. The inner region is widely "dead" tissue while the rapidly growing outer region is intensively supplied with blood. Discrepancies are observed for sample #8 and #12, which give strongly deviating results to higher or lower values, respectively. These effects are ascribed to the process of taking the micro samples of less than 1 mm³ size of soft biological tissue manually. It is fairly complicated to ensure proper correlation between the actual material and the location in the MRI image and hence the origin of the tissue under study is not clear for all cases. Further improvement on this part of the studies concerning the preparation and extraction of the tissue is under development. Skin samples (#2, #14 and #15), as well as those taken from spleen (#1) show an intermediate Gd level. For sample #10 complete sample loss occurred due to technical problems. For detailed and conclusive biomedical measurement of Gd concentration in tissue further refined sample collection procedures will be developed. Additional measurements are under way. These will be published elsewhere [6].

CONCLUSION AND OUTLOOK

Selective ultra trace determination of Gd for bio-medical investigations has been carried out by diode laser based RIMS. Using our compact RIMS instrument a set of more than 60 analytical samples, including blank and calibration samples, has been analyzed with high precision and very low sample loss. Detection limits of about 10^9 atoms have been demonstrated, far above the results for conventional techniques, which are limited by isobaric interferences.

Further experimental developments on the RIMS side include the upgrade of the diode laser system for higher power in the ionization step and the development of refined atomization procedures. These should lead to an increased overall efficiency in Gd ultra trace determination of $>10^{-5}$ with detection limits of $<10^7$ atoms corresponding to about 1 fg of Gd, which are of particularly high importance for applications of the technique to cosmo-chemical studies. Future bio-medical investigations by RIMS are foreseen after optimization of the sample taking process.

REFERENCES

1. P. Müller et al., *Ultratrace Determination of the Long-lived Isotope ^{41}Ca by Narrowband CW-RIMS*, RIS 98, AIP Conference Proceedings 454, 73-78 (1998)
2. P. Müller et al., Trace Detection of ^{41}Ca in Nuclear Reactor Concrete by Two-Step Resonant Laser Ionization Mass Spectrometry, *Radiochim. Acta*, (in press)
3. K. Blaum et al., *Diode laser resonance ionization mass spectrometry (RIMS) of Gadolinium*, RIS 98, AIP Conference Proceedings 454, 275-278 (1998)
4. K. Blaum et al., Isotope shifts and hyperfine structure in the [Xe] $4f^7$ 5d $6s^2$ $^9D_J \rightarrow$ [Xe] $4f^7$ 5d 6s 6p $^9F_{J+1}$ transitions of gadolinium, *Eur. Phys. J. D* **11**, 37-44 (2000)
5. W. Nörtershäuser et al., Double-resonance measurements of IS and HFS in Gd I with hyperfine-state selection in an intermediate level, *Phys. Rev. A* **62**, 022506 1-4 (2000)
6. W. Schreiber et al., *Magnetic Resonance in Medicine* (in preparation)

Determination of Trace Amounts of Plutonium in Environmental Samples by RIMS Using a High Repetition Rate Solid State Laser System

[*]C. Grüning, [†]G. Huber, [*]J.V. Kratz, [†]G. Passler, [*]N. Trautmann,
[*]A. Waldek, and [†]K. Wendt

[*]Institut für Kernchemie, Johannes Gutenberg-Universität Mainz
[†]Institut für Physik, Johannes Gutenberg-Universität Mainz
D-55099 Mainz, Germany

Abstract. A reliable and easy to handle high repetition rate solid state laser system has been set up for routine applications of Resonance Ionization Mass Spectrometry (RIMS). It consists of three Titanium-Sapphire (Ti:Sa) lasers pumped by one Nd:YAG laser, providing up to 3 W of tunable laser light each in a wavelength range from 725 nm to 895 nm. The isotope shifts for ^{238}Pu to ^{244}Pu have been measured in an efficient ionization scheme with $\lambda_1 = 420.76$ nm, $\lambda_2 = 847.28$ nm and $\lambda_3 = 767.53$ nm. An overall detection efficiency of the RIMS apparatus of $\varepsilon = 1 \times 10^{-5}$ is routinely reached, resulting in a detection limit of 2×10^6 atoms (0.8 fg) of plutonium. The isotopic compositions of synthetic samples and the NIST standard reference material SRM996 were measured. The content of ^{238}Pu to ^{242}Pu has been determined in dust samples from the surroundings of a nuclear power plant and ^{244}Pu was determined in urine samples for the National Radiation Protection Board (NRPB), U.K. Routine operation of plutonium ultratrace detection could thus be established.

INTRODUCTION

Trace amounts of plutonium are present in the environment mainly as a result of nuclear weapons tests, accidents and releases from nuclear facilities. A fast and sensitive detection method is required for low-level surveillance of the environment, studies of the migration behavior, personnel dose monitoring and nuclear safeguards. The isotopic composition of a plutonium contamination is directly related to its production method and thus to its origin. α-spectroscopy as the standard method of plutonium detection is limited in its ability to distinguish between ^{239}Pu and ^{240}Pu due to their similar α-energies, in its detection limit for the long living isotopes and the β-emitter ^{241}Pu is not even detected in this manner.

Resonance ionization mass spectrometry (RIMS) has been extensively applied by our group for ultratrace analysis and spectroscopy of plutonium, other actinides and technetium (1-4). The laser system used previously consisted of three dye lasers pumped by two powerful Cu - vapor lasers. Its important properties are a high repetition rate of 6.6 kHz, a broad tuning range through the usage of suitable dyes and a linewidth of 1.5-6 GHz (1). The drawbacks of this system such as size, high maintenance efforts and costs lead to the development of a new, powerful and easy to

CP584, *Resonance Ionization Spectroscopy 2000: 10th Int'l. Symp.*, edited by J. E. Parks and J. P. Young
© 2001 American Institute of Physics 0-7354-0024-5/01/$18.00

handle solid state laser system. This facilitates the application of RIMS for routine analysis of actinides (5). The characteristics of this laser system, spectroscopic measurements with different plutonium isotopes and isotope selective plutonium determination in synthetic samples, urine and dust samples are presented in the following.

SOLID STATE LASER SYSTEM AND RIMS - SETUP

A primary requirement for the application of RIMS for routine analysis is a reliable and easy to handle laser system. In combination with a time-of-flight mass spectrometer, the essential characteristics of pulsed lasers are high power, a high repetition rate, a broad tuning range and a rather narrow linewidth which are important for efficient resonance ionization.

A commercially available, Q-switched and intra-cavity doubled Nd:YAG pump laser (Clark-MXR ORC-1000) with a repetition rate of 1 - 25 kHz, a power of up to 50 W at 532 nm and a pulse length of approximately 400 ns is used to simultaneously pump three titanium-sapphire (Ti:Sa) lasers (5). In their tuning range from 725 nm to 895 nm each laser provides a power of up to 3 W with a linewidth of 5 GHz and a pulse duration of 60 – 150 ns. The three lasers are synchronized with intracavity Pockels cells used as Q-switches. As the ionization energies of the actinides are approximately 6 eV, an external, single pass frequency doubling setup applying a BBO crystal is required in order to apply a three step ionization scheme.

Measurements with plutonium were carried out using our standard TOF mass spectrometer (1) with the following procedure: Plutonium is atomized by evaporation from a sandwich filament. The atoms are excited into a Rydberg state by three step resonant excitation and subsequently field ionized. The ions are then mass separated in a reflectron time-of-flight mass spectrometer and counted. The TOF mass resolution of $m/\Delta m = 600$ is sufficient to discriminate different isotopes in the plutonium region around A = 240 [amu] and it serves for effective background suppression.

SPECTROSCOPIC MEASUREMENTS WITH PLUTONIUM

Excitation Scheme

Spectroscopic investigations in the spectrum of Pu I have been carried out to search for an efficient three step ionization scheme using wavelengths of 725 - 895 nm and 365 - 445 nm available from the Ti:Sa laser system.

Starting from the $5f^6\ 7s^2\ ^7F_0$ ground state, one favorable excitation scheme uses a wavelength of $\lambda_1 = 420.76$ nm in the first step to the $5f^6\ 7s\ 7p\ ^7D_1$ state. The second step leads to a J = 2 state with $\lambda_2 = 847.28$ nm. Scans of the ionization laser showed several autoionising states in the range from $\lambda_3 = 750$ to 758 nm and an especially strong transition to a high lying Rydberg state with $\lambda_3 = 767.53$ nm from which field ionization subsequently occurs with high efficiency. Measurements of the saturation for all three transitions of the excitation scheme have been carried out and showed saturation powers of 5 mW for the first step, 35 mW for the second and 410 mW for the third one. During analytical measurements, typical laser powers at the interaction region of the laser light with the atomic beam were 90 mW, 700 mW and 900 mW. This leads to power broadening in the first two steps making the system less sensitive against drifts in laser frequency.

Isotope Shifts

For isotope ratio measurements the isotope shifts in the excitation scheme for all isotopes have to be known precisely. Measurements with the plutonium isotopes 238, 239, 240, 241, 242 and 244 showed a shift of −19.5 GHz for ^{238}Pu relative to ^{244}Pu in the first transition, a smaller one of 11 GHz in the second and no significant isotopic shift in the third transition (figure 1).

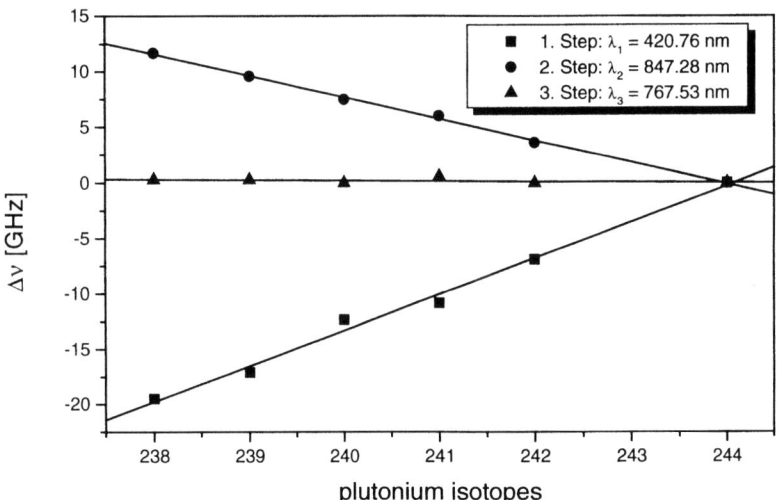

FIGURE 1. Isotope shifts of ^{238}Pu to ^{242}Pu with respect to ^{244}Pu in the excitation scheme discussed in the text.

ANALYTICAL MEASUREMENTS

Efficiency measurements were carried out to demonstrate the effectiveness of the excitation scheme, the ability of the Ti:Sa laser system to perform ultratrace analysis of plutonium and to compare it to the old copper vapor / dye laser system (6). Efficiencies of $\varepsilon = 1 \times 10^{-5}$ are routinely reached for a single isotope, resulting in a detection limit of 2×10^6 atoms with a confidence level of 99,7% (3σ) by taking the background counts into account.

To demonstrate the precision and accuracy of plutonium isotope ratio measurements, synthetic isotope mixtures of plutonium 238 to 244 have been analyzed. In figure 2, the time-of-flight spectrum of a sample of the NIST standard reference material SRM996 with some 10^{12} atoms of ^{244}Pu is shown.

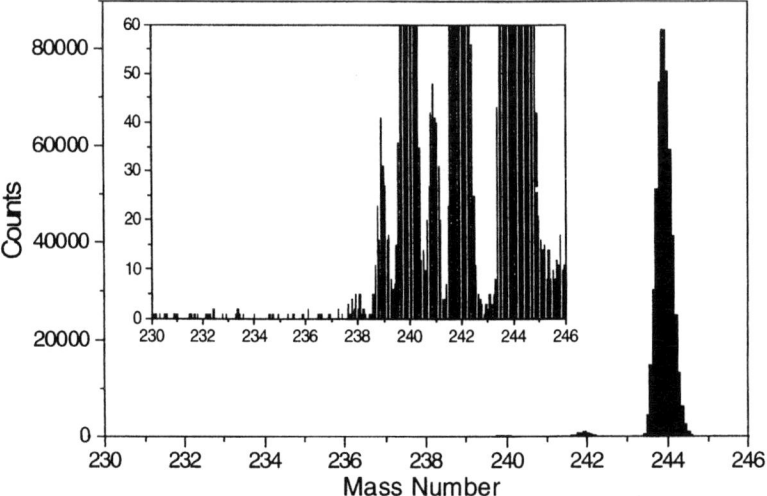

FIGURE 2. Time-of-flight spectrum of some 10^{12} atoms of the NIST standard SRM996. Inset: same spectrum with enlarged y-scale.

The measured values agree well with the certified sample composition shown in table 1. The dynamic range of the apparatus stretches over 4 orders of magnitude from 17 counts of ^{238}Pu to 485000 counts of ^{244}Pu.

TABLE 1. Isotopic composition of the NIST standard SRM996; errors are 3σ.

Pu – Isotopes	Measured Ratios	Certified Ratios
238	0.00004(3)	0.00004(1)
239	0.00039(9)	0.00035(2)
240	0.0065(4)	0.0069(1)
241	0.00054(10)	0.00037(2)
242	0.0137(5)	0.0135(1)
244	1	1

^{244}Pu Determination in Urine

The 'National Radiation Protection Board' (NRPB), U.K., investigates the uptake and urinary excretion of plutonium in human volunteers after an ingestion of 10^{14}

atoms of ^{244}Pu and an injection of 2×10^{12} atoms (7). A total of 49 urine samples have been measured in the ongoing collaboration so far for their ^{244}Pu content (6), 11 with the Ti:Sa – laser system. In the latter samples, 1×10^{8} to 3×10^{8} atoms of ^{244}Pu per sample have been detected.

Dust Samples

A total of 24 dust samples from the surroundings of a nuclear power plant have been analyzed for their total plutonium content and their isotopic composition from ^{238}Pu to ^{242}Pu. A typical time-of-flight mass spectrum of such a dust sample is shown in figure 3. On average 5×10^{8} atoms of ^{239}Pu have been found per gram dust. The ratio of ^{240}Pu/^{239}Pu has been measured to be 0.16(4).

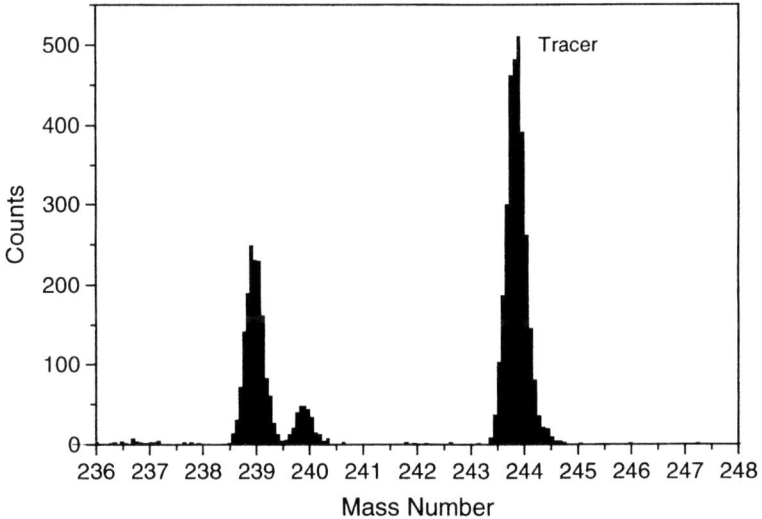

FIGURE 3. Time-of-flight spectrum of a dust sample.

CONCLUSION AND OUTLOOK

The capabilities of the new Ti:Sa laser system in the routine analysis of isotope mixtures of plutonium have been demonstrated for different sample matrices. The detection limit of 2×10^{6} atoms of plutonium is comparable to the one obtained with the copper vapor / dye laser system (1) and two orders of magnitude better than the detection limit of conventional α-spectroscopy for ^{239}Pu.

A compact diode laser system in combination with a quadrupole mass spectrometer is under construction to further reduce the size, costs and maintenance efforts for the RIMS apparatus.

REFERENCES

1. Passler, G., et al., *Kerntechnik* **62**, pp. 85 (1997).
2. Köhler, S., et al., *Spectrochimica Acta* **B52**, pp.717 (1997).
3. Nunnemann, M. et al., *Journal of Alloys and Compunds* **271-273**, pp. 45 (1998).
4. Waldek, A., et al., "RIMS Measurements for the Determination of the First Ionization Potential of the Actinides Actinium up to Einsteinium," in *this volume*.
5. Grüning, C., et al., "A High Repetition Rate Solid State Laser System for Resonance Ionization Mass Spectrometry of Actinides" in *Resonance Ionization Spectroscopy*, edited by J. C. Vickerman et al., AIP Conference Proceedings 454, New York: American Institute of Physics, 1998, pp. 285.
6. Erdmann, N., et al.," Trace Analysis of Plutonium by Resonance Ionization Mass Spectroscopy (RIMS)" in *Resonance Ionization Spectroscopy*, edited by J. C. Vickerman et al., AIP Conference Proceedings 454, New York: American Institute of Physics, 1998, pp. 279.
7. Ham, G.J., and Harrison, J.D., *Radiation Protection Dosimetry* **87,** No. 4, pp. 267 (2000).

Two-color resonance ionization mass spectrometry of hafnium

Shuichi Hasegawa, Junji Nakagawa, Hitoshi Kurosawa,
Susumu Tanji, and Atsuyuki Suzuki

Department of Quantum Engineering and Systems Science
University of Tokyo
7-3-1 Hongo, Bunkyo-ku, Tokyo 113-8656, JAPAN

Abstract.
We have observed high-lying Rydberg series converging to the first ionization limit of Hf atoms ($5d6s^2$ $^2D_{3/2}$) by two-color resonance ionization spectroscopy. The observed series were excited through three different intermediate states, $5d^26s6p$ $^5D_1$, $^3F_2$, $^3F_3$. This suggests the correlation between s and d electrons is important. Using the quantum defect theory, the converging limit of one of the observed series was in good agreement with the established value.

INTRODUCTION

The absorption cross sections for thermal neutron by hafnium are larger three orders of magnitude than those of zirconium isotopes. This property is utilized in the field of nuclear engineering. For instance, the nuclear fuel clad consists of zirconium and hafnium is used for the neutron absorber. It is necessary to separate two elements to make use of the property. However, the separation is one of the most difficult chemical processes because the chemical properties of the two elements are very similar. In order to develop a new process or investigate the quality of the separation, the better quantitative analysis method is required. Hafnium is contained in zircon sand which mainly consists of zirconium oxide. For the nuclear application, hafnium in zirconium sponge is eliminated less than 100 ppm. To investigate the separation process, one needs the trace analysis method. We have studied the possibility to apply resonance ionization mass spectrometry to the problem. However, the ionization scheme of hafnium has not been fully investigated because hafnium is a refractory element and the ionization limit of hafnium is as high as 55000 cm^{-1}. Callendar et al. investigated the first ionization limit of hafnium using two color RIS [1]. They determined the first ionization limit as 55047.9 cm^{-1} by fitting the observed Rydberg series to the Ritz formula. To

CP584, *Resonance Ionization Spectroscopy 2000: 10th Int'l. Symp.*, edited by J. E. Parks and J. P. Young
© 2001 American Institute of Physics 0-7354-0024-5/01/$18.00

our knowledge, there is no other work to ionize hafnium atoms by means of laser resonance ionization spectroscopy.

The electronic configuration of hafnium is different from that of the elements in the same column on the periodic table such as zirconium and titanium [2]. The ground state of Zr I is $4d^2 5s^2\ ^3F_2$ and that of Zr II is $4d^2 5s^4 F_{3/2}$. This indicates that the electronic configuration of zirconium can be considered as two-electron system for the first order approximation. However, the ground state of Hf I is $5d^2 6s^2\ ^3F_2$ and that of Hf II is $5d6s^2\ ^2D_{3/2}$. In Ref. [1], $5d^2 6s6p\ ^3F_2$ is chosen as the intermediate state so that they didn't expect to observe the Rydberg series converging to the first ionization limit. Actually, they measured the Rydberg series converging to the excited states of Hf II, such as $5d^2 6s\ ^4F_{3/2,5/2,7/2}$ and subtracted the known excited energy from the ground state of HfII to determine the first ionization limit using quantum defect theory. These indicate that high-lying Rydberg series can give the information on the correlation between s and d electrons.

From the view point of nuclear physics, hafnium has been attractive because it has the longest chain of 31 isotopes. Measurements of isotope shift and hyperfine structure give the insight into the nuclear property. High resolution spectroscopic methods such as saturated absorption spectroscopy [3], laser induced fluorescence [6] have been used by several groups. Resonance ionization spectroscopy is one of the powerful and sensitive methods to investigate the isotope shift and the hyperfine structure. The investigation of the ionization scheme can provide the opportunity to measure the states in broader energy range and opposite parity, especially highly excited states which is only accessible by means of RIS. We have performed two color resonance ionization spectroscopy to investigate high-lying Rydberg series through several intermediate states.

EXPERIMENTAL SETUP

The experimental apparatus consists of a laser system, a vacuum chamber and a measurement system. The schematic setup is shown in Fig. 1. Two dye lasers (Lambda Physik FL3002, LPD3000) pumped by a XeCl excimer laser (Lambda Physik COMPex201) were used to resonantly excite Hf atoms. The wavelength to reach the first ionization limit by two photons was generated by the laser dye, DMQ (346 nm $\sim$ 377 nm). The irradiation of the second step laser beam was delayed for 15 ns by the longer path length to assure the stepwise excitation. To calibrate the scanning wavelength, an optogalvano cell was used for the absolute wavelength and a Fabry-Pérot etalon for the interpolation. The optogalvanic signal and the interference signal were integrated by boxcar integrators. The vaporization of hafnium was achieved by an electron beam in the vacuum chamber whose pressure was ca. 10^{-6} Torr. The vapor was collimated to obtain the atomic beam, which the two laser beams intersected perpendicularly. The two laser beams counter-propagated collinearly. The excited atoms were ionized by a pulsed electric field, which ex-

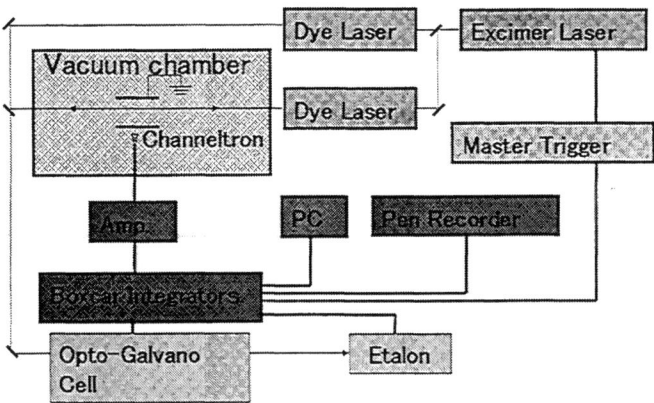

FIGURE 1. Experimental setup for two-color resonance ionization spectroscopy

tracted the ions to a channeltron to detect. The signal was amplified and fed into a boxcar integrator. The outputs from the boxcar integrators were digitized and recorded on a computer.

RESULTS AND DISCUSSION

Figure 2(a) shows the intermediate states obtained by scanning the first step laser wavelength from 360 nm to 370 nm. The ionization was achieved by the second laser beam whose wavelength was fixed at 360 nm. The ion signal corresponding to the transition from the ground state to $5d^2 6s6p \, ^3F_3$ was very strong. After scanning the second step laser wavelength to excite and ionize the atoms via the intermediate state, it turned out that there existed the autoionization state very close to the position which corresponded to the energy at which the first laser photon energy doubled. So, the one color double resonant ionization to the autoionization occurred, which made the ion signal very strong. From the states observed in Fig. 2 (a), three states were chosen as the intermediate states, which were assigned as $5d^2 6s6p \, ^3F_2$, $^3F_3$, and $^5D_1$ [2]. The excitation scheme was shown in Fig. 2 (b). The first step laser wavelength was fixed to resonantly excite the intermediate state and the second step laser wavelength was scanned just below the first ionization limit to excite the high-lying states. The excited atoms were ionized by the pulsed electric field. The ionization spectra through the three different intermediate states were shown in Fig. 3. The difference of the J values of the intermediate states should reveal the J values of the each high-lying state. However, because of the laser bandwidth (0.2 cm^{-1}) all peaks could not be resolved. So the complete assignment of the peaks was not possible for these spectra.

A part of the middle spectrum in Fig. 3 was magnified in Fig. 4 (a). This spectrum shows the ion signal obtained through $5d^2 6s6p \, ^3F_2$ (27149.64 cm^{-1}). The

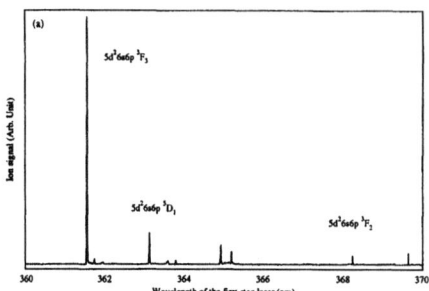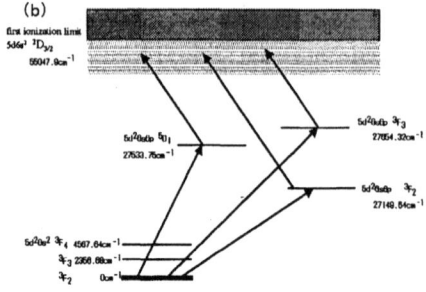

FIGURE 2. (a)The spectrum of the intermediate states scanning the first step laser wavelength between 360 nm and 370 nm. The second laser wavelength was fixed at 360 nm. (b)The intermediate states used in this study. Through these states, the atoms were excited to the Rydberg series.

periodic series are clearly observed. Three dominant peaks were designated as the first, second, and third series as shown in Fig. 4 (a). Reference [2] assigned the intermediate states as $5d^2 6s6p$. So, the expected Rydberg series are $5d^2 6sns$ or nd, which should converge to the ionization limit of $5d^2 6s$. Actually the authors in Ref. [1] didn't expect they could observe the Rydberg series converging to the first ionization limit. In order to verify that the observed series in this study converge to the first ionization limit, we used the quantum defect theory. The quantum defect is defined as,

$$E = I - \frac{R_{Hf}}{\nu^2},$$

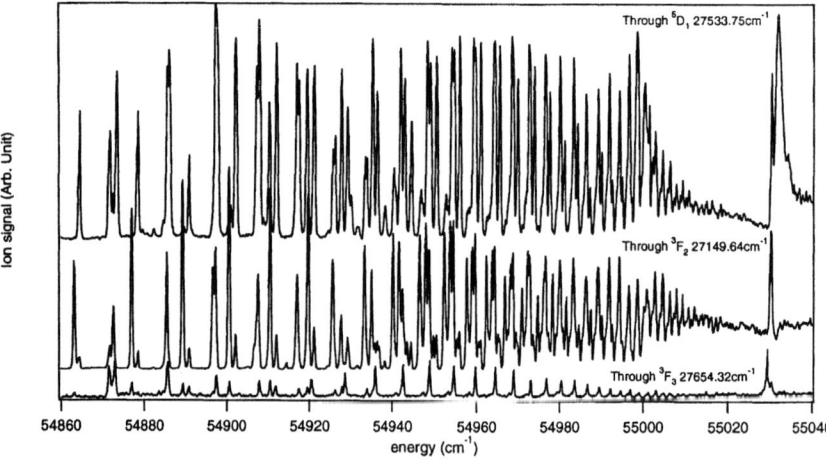

FIGURE 3. The high-lying Rydberg series through the three different intermediate states. The energy was calibrated by the optogalvano cell and the Fabry-Pérot etalon.

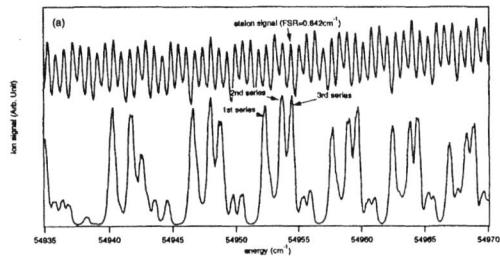 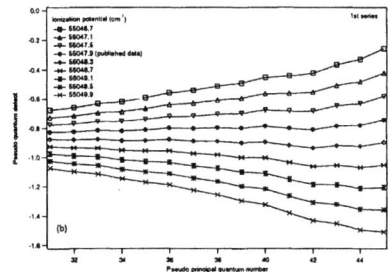

FIGURE 4. (a)Magnified spectrum of the Rydberg series excited through $5d^2 6s6p\ ^3F_2$ (27149.64 cm^{-1}). The spectrum mainly consisted of three peaks which was designated as the first, second, and third series in the Figure. (b)The pseudo quantum defect of the first series in Fig. 4 (a) as a function of the pseudo principal quantum number. The established value of the ionization limit (55047.9 cm^{-1}) are in good agreement to the observed Rydberg series.

$$\mu = n - \nu,$$

where E is the energy value of the state, I is the ionization limit, R_{Hf} is the Rydberg constant for Hf (109715 cm^{-1}), ν is the effective quantum number, n is the pseudo principal quantum number, and μ is the pseudo quantum defect. According to the theory, the quantum defect of the states belonging to the identical Rydberg series should be constant unless there is perturbation to the series. The pseudo quantum defects of the first series as a function of the pseudo principal quantum number were plotted in Fig. 4 (b). The ionization limit I was found by adjusting it to make the pseudo quantum defect constant as much as possible. As shown in Fig. 4 (b), the value of the ionization limit was changed by 0.4 cm^{-1} (the laser bandwidth ×2) and the established value by Ref. [1] was best fitted to the observed data in this study. This result shows that the Rydberg series via the $5d^2 6s6p$ intermediate state converged to the first ionization limit $5d6s^2\ ^2D_{3/2}$. The ionic core of the series should be $5d6s^2\ ^2D_{3/2}$. However, the intermediate states used in this study were all assigned as $5d^2 6s6p$ in Ref. [2]. According to the assignment, the observed series through these intermediate states should be designated as $5d^2 6sns$ or $5d^2 6snd$. But the series of the recent works by Bouazza et al. [3–5] show that the components of $(d + s)^3 p$ electrons are considerably mixed. Assuming that the intermediate states used in this study include $5d6s^2 6p$, the one photon transition could induce the Rydberg series of $5d6s^2 ns$ or $5d6s^2 nd$, which converge to the first ionization limit $5d6s^2$. The interaction between the ionic core and the excited electron decreases as the excited electron leaves away from the core. In this case, jj coupling scheme can be applied better than LS coupling scheme [7]. The J value of the ionic core $^2D_{3/2}$ is $3/2$, and the J values of the ns electron and nd electron are $j_s = 1/2$ and $j_d = 3/2, 5/2$. The total J values of the excited states are $(3/2, 1/2)_{1,2}$ for ns series and $(3/2, 5/2)_{1,2,3,4}$ and $(3/2, 3/2)_{0,1,2,3}$ for nd series. There can exist ten fine structure multiplets. The J selection rule limits the number of the series to seven multiplets from the $J = 1$ intermediate state, eight from $J = 2$, and six from $J = 3$. In Ref. [2], the states of 39546.34, 40513.53, 42061.60 and 46183.08 cm^{-1} are assigned as $5d6s^2(a^2D)7s$ $d^3D_{1,2,3}$ and e^1D_2, respectively. The effective quantum numbers of these states are 2.66, 2.75, 2.91, and 3.52, respectively. Meanwhile, there is no assignments for $5d6s^2nd$ states in Ref. [2]. In the experiment, the multiplets could not be resolved so fully as the theory predicted. Measurements of the Rydberg series $5d6s^2ns$ and nd of small n numbers are necessary to extend the assignment to the high-lying series and analyze the configuration interaction of the electrons.

In summary, we have performed two-color resonance ionization spectroscopy to investigate the high-lying states of Hf atoms. Through the intermediate state assigned as $5d^26s6p$, the Rydberg series were observed and identified that they were converging to the first ionization limit $5d6s^2$ $^2D_{3/2}$.

REFERENCES

1. C.L. Callendar, P.A. Hackett and D.M. Rayner, J. Opt. Soc. Am. B5, 1341 (1988).
2. W.F. Meggers and C.E. Moore, NBS Monograph 153 (1976).
3. S. Bouazza and M. Wilson, J. Phys. B33, 933 (2000).
4. S. Bouazza, M. Fienhold, G.H. Guthöhrlein, H.O. Behrens, and J. Dembczynski, Eur. Phys. J. D6, 303 (1999).
5. S. Bouazza, H.O. Behrens, M. Fienhold, J. Bembczynski, and G.H. Guthöhrlein, Eur. Phys. J. D6 311 (1999).
6. D. Zimmermann, P. Baumann, D. Kuszner, and A. Werner, Phys. Rev. A 50, 1112 (1994).
7. R.D. Cowan, *The theory of atomic structure and spectra*, Berkeley, University of California Press, 1981.

Measurement of the ionization potential of highly excited Rydberg states

Xinghua Li, Jens Lassen and Hans A. Schuessler

Dept. of Physics, Texas A&M University
College Station, TX 77843-4242

Abstract. We have measured the critical field strength of the high-lying Rydberg states $31s^2[3/2]_2$ and $29d^2[3/2]_2$ of krypton using mass-separated collinear fast atom beam laser spectroscopy. The results for the critical field strengths are F_c=0.525kV/cm and F_c=0.620kV/cm, respectively. A comparison with results from the saddle-point ionization model is made.

INTRODUCTION

Ultra-sensitive detection of rare krypton isotopes calls for particle- instead of photon- detection. Therefore field-ionization of Rydberg atoms is the most efficient way to produce charged particles for detection in fast beam experiments. This requires the efficient population of a suitable Rydberg-state. The ionization efficiency can be made close to unity and the ionization process can also be well defined in space. Moreover, field-ionization can be done state-selective, since the critical field-strengths of individual Rydberg-states differ between different states.

We investigated the critical field-strengths of the Kr $31s^2[3/2]_2$ and $29d^2[3/2]_2$ Rydberg levels. Metastable krypton atoms were produced through charge-exchange of a mass-separated krypton ion-beam in Cs or Rb vapor. Through two-step laser excitation the metastable krypton atoms were excited to the Rydberg levels investigated. The critical field-strengths were determined using a longitudinal field-ionizer by increasing the ionizer voltage until the ion-signal was observed. The results are compared with the saddle-point ionization model.

EXPERIMENTAL SETUP

A hollow-cathode discharge ion source (Danfysik 911A) generates a beam of Kr ions, which were accelerated to 5-12keV and mass-separated with a low-resolution bending magnet ($m/\Delta m$~200). The mass-separated ion-beam is deflected a few degrees to co-propagate with beams from a Ti:Sapphire ring-laser and a frequency-stabilized single-mode Ar^+ laser. The ion-beam enters an alkali-vapor charge-exchange cell for neutralization. A large fraction of the ions is thereby transferred into the metastable levels of krypton. The remaining ions are deflected and the fast neutral atoms enter the interaction region with the laser beams of about 0.8m in length. Thereafter the atoms pass through an axial field-ionizer. The Rydberg atoms are field-ionized and the product ions are detected with a Faraday cup.

CP584, *Resonance Ionization Spectroscopy 2000: 10th Int'l. Symp.*, edited by J. E. Parks and J. P. Young

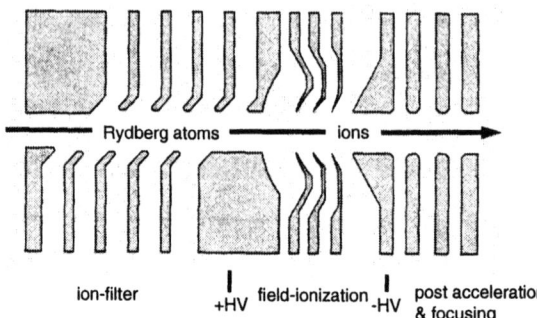

ion-filter   +HV   field-ionization   -HV   post acceleration & focusing

Fig. 1: Optimized axial field-ionizer for fast beam spectroscopy. The field-ionizer has three functional sections: ion filter, field ionization, and the post acceleration & focusing section.

The field ionizer geometry is chosen to preserve the high beam quality after ionization. This field-ionizer naturally has a longitudinal electric-field and allows for state selective field-ionization. The design of the longitudinal field-ionizer [4] is shown in Fig. 1. The set of concentric electrodes with a voltage divider chain is used to raise the axial potential slowly, before rapidly dropping back to a low potential. This creates a steep potential gradient and necessarily a large electric field. The ion-

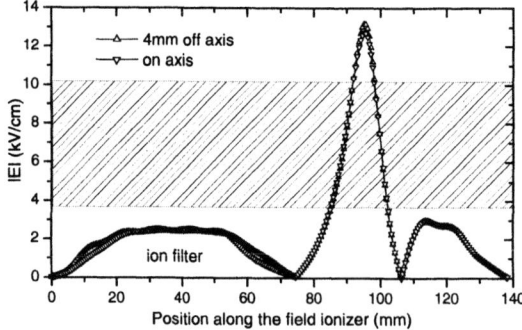

Fig. 2: Absolute electric field strength and potential of the field ionizer. These field strengths and potentials have been calculated using SIMION 6 on axis and 4mm off-axis. The dashed area is the operating range for proper operation of the field ionizer.

filter section in the front is composed of tilted electrodes as to deflect residual ions out of the beam. The electrodes in the ionization-region are shaped to produce a spherical potential such that surfaces of equal electrical field strength coincide with equipotential surfaces. This assures equal energy-offset for Rydberg-atoms ionized at characteristic field-strength and energy-labels the signal-ions. An example for the absolute value of the field-strength along the field-ionizer used for field-ionizing the *31s* and *29d* Kr Rydberg-states is shown in Fig. 2. These values were derived from SIMION 6 simulations of the field-ionizer setup for typical operating parameters. The critical field-strength of the Rydberg-atoms to be field-ionized and analyzed must be chosen to be above the field-strength values in the ion-filter section (the dashed region) in order to avoid ionization in the the ion-filter section.

RESULTS AND DISCUSSION

Metastable atoms in the $5s^2[3/2]_2$ level are excited to the $5p^2[5/2]_3$ level with a Ti:Sapphire laser. Once the resonance is found, the frequency of the Ti:Sapphire laser is tracked with respect to Doppler-tuning of the atom-beam energy. Excitation of the atoms in the $5p^2[5/2]_3$ level further up to high-lying Rydberg levels is observed by scanning the beam energy while simultaneous tracking of the first-step resonance. In Fig. 3 we present the spectra of the $5p^2[5/2]_3$-$31s^2[3/2]_2$ transition of ^{84}Kr. The upper trace shows the spectrum of the $31s^2[3/2]_2$ Rydberg level of Kr detected by collisional ionization. Due to the ionization of metastable atoms by collisional ionization, a constant background ion current is present. Excitation of the metastable atoms results in an increase of the ion current since excited levels possess increased collisional ionization cross-sections. This is illustrated by the dips on either side of the $5p^2[3/2]_2 \rightarrow 31s^2[3/2]_2$ peak produced by blocking the Ti:Sapphire laser light. The lower trace shows the same transition detected by field ionization. Since only the Rydberg-atoms are ionized via field-ionization, practically no background current is present in the spectrum.

Once the excitations to the Rydberg levels are found, the laser frequencies are kept on resonance and the critical field-strengths are determined. In Fig. 4 we present the critical field strength determination for the Kr $29d^2[3/2]_2$ Rydberg level. The sharp rise in ion signal observed indicates the threshold behavior of the field-ionization

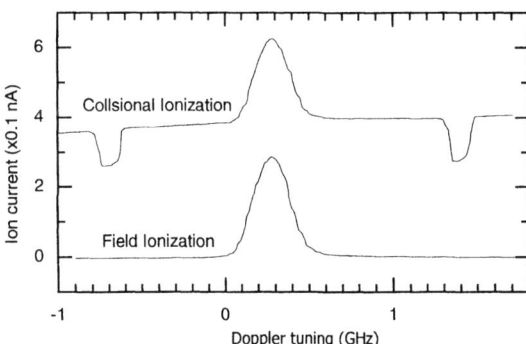

Fig. 3: Detection of the Kr $31s^2[3/2]_2$ Rydberg atoms by collisional- (upper trace) and field- (lower trace) ionization. The atoms were excited to the Rydberg-level via charge exchange and two-step laser excitation Kr $5s^2[3/2]_2 \rightarrow 5p^2[5/2]_3 \rightarrow 31s^2[3/2]_2$. The target gas pressure for collisional ionization is 100 mtorr; the voltage settings on the ion-filter and field-ionizer are 0 kV and −2.0 kV, respectively.

process. The critical field-strength is determined as the point where the signal rises to 10% of the maximum field-ionization signal.

A number of critical field-strengths have been published for Kr Rydberg levels [5]. With these values and the critical field strength F_c from the saddle point model the field-ionizer performance can be analyzed. According to the saddle point model [6], the critical field strength for field ionization F_c is related to the effective quantum number $n^*=n-\delta_{nlj}$ of a particular Rydberg state according to:

$$F_c = \frac{1}{16(n^*)^4} = 3.214 \cdot 10^8 \frac{1}{(n^*)^4} \frac{Volts}{cm},$$

where δ_{nlj} is the quantum defect of the Rydberg level. In Tab. 1 some calculated and observed critical field-strengths for the $29d^2[3/2]_2$ and $31s^2[3/2]_2$ levels are given.

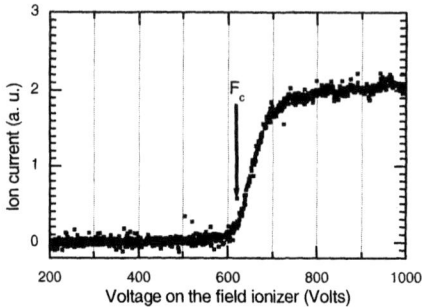

Fig. 4: Critical field-strength determination for the Kr $29d^2[3/2]_2$ Rydberg level. The ion-filter section is operated at ground potential and the voltage on the ionizer section is scanned for the measurement. F_c is determined from the applied voltages.

TABLE 1. Calculated and experimental critical field-strengths for Kr $31s^2[3/2]_2$ and $29d^2[3/2]_2$ levels

State	Energy (cm^{-1})	δ_{calc}	$F_{c\,cal}$ (V/cm)	$F_{c\,exp}$ (V/cm)
$31s\,^2[3/2]_2$	112773.41	3.10	0.53	0.525
$29d\,^2[3/2]_2$	112770.72	1.36	0.55	0.620

CONCLUSION

We have investigated the critical field ionization strength of the Kr $29d^2[3/2]_2$ and $31s^2[3/2]_2$ levels. The metastable krypton atoms produced through charge-exchange of krypton ions in rubidium or cesium vapor were excited to the high-lying Rydberg levels by resonant two-step laser excitation. The Rydberg-atoms were field-ionized. By scanning the field-ionizer voltage the critical field-strength was obtained.

REFERENCES

1. R. Neugart, W. Klempt, and K. Wendt, *Nuc. Instrum. Methods* **B17**, 354-359 (1986).
2. W. Borchers, E. Arnold, W. Neu, R. Neugart, K. Wendt, G. Ulm and ISOLDE Corporation, *Phys. Lett.* **B216**, 7 (1989)
3. H. A. Schuessler, E. C. Benck, and F. Buchinger, in *Resonance Ionization Spectroscopy 92*, Parks and Miller, editors, Bristol, Inst. of Phys. Conf. Ser. **128**, 37-40 (1992).
4. K. Stratmann,R. Hohmann,H.J. Kluge,S. Kunze,J. Lantzsch, L. Monz, E. W. Otten, G. Passler, J. Stenner, K. Wendt , K. Zimmer, Rev. Sci. Instrum. **65**, 1847-1852 (1994)
5. C. Delsart, J.-C. Keller and C. Thomas, *J. Phys. B: At. Mol. Phys.* **14**, 4241-4254 (1981)
6. T. Gallagher, *Rydberg Atoms*, Cambridge, Cambridge University Press, 1994

Collisional ionization of krypton and xenon excited states

Xinghua Li, Jens Lassen and Hans A. Schuessler

Dept. of Physics, Texas A&M University
College Station, TX 77843-4242

Abstract. We investigated the collisional ionization of krypton and xenon atoms in a mass-separated atomic beam in collinear fast-beam laser spectroscopy. It is found that collisional ionization can be used for sensitive detection of recyclable and partially recyclable transitions. It can also be used to locate the strongest transitions for isotopes with *HFS* splitting. Incorporated into our trace-detection system, collisional ionization allows ultra-sensitive particle detection of rare krypton isotopes.

INTRODUCTION

Collisional ionization is a powerful and simple technique to detect laser excitation processes into low, intermediate and highly excited states. The detection of collisionally formed ions does not suffer from the solid angle detection disadvantage inherent in photon counting. Collisional ionization is a stripping process where the outer electron of a fast neutral atom is removed through collisions with a gas target. Other advantages of collisional ionization are that: ions can be collected with unity efficiency, and contrary to photon detection, scattered light does not affect the signal to noise ratio. In spite of the small cross-section (about 10^{-15} cm^2) collisional ionization can be used to efficiently detect optical pumping between atomic levels with a sensitivity superior to fluorescence detection, provided that the energy levels involved are well separated.

We present a detailed investigation on the collisional ionization of krypton and xenon atoms in a mass-separated fast atom beam. The features of collisional ionization in different situations are studied.

EXPERIMENTAL SETUP

A hollow-cathode discharge ion source (Danfysik 911A) generates a beam of Kr or Xe ions, which is accelerated to 5-12keV and mass-separated with a low-resolution bending magnet ($m/\Delta m \sim 200$). The mass-separated ion beam is deflected by a few degrees and co-propagates with laser light from a Ti:Sapphire ring-laser. For two-step resonant excitation an additional ring dye-laser or single frequency Ar$^+$ laser is used. The ion beam enters an alkali-vapor charge-exchange cell for neutralization. A large fraction of the ions is thereby transferred into the metastable levels of krypton or xenon. The remaining ions are deflected and a beam of fast neutral atoms enters the

CP584, *Resonance Ionization Spectroscopy 2000: 10th Int'l. Symp.*, edited by J. E. Parks and J. P. Young
© 2001 American Institute of Physics 0-7354-0024-5/01/$18.00

interaction region with the lasers of about 0.8 meters. Thereafter the atoms pass through a collisional ionization cell and a fraction of the excited atoms are collisionally ionized. These product ions are subsequently detected with a Faraday cup.

RESULTS AND DISCUSSION

Metastable krypton and xenon atoms can be ionized through collisional ionization. As a result a constant background ion current is always present due to collision ionization of the metastable krypton and xenon atoms. Optical excitation of atoms in the metastable level can change the distribution of excited atoms in different levels. If the only decay mechanism of the excited atoms is to radiate back to the initial level, the transition is called a recyclable transition. The Kr $5s^2[3/2]_2 \rightarrow 5p^2[5/2]_3$ transition is an example for a recyclable transition. Because excited atoms possess a larger collisional-ionization cross-section than the non-excited atoms in the metastable level, optical excitation increases the detectable ion current. On the other hand, if atoms in the excited level have additional decay channels leading to the net decrease of the number of excited atoms, a dip or flop-out signal is observed. For example, if metastable krypton atoms were excited to the $5p^2[5/2]_2$ level, atoms in the $5p^2[5/2]_2$ level can decay to either the metastable level or the $5s^2[3/2]_1$ level, from where they undergo fast radiative transitions to the ground state. Thus excitation of metastable atoms to the $5p^2[5/2]_2$ level will result in optical pumping and a net loss of atoms in the metastable level. Hence a decrease in the ion current is observed. Fig. 1 presents registration curves for the two different kinds of transitions.

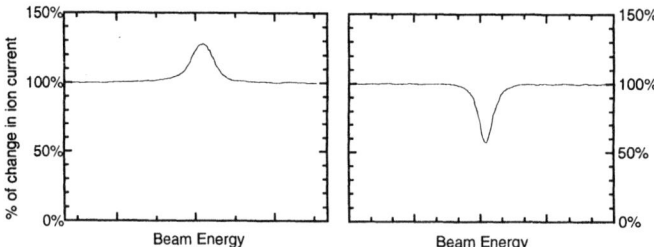

Fig. 1: Change of the observed collisional ionization current with respect to different type of transitions. Left panel: Kr $5s^2[3/2]_2 \rightarrow 5p^2[5/2]_3$, a recyclable transition; right panel: Kr $5s^2[3/2]_2 \rightarrow 5p^2[5/2]_2$, a partially-recyclable transition

The dependence of collsional ionization cross-section with respect to different target gases has been investigated by Neugart et al. [1,2] It was found that for electron-negative gas such as chlorine large collisional ionization cross-sections are obtained. To avoid the use of corrosive halogens such as Cl_2, we chose to use inert Ar instead. For our collisional ionization cell with an effective cell length of 50 mm the best signal depths were obtained for collisional target pressures ranging from 40 mTorr to 100 mTorr. In Fig. 2 the dependence of collisional ionization current versus target gas

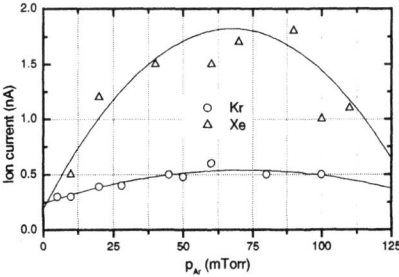

Fig. 2: Collisional ionization current versus argon gas pressure in the collisional ionization cell

pressure for metastable krypton and xenon atoms is presented. Next, two typical registration curves for measurements on Xe and Kr beams are given.

Fig. 3 presents composite spectra of the Xe $6s^2[3/2]_2 \rightarrow 6p^2[3/2]_2$ transition detected by collisional ionization. The spectra were obtained in the mass-separated atom beam with excited at 823 nm. The transition is partially recyclable, since part of the atoms in the excited $6p^2[3/2]_2$ can make cascade transitions to the ground state. Thus optical pumping results in a loss of excited atom population in the metastable and $6p^2[3/2]_2$ levels. It can be seen that collisional ionization can be used to efficiently detect optical transitions.

For the odd-isotopes of krypton/xenon, there is hyperfine structure (*HFS*) splitting.

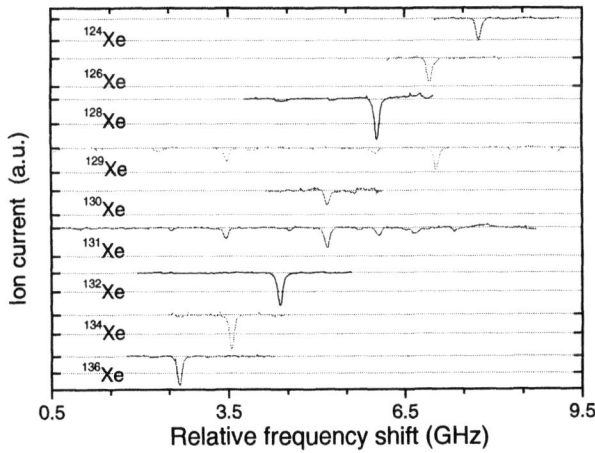

Fig 3: Optical excitation of Xe $6s^2[3/2]_2 \rightarrow 6p^2[3/2]_2$ transition in a mass-separated collinear fast atom beam detected by collsional ionization (target gas (Ar) pressure 60 mTorr)

The population is distributed among several *HFS* sublevels according to their statistical weights. Thus for odd-isotopes the fraction of population that can be excited through optical pumping is smaller than that for even-isotopes without *HFS*. As a result the visibility of an optical transition is comparatively smaller for odd- than for even-isotopes.

Collisional ionization can also be used to detect optical excitations to the high-lying Rydberg levels of krypton/xenon. In Fig. 3 we present spectra of the two-step ^{83}Kr $5s^2[3/2]_2 \rightarrow 5p^2[5/2]_3 \rightarrow 10s^2[3/2]_2$ transition. In the left panel the spectrum of the ^{83}Kr $5s^2[3/2]_2 \rightarrow 5p^2[5/2]_3$ transition is shown. The strongest transition, $F''=15/2 \rightarrow F'=13/2$, is marked with an arrow. Once the transition is located, the Ti:Sapphire laser frequency and the atom beam energy are kept at the resonance (see arrow in Fig. 3). The output from an Ar$^+$-pumped ring dye-laser is used to further excite the atoms in the intermediate $5p^2[5/2]_3$ level to the $10s^2[3/2]_2$ level. Through tuning of the dye laser, an increase in ion current is observed as the dye laser is tuned into resonance with the $5p$-$10s$ transition. The spectrum is shown in the right panel. The dips on either side of the spectrum are produced by blocking the Ti:Sapphire laser beam. This is used to check whether the first-step transition is still in resonance.

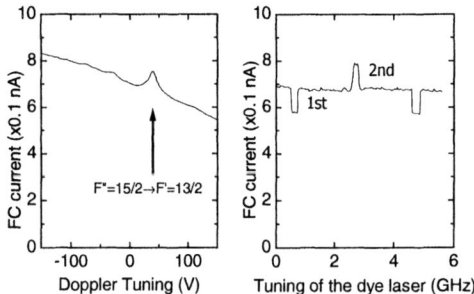

Fig. 4: Spectra of the two-step laser excitation of krypton atoms detected by collisional ionization. Left panel: spectrum of the ^{83}Kr $5s^2[3/2]_2 \rightarrow 5p^2[5/2]_3$ transition in the first step; Right panel: spectrum of the $5p^2[5/2]_3 \rightarrow 10s^2[3/2]_2$ transition in the first and second laser excitation step.

CONCLUSION

We have investigated the possibility of using collisional ionization to detect optical excitations of krypton/xenon atoms in a mass-separated collinear fast atom-beam laser spectroscopy. It is found that collisional ionization can be used efficiently to detect optical transitions under different excitation conditions. The collisional ionization technique has now been incorporated into our two-step excitation and field ionization scheme for trace detection of ^{85}Kr in ambient krypton.

REFERENCES

1. R. Neugart, W. Klempt, and K. Wendt, *Nuc. Instrum Methods* **B17**, 354-359 (1986).
2. W. Borchers, E. Arnold, W. Neu, R. Neugart, K. Wendt, G. Ulm and ISOLDE Corporation, *Phys. Lett.* **B216**, 7 (1989)
3. H. A. Schuessler, E. C. Benck, and F. Buchinger, in *Resonance Ionization Spectroscopy 92*, Parks and Miller, editors, Bristol, Inst. of Phys. Conf. Ser. **128**, 37-40 (1992).

Isotope shifts of Kr Rydberg-levels by resonance ionization

Xinghua Li,[1] Jens Lassen,[1] M. I. Bhatti[2] and Hans A. Schuessler[1]

[1]Physics Department, Texas A&M University, College Station, TX 77843-4242
[2]Physics and Geology Department, University of Texas Pan American, Edinburg, TX 78539

Abstract. Accurate spectroscopic data on isotope shifts of high-lying states, in particular Rydberg-states are rare, because of their small values. We report results on transitions involving an intermediate $5p^2[5/2]_3$ level and high-lying Rydberg-levels. The excitation scheme utilizes two-step excitation of metastable krypton atoms in a mass-separated collinear fast atom beam followed by field-ionization. It is found that the specific mass shift is of the order of a few tens of MHz.

INTRODUCTION

There exists few measurements on isotope shifts (IS) involving high-lying Rydberg-levels. In this paper we present measurements on krypton atoms involving the $5p^2[5/2]_3$ level and high-lying levels of the $29d^2$ and $31s^2$ manifolds. The study was motivated by the search for ultra-sensitive detection schemes capable of counting a few long-lived radioactive krypton isotopes in samples containing much larger amounts of the stable isotopes. The detection schemes are of interest in many applications, including radioactive dating and nuclear non-proliferation monitoring.

EXPERIMENTAL SETUP

The experimental setup for the two-step excitation and field-ionization is shown in Fig. 1. A hollow-cathode discharge ion source (Danfysik 911A) generates a beam of Kr ions, which were accelerated to 5-12 keV and mass-separated with a low-resolution bending magnet ($m/\Delta m{\sim}200$). The mass-separated ion-beam is deflected by a few degrees and co-propagates with laser light from a Ti:Sapphire ring-laser and a single-mode Ar$^+$ laser. The beam-line is evacuated by turbo-molecular pumps to better than 10^{-6} Torr. The ion beam crosses an alkali-vapor charge-exchange cell for neutralization. A large fraction of the ions is thereby transferred into the metastable $5s^2[3/2]_2$ level. The remaining ions are deflected and the fast neutral atoms enter the interaction region with the laser light of 0.8m length. Thereafter the atoms pass through a collisional ionization [1] cell and a field-ionizer. The Rydberg-atoms can be either collisionally- or field- ionized. The product ions are detected on a Faraday cup.

CP584, *Resonance Ionization Spectroscopy 2000: 10th Int'l. Symp.*, edited by J. E. Parks and J. P. Young
© 2001 American Institute of Physics 0-7354-0024-5/01/$18.00

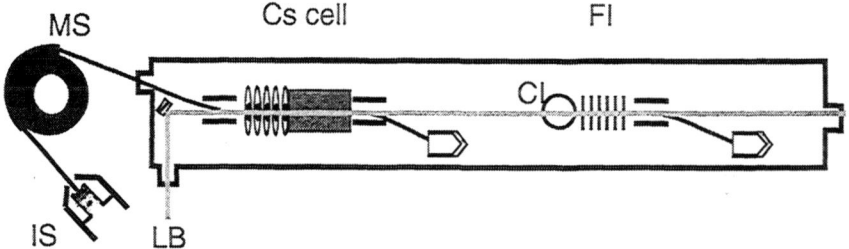

MS   Cs cell   FI

IS   LB

Fig. 1: Schematic beamline setup: CI-collisional ionization cell; FI: field-ionizer; LB: laser beams; IS: ion source, MS: mass separator.

The metastable krypton atoms produced by charge-exchange in Rb-vapor from a fast Kr ion-beam are excited by a Ti:Sapphire laser at 811 nm to the intermediate $5p^2[5/2]_3$ level in the first step. Because the only decay channel for atoms in the intermediate level is back to the metastable level, the transition is a recyclable, e.g. closed transition. By means of Doppler-tuning of the atom-beam velocity the atoms can be further excited to various high-lying Rydberg-levels with a single-mode Ar^+ laser at 488 nm.

Since field-ionization can only be applied to high-lying Rydberg-atoms, collisional ionization [1] is used for monitoring the resonant excitation of atoms from the metastable level into the intermediate $5p^2[5/2]_3$ level during system alignment.

The experimental procedure is as follows: first the $5s$-$5p$ transition is located using the collisional ionization technique. Since it is a closed transition, an enhancement in ion current is observed whenever the metastable atoms are excited to the intermediate level. In the next step a simultaneous tracking of the acceleration voltage with respect to the laser wavelength of the Ti:Sapphire laser is established. Transitions of the atoms in the intermediate level to the Rydberg-levels are detected by Doppler tuning the atom-beam energy through resonance with the fixed frequency 488nm Ar^+ laser light.

RESULTS

In the energy range 5-12keV the following Kr Rydberg-levels are accessible with excitation from the $5p^2[5/2]_3$ level using an Ar-ion laser: $31s^2[3/2]_2$, $29d^2[5/2]_3$, $29d^2[5/2]_3$, $29d^2[7/2]_3$, $29d^2[7/2]_4$ and $29d^2[3/2]_2$. The strongest transition among theses is the $5p^2[5/2]_3 \rightarrow 29d^2[7/2]_4$ transition. It is to be used for RIS trace-detection. In Fig. 2 we present composite spectra of the Kr $5p^2[5/2]_3 \rightarrow 29d^2[3/2]_2$ transition for the even-isotopes. The relative total frequency-shift of the transition is obtained by subtracting the known instrumental artificial Doppler-shifts which are shown in Fig. 3. For clarity the peak height of each isotope were normalized. A similar procedure was carried out for all other transitions. The isotope-shifts measured are given in Tab. 1.

The contribution from the field-shift (FS) to the total isotope-shift (IS) is usually small and within the experimental error in the IS measurement. Thus it is omitted from the analysis. The normal mass-shift (NMS) can be calculated according to $\delta v_{NMS}^{AA'} = (v/1836.15)(A' - A)/A'A$, with v being the atomic transition frequency. The normal mass-shift (NMS) accounts for approximately 100 MHz per two neutrons and

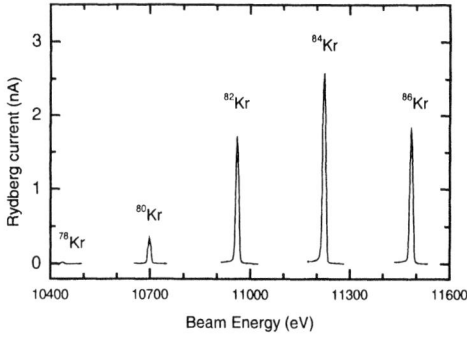

Fig 2: Composite spectrum of the Kr $5p^2[5/2]_3 \rightarrow 29d^2[3/2]_2$ transition for the even krypton isotopes. The single-mode Ar$^+$ laser is locked to 20486.65 cm^{-1}.

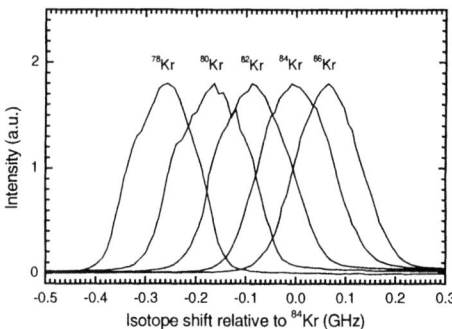

Fig 3: Total isotope shift relative to ^{84}Kr for the $5p^2[5/2]_3 \rightarrow 29d^2[3/2]_2$ transition

is listed in Tab. II. To obtain the specific mass-shift (SMS), the calculated NMS was subtracted from the observed total IS. The obtained SMS are shown in Tab. III.

TABLE I. Total isotope shifts (in MHz) measured for various transitions.

Isotope Pair	$5p^2[5/2]_3 \rightarrow$ $31s^2[3/2]_2$	$5p^2[5/2]_3 \rightarrow$ $29d^2[5/2]_3$	$5p^2[5/2]_3 \rightarrow$ $29d^2[5/2]_2$	$5p^2[5/2]_3 \rightarrow$ $29d^2[7/2]_3$	$5p^2[5/2]_3 \rightarrow$ $29d^2[7/2]_4$	$5p^2[5/2]_3 \rightarrow$ $29d^2[3/2]_2$
80-78	47	103	57	95	52	96
82-80	69	86	80	100	80	80
84-82	98	77	57	66	83	86
86-84	90	67	98	74	72	65

TABLE II. Calculated normal mass shifts (in MHz) for transition at ~20476 cm^{-1}.

Isotope Pair	Normal mass shift (in MHz)
80-78	107.2
82-80	102.0
84-82	97.2
86-84	92.7

TABLE III. Specific mass shifts (MHz) evaluated for various transitions into high lying Kr Rydberg states.

Isotope Pair	$5p^2[5/2]_3 \rightarrow 31s^2[3/2]_2$	$5p^2[5/2]_3 \rightarrow 29d^2[5/2]_3$	$5p^2[5/2]_3 \rightarrow 29d^2[5/2]_2$	$5p^2[5/2]_3 \rightarrow 29d^2[7/2]_3$	$5p^2[5/2]_3 \rightarrow 29d^2[7/2]_4$	$5p^2[5/2]_3 \rightarrow 29d^2[3/2]_2$
80-78	-60	-4	-50	-12	-55	-11
82-80	-33	-16	-22	-2	-22	-22
84-82	1	-20	-40	-31	-14	-11
86-84	-3	-26	5	-19	-21	-28

CONCLUSION

We have measured the isotope shift for transitions involving the intermediate $5p^2[5/2]_3$ and the $31s^2[3/2]_2$, $29d^2[5/2]_3$, $29d^2[5/2]_3$, $29d^2[7/2]_3$, $29d^2[7/2]_4$, $29d^2[3/2]_2$ Rydberg-levels using two-step excitation in a mass-separated collinear fast atom beam and a field-ionization technique. It is found that the isotope shifts for all transitions are similar, showing that the transition isotope shifts into Rydberg-states are dominated by the shift of the common intermediate level $5p^2[5/2]_3$. After subtracting the normal mass shift, it is found however, that there is a remaining small, yet for cw-RIS techniques important specific mass shift. A detailed analysis including Hartree-Fock calculations is in progress.

REFERENCES

1. R. Neugart, W. Klempt, and K. Wendt, *Nuc. Instrum. Methods* **B17**, 354-359 (1986).

Formation and Reactions of Cluster Ions From Aromatic Carboxylic Acids Together with Amino Acids and Small Peptides

Andreas Uphoff, Anja Meffert and Jürgen Grotemeyer

*Institut für Physikalische Chemie, Christian-Albrechts Universität zu Kiel,
Olshausenstraße 40, D-24098 Kiel, Germany*

Abstract : The cluster formation of several aromatic carboxylic acids is investigated by means of laser desorption into a supersonic beam followed by multi photon ionization - time-of-flight mass spectrometry. The formation of not only homogeneous clusters but also that of heterogeneous clusters with some small amino acids and peptides is studied by employing either nano- or femtosecond laser pulses.

INTRODUCTION

The formation of gas phase bimolecular clusters can be used for the understanding of intra molecular hydrogen bonding [1], hydrophobic and hydrophilic effects like in water or alcohols [2] or as well as other applications [3]. Different theoretical and experimental publications have discussed the formation of bimolecular cluster - neutral and ionic- as a source for the ionization of large biomolecules by the technique of matrix assisted laser desorption ionization (MALDI) [4, 5]. In this work we have investigated the cluster formation between typical MALDI-matrices, amino acids and small peptides by means of laser desorption into a supersonic beam followed by multi photon ionization [6]. This mass spectrometric technique allow the separation of desorption and ionization processes in time and space. Our interest is focussed here on the question whether cluster formation is been observed, and possibly even reactions in the ionized clusters can be detected. The results are compared to simple MALDI mass spectra.

EXPERIMENTAL SETUP

The experiments were performed in a reflectron time-of-flight (RETOF) mass spectrometer which is described in detail elsewhere [7]. Desorption of the sample molecules

CP584, *Resonance Ionization Spectroscopy 2000: 10th Int'l. Symp.*, edited by J. E. Parks and J. P. Young
© 2001 American Institute of Physics 0-7354-0024-5/01/$18.00

from a matrix into a supersonic beam of argon was performed with the help of a pulsed CO_2 laser (10.6 µm). In our investigations concerning heterogeneous cluster formation the MALDI matrix substance and the respective amino acid were mixed in equal molar parts. The resulting probe mixtures were pressed together with polyethylene to pellets and applied to the probe tip. The probe tip was placed in the desorption chamber in a distance of approximately 1mm in front of the nozzle of a pulsed valve which produces a supersonic beam of argon. In the experiments a back pressure of about 3bar argon is applied. The neutral sample molecules and neutral clusters were cooled in it degrees of freedom and transported through a skimmer into the ion source of the RETOF mass spectrometer. Here the multi photon ionization (MUPI) by either nanosecond or femtosecond laser pulses is performed. It should be noted that any preformed ions originating from the desorption process are prevented from getting into the ion source. The femtosecond laser pulses were generated with the help of a dye laser system which was pumped by a XeCl excimer laser ($\lambda \sim 260$ nm, pulse length 500 fs, energy / pulse 10-15 µJ). For the ionization with nanosecond laser pulses the frequency doubled output of a Nd:YAG pumped dye laser system was used ($\lambda = 260$ nm, pulse length 8 ns, energy / pulse ~ 500 µJ). The spectra were recorded using a 200 MHz transient digitizer and the evaluation of the registered data was performed on a VME bus computer. Each spectrum represents at least the sum of 25 laser shots. The MALDI mass spectra were obtained with home build reflectron time-of-flight mass spectrometer. The samples were prepared in according to standard procedures.

All matrix compounds as well as the employed amino acids and peptides were purchased from Sigma Aldrich Chemistry and used without further purification.

RESULTS AND DISCUSSION

In figure 1 the cluster mass spectra of different amino acids with sinapinic acid taken with 500 fs laser pulse duration are shown. As seen from these mass spectra strong signals for the monomeric unit of the sinapinic acid as well as the homogenous clusters from this acid are observed. Beside these intense signals the cluster formation of the sinapinic acid with the different amino acids alanine, valine and lysine is seen. In case of the alanine and the lysine only the heterogenous dimer can be found, while the clustering with valine yields also higher order heterogeneous clusters like the trimer and other. This different behavior is accounted to the formation process and the actual intensity of the femtosecond laser pulse. Apparently the intensity of the signal for the protonated amino acid is for the three cases different. While alanine show the smallest intensity for the transfer of the proton, this signal reaches over 40 % with lysine. Using the amino acid glycine for clustering together with the sinapinic acid results in the observance of the heterogenous cluster but not in any signal for the proton transfer to the amino acid.

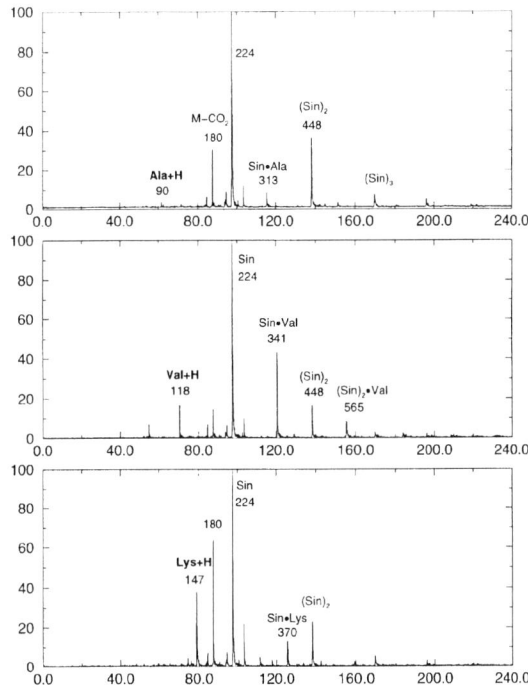

As already shown in previous papers [8] this is due to a process of a dissociative proton transfer. Therefore for the formation of the protonated species is controlled by the proton affinity (PA) of the amino acid as well as of the deprotonated aromatic carboxylic moiety.

In table 1 the complete results of these investigations are summarized. Here the intensities for the signal of the protonated amino acids are compared. Clearly the most acids used as matrices yield not in a signal for any proton transfer in case of the amino acid glycine. Both the di-hydroxy-benzoic acids as well as the sinapinic acid show no signal at all, while the vanillic and the ferulic acid can undergo this dissociative proton transfer reaction with medium to high intensities. Using alanine as a proton receptor again the 2,5 DHB yield no signal. In this case not even a signal for the homogenous cluster with femtosecond laser pulses could be observed. The other isomer of this benzoic acid shows a small signal for the protonated alanine like the sinapinic acid. Increasing the size of the non-aromatic residue of the

Table 1. Intensities of the protonated amino acid from cluster reactions with different carboxylic acids.

	Glycine	Alanine	Valine	Leucine	Lysine
2,5 DHB	-	-	o	+	++
3,4 DHB	-	o	+	+++	+++
Sinapinic acid	-	o	+	++	+++
Vanillic acid	-	+	++	++	+++
Ferulic acid	+	+	++	+++	+++

Legend :   - not observed; o appearing, but weak; ı intensity < 15 %, ++ intensity < 20 %; +++ intensity < 30 %, ++++ intensity > 30 %

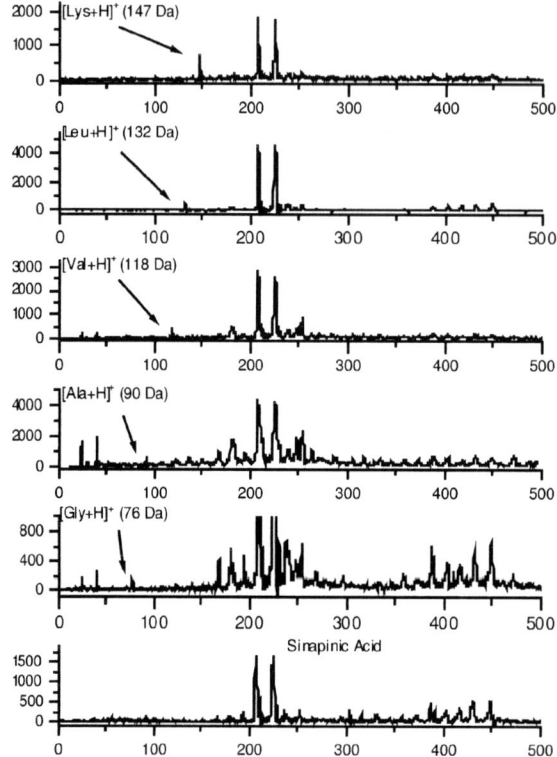

amino acid or the introduction of a functionalized amino acid yield in strong signals for the protonated amino acid.

At this point the question should be addressed whether these results obtained from investigation in a supersonic expansion followed by multi photon ionization can be compared to the MALDI method. In figure 2 some MALDI mass spectra of the simple amino acids together with the matix substance sinapinic acid are shown. Obviously strong signals for the matrix are observed throughout in all 6 presented mass spectra. By comparison between the mass spectrum of the pure sinapic acid and the mixture mass spectra these signals can be unambiguously identified. Although these mass spectra are taken with a simple linear time of flight mass spectrometer, the mass resolution is high enough to distinguish between these signals. As seen also from the multi photon ionization mass spectra the signal at mass 225 is due to the protonated sinapinic acid. At mass 207 the loss of water is observed. This signal has no equivalent in the multi photon ionization.

Comparing the other signals in the different MALDI mass spectra clearly different intensities of the protonated amino acids can be observed. While the signal for the protonated lysine has the highest intensity the other signals are decreasing with the size of the side chain. As demonstrated previously [13] this is a direct result of the proton affinity. As seen from table 1 the proton affinity of the glycine must have the lowest value of all the presented amino acids. Therefore this amino acid yields nearly no signal for its protonated species. In figure 2 the corresponding MALDI mass spectrum is shown with increased sensitivity thus displaying a very small signal for the mass at m/z 76. It should be noted that in the ionization procedure a large number of different other processes can lead to the formation of this ion. Therefore the small intensity can be explained sufficiently by these other processes. Obviously the formation of the protonated amino acids can easily and sufficiently explained through the cluster formation followed by a dissociative proton transfer after ionization of the cluster. It should be noted that a necessary condition for the proton transfer is the proton affinity as already discussed above.

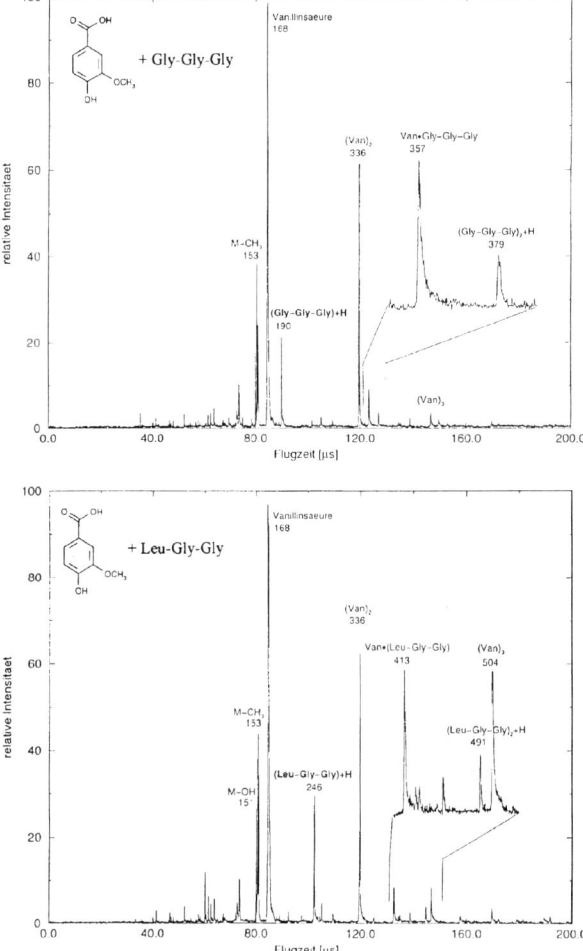

A further condition to be fulfilled prior this transfer process can be established is a specificity of the matrix molecule. It has been shown by labeling studies that an intense protonated amino acid signal is only observed if the matrix has a phenolic proton. In the dissociative proton transfer only this proton can be used. The acidic proton from the carboxylic group is invariant and not part of the process.

In figure 3 the multi photon mass spectra of some smaller tripeptides are shown. Again these peptides have only non-aromatic side chains, therefore the multi photon ionization take place only in the aromatic ring system of the matrix substance. The most striking difference to the cluster mass spectra of the simple amino acids is the intense signal for the protonated peptides. As seen from figure 3 the tripeptide Gly-Gly-Gly yield a signal for its protonated analog. It should be noted that the amino acid glycine does not show any signal for the protonated compound with all investigated matrices. This is also seen by the replacement of a glycine by a leucine. Again

283

the protonated species show a strong signal for the dissociative proton transfer. Furthermore both mass spectra display a variety of different other clusters. Most notably is the formation of a proton bound dimer as well as the attachment of the matrix to the small peptide yielding intense signals. This is also well known from the MALDI process.

Conclusions

The gas phase behavior of some aromatic carboxylic acids were investigated by means of laser desorption multi photon ionization time-of-flight mass spectrometry. Using femtosecond laser pulses for the ionization process intense homogeneous clustering of these compounds was observed and, when codesorbed with small non-polar amino acids and peptides into the supersonic beam, also the formation of homogeneous clusters, and especially heterogeneous dimers, was detected. In addition protonated amino acids as well as the protonated peptides were detected, which could not have been formed by direct ionization as these substances do not absorb laser light under the chosen ionization conditions. Most important the experimental results disclose a relation between the formation of heterogeneous dimers and protonated amino acids. On the basis of these experimental findings the protonated amino acids and peptides must be formed via proton transfer in the heterogenous dimer ions from the carboxylic acid to the analyte molecule and subsequent decay of the dimer releasing the respective protonated amino acid, also by dissociative proton transfer. This proposed mechanism is promoted by the observed correlation of the signal intensity of the detected protonated amino acids with their gas phase proton affinities.

Acknowledgment

This work is supported by the Deutsche Forschungsgemeinschaft (Project Gr 917/9-3) and in part by the Volkswagen Stiftung. Support from the Fonds der Chemischen Industrie is gratefully acknowledged.

References

1. a) M.S. de Vries, *J.Am.Chem.Soc* **122**, 8091 (2000).
    b) M. Dey, J. Grotemeyer, E.W. Schlag; *J.Am.Chem.Soc.* **116**, 9211 (1994).
2. U. Boesl, J. Grotemeyer, K. Müller-Dethlefs, H.J. Neusser, H.L. Selzle, E.W. Schlag; *Int.J.Mass Spectrom.Ion Proc.* **118**, 191 (1992).
3. E.R Bernstein, *J.Phys.Chem.***96**, 10105 (1992).
4. M. Karas, D. Bachmann, U. Bahr and F. Hillenkamp, *Int. J. Mass Spectrom. Ion Processes* **78**, 53 (1987).
5. P.C. Liao and J. Allison, *J Mass Spectrom.* **30**, 408 (1995).
6. J. Grotemeyer, E.W. Schlag; *Angew.Chem.*, **100**, 461 (1988).
7. U. Boesl, K. Walter, J. Grotemeyer, E.W. Schlag; Anal. Instrum. **16**, 151 (1987).
8. A. Meffert, J. Grotemeyer; *Ber.Bunsenges.Phys.Chem.* **102**, 459 (1998).

A new method to obtain a 212 nm laser source for Lyman-α light with CLBO crystals

Y. Miyake[1], K. Kato[2], S. Makimura[1], K. Shimomura[1], Y. Matsuda[3], P. Bakule[3], R.J. Scheuermann[3], K. Nagamine[1,3]

[1] *Meson Science Laboratory, High Energy Accelerator Research Organization (KEK-)MSL), Oho, Ibaraki-305-0801, Japan*
[2] *Chitose Institute of Science and Technology, 758-65, Bibi, Chitose-city, Hokkaido-066-8655, Japan*
[3] *Muon Science Laboratory, Institute of Physics and Chemical Research (RIKEN), Saitama 351-01, Japan*

Abstract. We have been generating Lyman-α laser light by the sum-difference frequency mixing method using two 212.5 nm photons for two-photon resonant excitation of the $4P^5 5P[1/2]_0$ state in Kr, subtracted by a photon with a tunable difference wavelength. To date, the 212.5 nm emission has been obtained by quadrupling single-mode 850 nm light by using two β-Ba_2BO_4 crystals. However, because of the low thermal conductivity of β-Ba_2BO_4 crystal, the generated 212.5nm beam became in an annular form. In this report, we will report on a new method for obtaining the Lyman-α light with $CsLiB_6O_{10}$ crystals together with the experimental results of the resonance spectra of tritium and deuterium generated by nuclear reactions.

INTRODUCTION

For the efficient ionization of a rare species such as muonium (designated as Mu; consisting of a μ^+ and an e^-, can be considered to be a light isotope of hydrogen), we require an intense pulsed laser system having the following properties:

- More than 18 kW/cm^2 of the Lyman-α light is needed in order to saturate the transition of Mu from the 1s state to the 2p state, and more than 15 mJ/pulse/cm^2 of $\leq$ 366 nm light is needed to ionize the 2p state of Mu.

- The band width of the laser should be around 200 GHz because the 1s-2p transition of Mu has a Doppler broadening of 237 GHz at 2300 K, a temperature at which thermal Mu can most efficiently be emitted into vacuum from a tungsten surface [1].

- The laser beam quality should be very good, not only to allow efficient frequency doubling and non-linear mixing, but also to transport the laser beams

CP584, *Resonance Ionization Spectroscopy 2000: 10th Int'l. Symp.*, edited by J. E. Parks and J. P. Young
© 2001 American Institute of Physics 0-7354-0024-5/01/$18.00

into the thermal Mu production target area in the primary proton beam line, which is located at a distance of around 11 m from the pulsed laser system. This distance is determined by the thickness of the radiation shield required for the primary beam line.

VUV GENERATION

At present, there exists no laser system which emits the Lyman-α laser light directly. Laser generation in the VUV (vacuum ultra violet) region has been investigated by many groups (see e.g. [2–4]) employing non-linear wave mixing techniques. The methods have been based upon non-linear processes, such as:

(a) Non-Resonant Tripling: $\omega_{VUV}=3\omega_1$,

(b) Resonant Sum Frequency Mixing: $\omega_{VUV}=\omega_1+\omega_2+\omega_3$ (or $2\omega_1+\omega_2$),

(c) Resonant Sum-Difference Frequency Mixing: $\omega_{VUV}=\omega_1+\omega_2-\omega_3$ (or $2\omega_1-\omega_3$) or

(d) Anti-Stokes stimulated Raman scattering.

Among these methods, the first (a) is the most general technique for the generation of the hydrogen Lyman-α emission (around 121.6-122.1nm) [3]. This method is technically the easiest method, where we can expect the original 366 nm laser light to also be used for ionizing the 2p state of Mu. However, even under a Kr-Ar environment that enhances phase matching, the conversion efficiency to generate VUV is too low to satisfy our power requirements unless its time width is of the order of pico-seconds (see A.H. Kung et al. [4]). One of the most successful experiments in the VUV region utilizing sum frequency mixing (method (b)) was carried out by C.H. Muller et al. in the so-called "SANDIA" system [5], where excitation of 6S-6P-7S-8P in Hg was used. Although this method showed the highest conversion efficiency of VUV light (up to 5 %) and generated several mJ/pulse, it requires not only three independent laser wavelengths of 555, 405 and 255 nm, but also a mercury cell with a heat pipe, requiring frequent maintenance. Thus, it is impractical for our application.

In the past, sum-difference frequency mixing (method (c)) was investigated with metal vapors, such as Sr, Mg or Hg [6,7]. Recently, rare gases such as Xe and Kr have been found to be appropriate media for non-linear phenomena in the sense that a reasonable conversion efficiency can be achieved while retaining ease of handling. G. Hilber et al. [8] adopted the use of the $4P^55P[5/2]$ two-photon resonance in Kr, achieving a power efficiency as high as 10^{-5}. Recently, Marangos et al. [9] have applied $4P^55P[1/2]_0$ two-photon resonance in Kr, and by employing buffer gases for phase matching, have obtained a conversion efficiency of $\sim 10^{-3}$ in the Lyman-α region. The Kr $4P^55P[1/2]_0$ resonance state was found by Miyazaki et al. [10] to provide the highest yield of XUV generation in sum frequency mixing compared to any other two-photon excitation level in Kr and Xe.

Recently, H. Wallmeier *et al.* [11] reported the generation of a tunable laser radiation with a range of 129 to 210nm using anti-Stokes Raman scattering of tunable dye laser radiation in cold hydrogen(method (d)).

After taking into account the conversion efficiency, a simplicity of the mixing cell, the number of photons required and possibility to obtain about 200 GHz of the band width to match the Doppler broadening of Mu at 2000 K, we decided to adopt method (c). In our experiments, following Magangos *et al.*, the Lyman-α wavelength has been achieved via the sum-difference frequency mixing method using two photons of 212.5 nm (ω_r) for two-photon resonant excitation of the $4P^5 5P[1/2]_0$ state in Kr, subtracted by a photon with a tunable difference wavelength (ω_t; 820-848 nm). The scheme is shown in the right top corner of Fig.1 [1] .

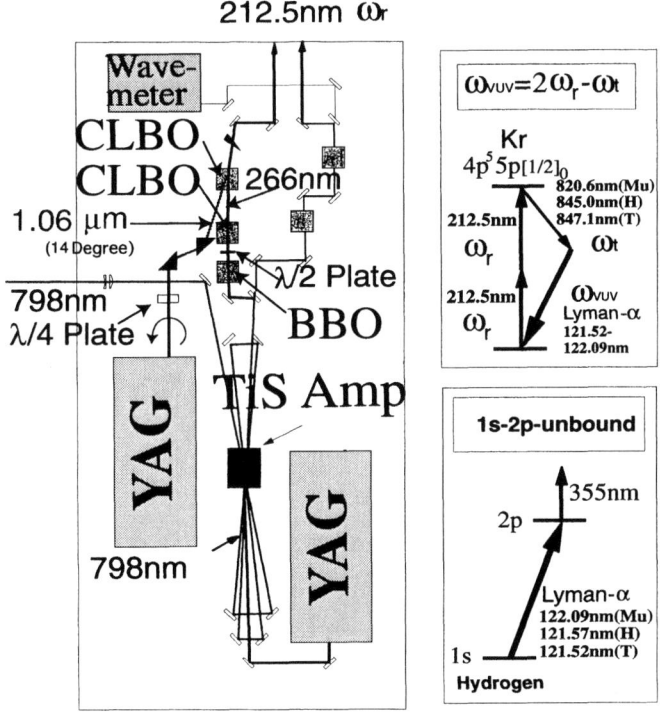

FIGURE 1. Scheme for generating an intense resonant frequency (ω_r) with a good beam quality by adopting CLBO crystals (left figure). Resonant ionization scheme via the 1s $\rightarrow$ 2p $\rightarrow$ unbound transition for the hydrogen isotopes, and the scheme for the Lyman-α generation by a sum-difference frequency mixing method using two 212.5 nm photons for two-photon resonant excitation of the $4p^5 5p[1/2]_0$ state in Kr, subtracted by a photon with a tunable difference wavelength (right figure).

GENERATION OF THE ω_R(212 NM) LASER BEAM

To date, ω_r has been obtained by quadrupling single-mode 850 nm light with a band width of 0.5-1.0 GHz. The 850 nm light with an intensity as high as 300-600 mJ/pulse is frequency-doubled by first using a β-Ba$_2$BO$_4$ crystal and generating 425 nm light with an intensity of 100-150 mJ/p with an excellent beam quality. The 425 nm light is then frequency-doubled by using a second β-Ba$_2$BO$_4$ crystal and generating ω_r with an energy of 5-10 mJ/pulse. Unexpectedly, the output of ω_r was found to have a very poor beam quality with a ring pattern. According to Kato, the phase matching angle of the β-Ba$_2$BO$_4$ crystal has a sharp temperature dependence in the wavelength regime of our interest. Since the absorption of 212.5 nm by the β-Ba$_2$BO$_4$ is very high to be 18 %/cm, the temperature difference between the center and edges of the β-Ba$_2$BO$_4$ crystal, and also due to its low thermal conductivity, there causes a significant change in the phase matching condition, consequently distorting the generated beam to an annular form [12].

In order to solve this problem, such an annular form induced by the thermal de-phasing effect due to the self-heating, we are now investigating a new method to generate ω_r, in which we are mixing Nd:YAG's fundamental 1.06 μm laser light and a single mode of the third harmonic of 798 nm by adopting a new non-linear CsLiB$_6$O$_{10}$ (CLBO) crystal developed by Mori et al. [13]. Since CLBO possesses a smaller walk-off angle and larger angular and spectral bandwidths than BBO, CLBO is suitable for the fourth (266 nm) and fifth (213 nm) harmonic generation of the fundamental of Nd:YAG laser (1.06 μm) [13,14]. In the case of CLBO, the fifth harmonic can be achieved by the sum frequency mixing of the fundamental and the fourth harmonics of 266 nm light [15].

Since we must have a tunable ω_r, a single mode 798 nm light (Continuum Mirage 800 OPO laser) with a band width of 0.5-1.0 GHz and 200-230 mJ/pulse is frequency-tripled by the combination of β-Ba$_2$BO$_4$ ($\theta = 27.4$ degrees, $\phi = 0$ degrees) and CLBO ($\theta = 57$ degrees, $\phi = 45$ degrees) crystals, generating the 266 nm light. Then 266 nm light is mixed with the fundamental of an injection seeded Nd:YAG (1.06 μm) light with a band width of 3 GHz in the second CLBO ($\theta = 71$ degrees, $\phi = 45$ degrees) crystal with an incident angle of 14 degrees. A schematic configuration of the new method for generating ω_r by adopting CsLiB$_6$O$_{10}$ (CLBO) crystals is shown in Fig.1.

Finally, using the new method adopting CLBO crystals, ω_r=212 nm with an intensity of 6-9 mJ/pulse with very good beam quality was obtained.

ON LINE EXPERIMENT

The Lyman-α wavelength can be easily tuned by changing the difference wavelength ω_t in the present sum-difference frequency mixing method. Table 1 shows the Lyman-α wavelength and its corresponding difference wavelength ω_t in vacuum and ω_t in air, for Mu, H, D and T.

The validity of the present method was demonstrated by on-line experiments with the use of 500-MeV protons at KEK-MSL. From the process of nuclear reactions,

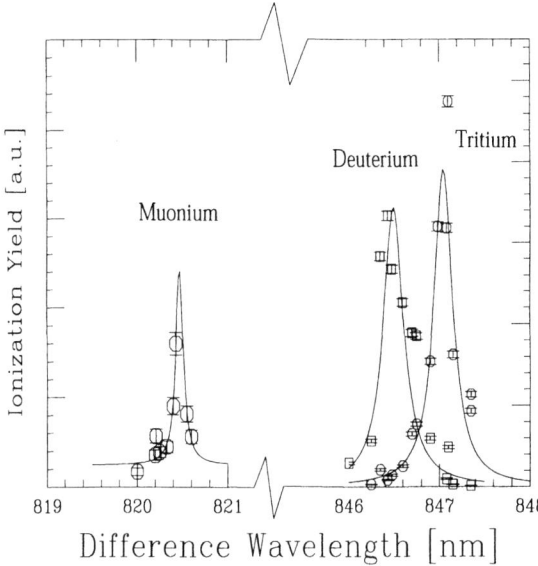

FIGURE 2. Resonance curve for Mu, D and T generated by nuclear reactions of 500 MeV proton

nearly all the elements, including short-lived particles, can be produced simultaneously. With use of the ω_r generated with CLBO, Lyman-α generation was achieved and measurements of the resonance spectra of Mu, tritium and deuterium generated by nuclear reactions were performed. Fig.2 shows the resonance spectra for the hydrogen isotopes Mu, D, and T, by changing the difference wavelength ω_t. The observed resonance peaks agree with the calculated values.

TABLE 1. The Lyman-α wavelength and its corresponding difference wavelength ω_t in vacuum and ω_t in air for Mu, H, D and T, in the case of a sum-difference frequency mixing method using two photons of 212.55 nm (ω_r) for two-photon resonant excitation of the $4P^55P[1/2]_0$ state in Kr.

	Muonium(Mu)	Hydrogen	Deuterium	Tritium
Lyman-α (nm)	122.09	121.57	121.53	121.52
ω_t (vacuum) (nm)	820.59	844.98	846.57	847.11
ω_t (air) (nm)	820.37	844.75	846.33	846.88

REFERENCES

1. Y. Miyake, K. Shimomura, A. P. Mills, and K. Nagamine, Hyperfine Interactions 106 (1997) 237.
2. J.F. Ward and G.H.C. New; Phys. Rev. 185 (1969) 57.
3. R. Mahon, T.J. McIlrath, V.P. Myerscough and D.W. Koopman; IEEE J. Quantum Electron. QE-15, 6 (1979) 444.
4. A.H. Kung, J.F. Young and S.E. Harris; Appl. Phys. Lett. 22 (1973) 301. and 28 (1976) 239.
5. G.H. Muller, D.D. Lowenthal, M.A. DeFaccio and A.V. Smith; Opt. Lett. 13, 81 (1988) 65.
6. R.T. Hodgeson, P.P. Solokin and J.J. Wynne; Phys. Rev. Lett. 32 (1974) 343., R. Hilbig and R. Wallenstein; IEEE J. Quantum Electron. QE-19 (1983) 12.
7. R. Hilbig and R. Wallenstein; IEEE J. Quantum Electron. QE-17 (1981) 8.
8. G. Hilber, A. Lago and R. Wallenstein; J. Opt. Soc. Am. B4, 11 (1987) 1753.
9. J.P. Marangos, N. Shen, H. Ma, M.H.R. Hutchinson, and J.P. Connerade; J. Opt. Soc. Am. B7 (1990) 1254.
10. K. Miyazaki, H. Sakai and T. Sato; Appl. Opt., 28, 4 (1989) 699.
11. H Wallmeier and H. Zacharias Appl. Phys. B45 (1988) 263.
12. K. Kato, The Review of Laser Engineering18,1 (1989)
13. Y. Mori, I. Kuroda,S. Nakajima, T. Sasaki and S. Nakai Jpn. J.Appl.Phys, 34, (1995), L296, Appl.Phys. Lett. 67(13)(1995)1818
14. N. Umemura, K. Yoshida, T. Kamimura, Y. Mori, T. Sasaki and K. Kato, Adv. Solid-State Lasers 26 (1999)715
15. Y.K. Yap, M Inagaki, S. Nakajima, Y. Mori and T. Sasaki, Opitcs Letters 21, 17 (1996), 1348

Analysis of Polycyclic Aromatic Hydrocarbons (PAHs) using Nanosecond Laser Desorption / Femtosecond Ionisation Laser Mass Spectrometry (FLMS)

L Robson[a], A D Tasker[a], S M Hankin[a], K W D Ledingham[a], R P Singhal[a], X Fang[a], T McCanny[a], C Kosmidis[b], P Tzallas[b], A J Langley[c], P F Taday[c], E J Divall[c]

[a] Dept of Physics and Astronomy, University of Glasgow, Glasgow, G12 8QQ, Scotland, UK
[b] Department of Physics, University of Ioannina, GR-45110, Ioannina, Greece
[c] Central Laser Facility, CLRC Rutherford Appleton Laboratory, Oxon, OX11 0QX, UK

Abstract. Nanosecond laser desorption / femtosecond ionisation laser mass spectrometry (LD/FLMS) allows ultra-sensitive detection and trace analysis of atoms and molecules. In this study, we have applied the LD/FLMS technique to the characterisation of polycyclic aromatic hydrocarbons (PAHs). Using high intensity femtosecond laser pulses (10^{13} - 10^{15} W/cm^2) at $\lambda \sim 395$nm and 790nm coupled to a reflectron time of flight mass spectrometer a series of PAHs have been investigated. In particular, anthracene, tetracene and pentacene are discussed. The spectra presented show intact parent ion at both wavelengths, with little fragmentation at lower ionisation laser intensities. This initial data suggests that the optimum wavelength to operate FLMS for PAHs may be 395 nm and not 790 nm for maximum parent ion production. Comparative studies adopting nanosecond ionisation are also discussed.

INTRODUCTION

The rapidly developing field of ultrafast laser-molecule interactions in combination with time-of-flight mass spectrometry is becoming an important tool for atomic and molecular analysis [1-5]. Coupling a laser desorption capability with the established technique of gas-phase femtosecond laser mass spectrometry allows ultra-sensitive detection and trace analysis of atoms and molecules in the solid-phase [5,6]. The technique involves two steps: desorption of intact neutral molecules from the sample surface and post-ionisation of the desorbed analyte using high intensity femtosecond laser pulses (10^{14}-10^{15} W/cm^2). The ions are mass-separated and detected according to their mass-to-charge ratio using a reflectron time of flight mass spectrometer. The intense, ultrafast laser pulses from the femtosecond laser are able to ionise virtually any class of molecule and almost eliminate the chance of dissociation prior to detection. The temporal and spatial decoupling of desorption and ionisation allows each step to be optimised independently yielding improvements in system

CP584, *Resonance Ionization Spectroscopy 2000: 10th Int'l. Symp.*, edited by J. E. Parks and J. P. Young
© 2001 American Institute of Physics 0-7354-0024-5/01/$18.00

performance. The generation of intact parent ions unambiguously identifies the analyte providing strong evidence for the potential of LD/FLMS as a fundamental analytical technique for the detection of labile environmental pollutants [2].

The analysis of PAHs is of increasing importance from the perspective of environmental analysis, considering their presence in atmospheric particles and their known carcinogenic, mutagenic and teratogenic effects [7], and from the perspective of studying fundamental molecular ionisation [1,8,9].

EXPERIMENTAL

Details of the reflectron time-of-flight (ToF) mass spectrometer used have been described in detail elsewhere [10]. Briefly, the source chamber and flight tube are pumped using rotary-backed turbomolecular pumps to a base pressure of 10^{-9} Torr. The sample for analysis is deposited on a stainless steel stub and transferred to the source chamber by means of a load-lock. Laser desorption of the solid samples was achieved using the fourth harmonic output (266 nm) from a nanosecond Nd:YAG laser (Minilite I, Continuum) operating from 30-150 µJ, focussed onto the sample stub using a 27 cm focal length lens. The ASTRA femtosecond laser at the Rutherford Appleton Laboratory produces pulses of ~9 nJ at a wavelength of ~790 nm with a pulse duration of ~50 fs. The low energy pulses are stretched to 300 ps and amplified in a second Ti:Sapphire rod pumped by 140 mJ in 20 ns of the second harmonic from a Nd:YAG laser. The beam is attenuated using a $\lambda/4$ waveplate and polariser and the amplified pulses are then recompressed to 50 fs in a grating pair and directed in to the source chamber of the mass spectrometer for post-ionisation of the desorbed molecules. When ionising radiation of wavelength ~395 nm is required, the fundamental laser beam is passed through a frequency doubling crystal (BBO) placed infront of the source chamber. Nanosecond ionisation of laser desorbed PAHs was carried out at the Glasgow laboratory. Laser desorption was achieved using the same method as above, while post ionisation was achieved using the fourth harmonic output (266 nm) of a second Nd:YAG laser (SL2Q/SL3A, Spectron Laser Systems). The 16 ns pulse was focussed using a 30 cm focal length lens generating intensities of up to 2.46×10^9 W/cm^2. Figure 1 displays a schematic of the desorption and post-ionisation events.

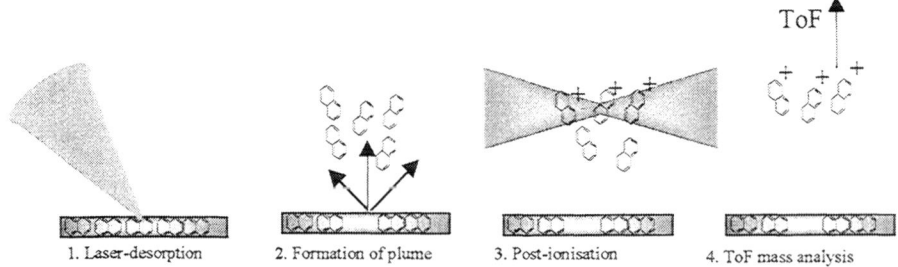

FIGURE 1. Two-step laser-desorption / post-ionisation time-of-flight mass spectrometry.

A variable delay from 1-37 µs was set up between desorption and ionisation using a delay generator (DG535, SRS). The ions, extracted from the source using purpose-designed ion optics, were guided into the reflectron after which they are detected by a multi-channel plate (Galileo). The signal output from the detector was coupled to a digital oscilloscope (9344C, LeCroy) for single-shot and averaged-data collection. A PC installed with GRAMS/32 software (Galactic) was connected to the oscilloscope through a GPIB interface and was used for data acquisition and analysis.

RESULTS AND DISCUSSION

Femtosecond Ionisation

The laser-desorbed PAH molecules were ionised using 50 fs laser pulses. The mass spectra for tetracene, recorded using wavelengths of 395 and 790 nm at the maximum and minimum intensities available, are shown in Figure 2.

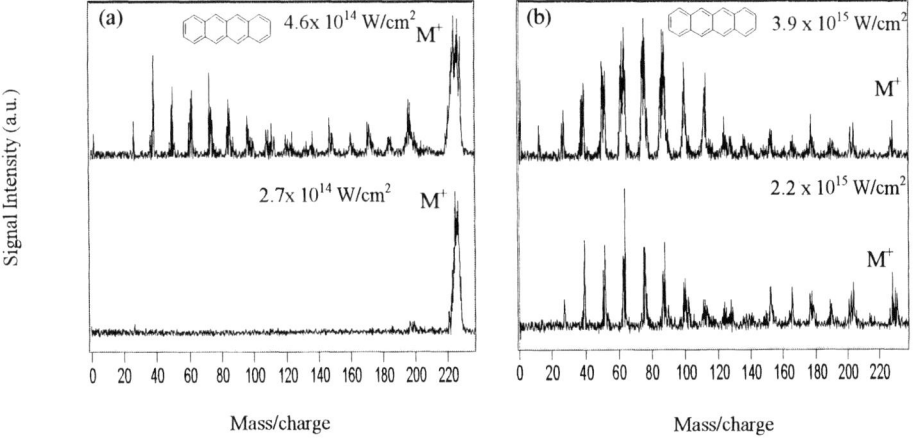

FIGURE 2. Time of flight mass spectra of laser-desorbed tetracene upon 50 fs laser post-ionistion using (a) λ = 395 nm and (b) λ = 790 nm.

The upper trace of Figure 2(a) shows extensive fragmentation; carbon clusters are dominant, with C_2^+ to C_{16}^+ readily resolved, but not at the expense of detecting the parent ion, denoted by M^+ at m/z=228 which is the largest peak in the spectrum. A notable feature of the spectra obtained for tetracene at 395 nm is that almost no fragmentation was observed and the parent ion dominated the spectra when the peak intensity of the ionising laser was reduced to 2.7×10^{14} W/cm². In contrast to this, at 790 nm ToF mass spectra for tetracene did not yield exclusively parent ions at the minimum ionisation intensity. As shown in Figure 2(b), extensive fragmentation occurs with a much smaller yield of parent ion at the maximum intensity of 3.9×10^{15} W/cm^2. Although the relative yield of parent ion increased as the intensity of the ionisation laser was lowered to 2.2×10^{15} W/cm^2, there was no intensity at which the spectra exhibited parent ions only, as was the case at 395 nm.

The results described above were common to all three molecules. At intensities of $2.2 - 4.0 \times 10^{15}$ W/cm^2 and a wavelength of 790 nm, it is believed that field ionisation is the predominant mechanism for molecular ionisation [11]. Using 395 nm, 50 fs laser pulses the intensity range is reduced to $1.9 - 3.7 \times 10^{14}$ W/cm^2. It is believed at these intensities, molecular ionisation could be attributed to a multiphoton and/or field ionisation mechanism. Also, the shape of the laser pulse at both wavelengths has to be considered when identifying the differences between the spectra, in addition to where spatially in the laser pulse the parent and low mass fragments are ionised. The high energy processes involving the fragmentation and ionisation of the carbon clusters indicates that ionisation takes place at the focus of the laser beam, where the intensity is at a maximum, whereas ion production of the parent is likely to occur in the lower energy 'wings' of the laser focus.

Nanosecond Ionisation

Comparative studies were also carried out to characterise laser-desorbed PAHs under nanosecond ionisation conditions. Figure 3 shows the typical pattern observed for tetracene using nanosecond laser post-ionisation at varying ionisation intensities.

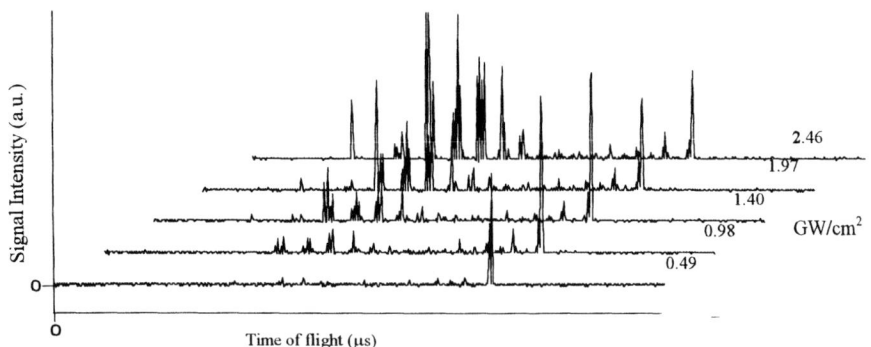

FIGURE 3. ToF mass spectra of laser desorbed tetracene post-ionised using 266 nm, 16 ns laser pulses as a function of ionisation intensity.

Figure 3 is representative of nanosecond ToF mass spectra for the three molecules studied: anthracene, tetracene and pentacene. At the highest laser intensity $(2.46 \times 10^9$ W/cm$^2)$ tetracene shows significant fragmentation, although there is still a detectable parent ion. Notably, as the ionisation intensity is lowered, a suppression of low mass fragments is observed, ultimately leading to parent only mass spectrum at an ionisation intensity of 0.49×10^9 W/cm^2. This is a similar trend to that observed for femtosecond laser ionisation at a wavelength of 395 nm. This suggests that a multiphoton mechanism may be playing a substantial role in ultrafast PAH molecular ionisation at 395 nm. Furthermore, initial studies of the fragmentation patterns have revealed differences in the overall and individual carbon cluster profiles. In the case of nanosecond ionisation, the overall profiles are peaked towards C3 for all PAHs studied, commensurate with an *in-pulse* ladder-switching mechanism [12]. In contrast to this, infrared femtosecond mass spectra exhibit carbon cluster profiles peaked towards C4-C6 and exhibit a lower abundance of small clusters, as a result of *post-pulse* explosive molecular fragmentation.

CONCLUSIONS

Laser-desorption has been coupled to femtosecond laser ionisation mass spectrometry for the characterisation of solid phase polycyclic aromatic hydrocarbons. This work has provided ToF mass spectra of high-mass PAH molecules that have not previously been analysed using this technique.

The PAHs anthracene, tetracene and pentacene have been studied under both femtosecond (395 and 790 nm) and nanosecond laser ionisation conditions. Analysis of the mass spectra has demonstrated that parent ions can be observed for each molecule under both fs and ns regimes. However, conditions could be found where parent ions only dominated the spectra at 395 nm (femtosecond) in each case, and is thought to be the result of a significant multiphoton component in the ionisation mechanism. Differing profiles of carbon clusters resulting from ns and fs ionisation highlight the significance of pulse duration on the formation of low-mass fragment ions.

Infrared femtosecond ionisation at 790 nm has been reported to be highly efficient for the analysis of molecules including aromatics, nitroaromatics and organometallics [13-15]. However, the data presented for these three molecules has shown that the optimum regime to operate FLMS for PAHs may be at 395 nm and not 790 nm in terms of parent ion production.

In conclusion, the combination of nanosecond desorption and with femtosecond post ionisation has been shown to be an effective way of analysing environmentally-sensitive, solid-phase materials. As part of an environmental analytical study, the technique will be applied to the detection of PAHs in atmospheric particles and to the detection of other pollutants including labile nitro- and oxy-PAHs, pesticides and explosives.

ACKNOWLEDGMENTS

The authors wish to express thanks to the Rutherford Appleton Laboratory for excellent facilities and assistance, and to EPSRC for funding and Studentships (LR and ADT). NERC arc also acknowledged for the award of a Post-doctoral Fellowship to SMH.

REFERENCES

1   Levis, R.J., and DeWitt, M.J., *J. Phys. Chem. A* **103,** 6493-6507, (1999).
2   Carsten, G., Heinicke, R., Weickhardt, C., Grotemeyer, J., *Int. J. Mass Spectrom.* **185/186/187,** 307-308, (1999).
3   Weickhardt, C., and Grotemeyer J, *J. Mass spectrom. Reviews* **15,** 139-162, (1996).
4   DeWitt, M.J., and Levis, R.J., *J. Chem. Phys.* **110,** 11368-11375, (1999).
5   Ledingham, K.W.D., and Singhal, R.P., *Int. J. Mass. Spectrom. Ion Proc.* **163,** 149-168, (1997).
6   Nicolussi, G.K., Pellin, M.J., Lykke, K.R., Trevor, J.L., Mencer, D.E., Davis, A.M., *Surface and Interface Anal.* **24,** 363-370, (1996).
7   *International Agency for Research on Cancer Monographs on the Evaluation of Carcinogenic Risks to Humans,* IARC, Lyon, France, 1983, Volume 32, part 1.
8   Dewitt, M.J., and Levis, R.J., *J. Chem Phys.* **102,** 8670-8673, (1995)
9   Hankin, S.M., Villeneuve, D.M., Corkum, P.B., Rayner, D.M., *Phys. Rev. Lett.* **84,** 5082-5085, (2000).
10 Borthwick, I.S., *"The Application of Resonance Ionisation Mass Spectrometry to Trace Analysis in Solids", PhD Thesis,* University of Glasgow, (1993).
11 Codling, K., and Frasinski, L.J., *J. Phys. B.* **26,** 783-809, (1993).
12 Grotemeyer, J., and Schlag, E.W., *Angew. Chem. Int. Ed. Engl.* **27,** 447-592, (1988).
13 Ledingham, K.W.D., Smith, D.J., Singhal, R.P., McCanny, T., Graham, P., Kilic, H.S., Peng, W.X., Langley, A.J., Taday, P.F., Kosmidis, C., *J. Phys. Chem. A* **103,** 2952-29963, (1999).
14 Kosmidis, K., Ledingham, K.W.D., Kilic, H.S., McCanny, T., Singhal, R.P., Langley, A.J., Shaikh, W., *J. Phys. Chem. A* **101,** 2264-2270, (1996).
15 Tzallas, P., and Kosmidis, *to be published.*

Quantitative Surface Analysis of Molecular Overlayers by Resonantly Enhanced Multiphoton Ionization of Sputtered Molecules

A. Schnieders, K. Rüschenschmidt, M. Schröder, A. Benninghoven,
H. F. Arlinghaus

Physikalisches Institut der Universtät Münster, Wilhelm-Klemm-Str. 10, D-48149 Münster, Germany

Abstract. Laser postionization of sputtered neutrals in combination with resonantly enhanced multiphoton ionization is a more quantitative analytical technique for organic surface analysis than secondary ion mass spectrometry due to the reduced matrix effect. Nevertheless, we observed a remaining sputter-induced matrix effect. We have investigated this matrix effect by comparing the secondary neutral and ion emissions from UHV-deposited molecular overlayers of adenine and β-alanine on Ag substrates under Ar^+, Xe^+, and SF_5^+ primary ion bombardment.

INTRODUCTION

Laser postionization of sputtered neutrals (Laser-SNMS) using resonant or non-resonant multiphoton ionization is a powerful tool for sensitive and quantitative analysis of inorganic surface species. Especially the reduction of the matrix effect (dependence of the secondary particle emission on the chemical environment on the surface), which often complicates quantification in secondary ion mass spectrometry, is advantageous in many applications. In recent years, many efforts have been made to also use these advantages of Laser-SNMS for the analysis of organic surfaces.

At present, resonantly enhanced multiphoton ionization (REMPI) is the most efficient ionization technique for organic molecules. Despite this effective ionization technique, however, a remaining matrix effect is observed, which has to be considered for quantitative surface analysis [1]. We have investigated this so-called sputter-induced matrix effect by comparing the secondary neutral emission and the corresponding secondary ion emission from UHV-deposited submonolayer and multilayer coverages of adenine and β-alanine on Ag substrates.

EXPERIMENTAL

All experiments were performed in a reflectron-type time-of-flight mass spectrometer equipped with a 10 keV electron impact gas ion source for sputtering. We chose Ar^+, Xe^+, and the polyatomic ion SF_5^+ as primary ion species to compare the

CP584, *Resonance Ionization Spectroscopy 2000: 10th Int'l. Symp.*, edited by J. E. Parks and J. P. Young
© 2001 American Institute of Physics 0-7354-0024-5/01/$18.00

influence of mass and the number of constituents of the primary ions on the yield.

The molecular overlayers were produced by Knudsen cell effusion on liquid nitrogen-cooled targets. Ag was chosen as substrate material since it shows the most significant effect. Starting with a sputter-cleaned substrate, the flux of photoions and secondary ions which are characteristic for the sputtered molecules was monitored online under static sputtering conditions until the substrate was covered by several monolayers (adenine: $M^{\oplus}$ and $(M + H)^+$; β-alanine: $(CNH_4)^{\oplus}$, which is formed via an α-cleavage process of the molecular radical ion after photoionization, and $(M + H)^+$).

REMPI of Molecules

REMPI is an efficient ionization technique for organic molecules. In principle, wavelength, laser intensity, and laser pulse width have to be adjusted to the spectro-scopic properties of the respective molecules. The use of short pulse lasers is often mandatory due to the necessity of (1) a broad bandwidth to cover all rovribronic states, which are excited by the sputter process, (2) a suitable laser intensity to prevent deple-tion of the excited state, and (3) low photon fluences to prevent photofragmentation of the molecular ions. Therefore we used a sub-picosecond excimer laser system at 248 nm for postionization of sputtered neutral adenine molecules. For postionization of β-alanine a conventional excimer laser ($\tau = 20$ ns) at 193 nm was used, since in this case a pulse shortening was not necessary (Figure 1) [2].

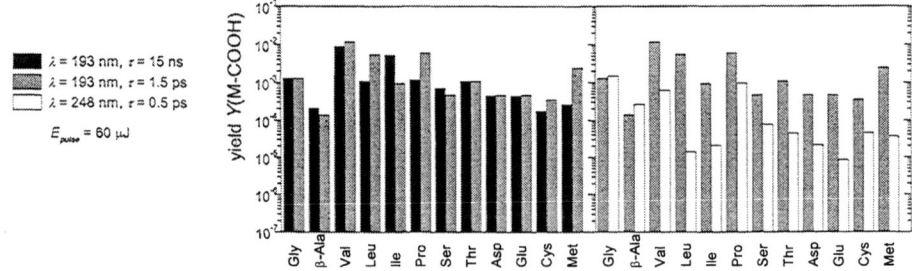

FIGURE 1. Photoion yields Y(M-COOH) of sputtered aminoacids for different laser parameters.

RESULTS AND DISCUSSION

Figure 2a shows the secondary particle signals in dependence on total surface coverage of adenine on Ag employing different primary ion species. Comparing the emission characteristics of the postionized neutral molecule $M^{\oplus}$ and the protonated molecule $(M + H)^+$ no significant difference could be observed. Both signals increased approximately in proportion to the surface coverage. After a distinct maxi-mum, the signals decreased despite still increasing layer thickness, reaching a plateau when a multilayer of the molecule had been adsorbed on the substrate. A characteris-tic quantity describing the dependence of the secondary particle yield on total surface coverage is the yield enhancement, which is defined as the ratio of maximum yield to

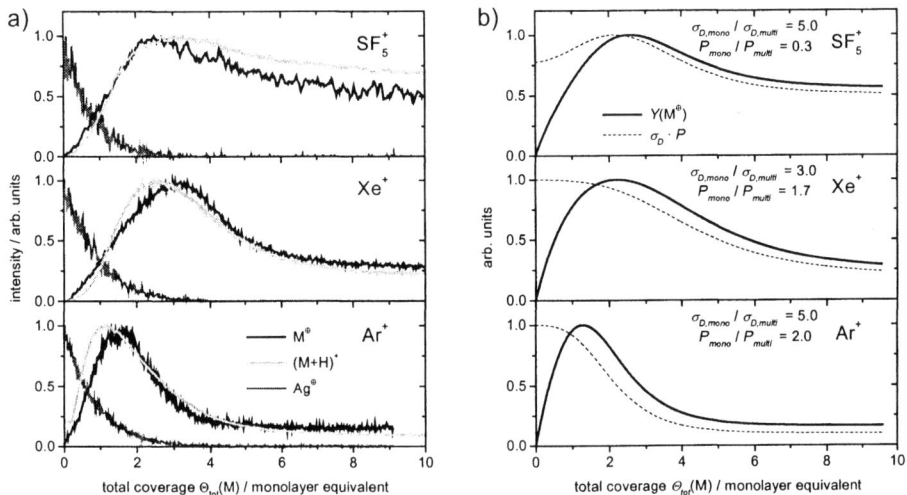

FIGURE 2. Secondary particle emission from molecular overlayers of adenine on Ag in dependence on total surface coverage: a) experimental results; b) calculated dependence using Eq. (1). The ratio $\sigma_{Dmono} / \sigma_{D,multi}$ was determined experimentally.

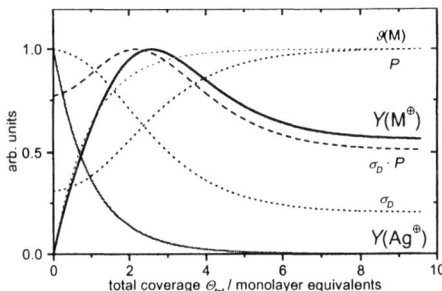

FIGURE 3. Model for the dependence of the yield on total surface coverage based on Eq. (1).

the yield obtained from a multilayer. It depends not only on the substrate material [1] but also on the primary ion used for sputtering. The influence of surface coverage on the sputtering process, which is reflected by the yield enhancement, is most pronounced by sputtering adenine using Ar^+ primary ions. It is reduced by changing to Xe^+. Using SF_5^+, the layer thickness has almost no influence on the yields.

The principal dependence of the yield on total surface coverage can be described by a simple model [1,3] based on yield equation (Figure 3):

$$Y(X_i^q) = \vartheta(M) \cdot \sigma_D \cdot P(M \to X_i^q) \cdot T \cdot D \qquad (1)$$

with surface density ϑ, disappearance cross section σ_D, transformation probability P, transmission T, and detection probability D. Assuming different values of σ_D and P in the submonolayer and multilayer range and a continuous transition between them, the yield dependence on total surface coverage can be calculated (Figure 2b).

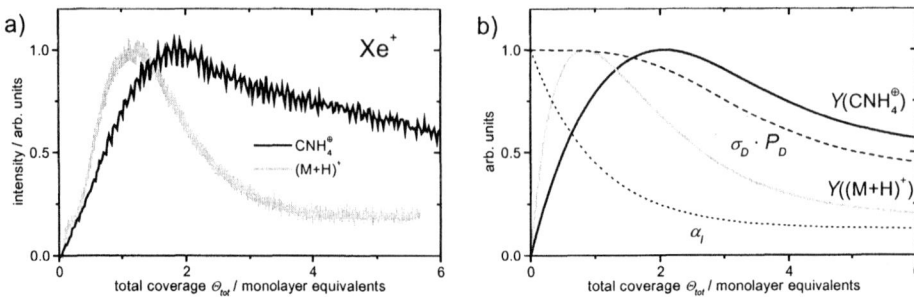

FIGURE 4. Secondary particle emission from molecular overlayers of β-adenine on Ag in dependence on total surface coverage: a) experimental results; b) calculated dependence using Eq. (1).

The behaviour of β-alanine is different from that of adenine as demonstrated in Figure 4. The secondary ion signal $I((M + H)^+)$ reaches its maximum before the maximum of the photoion signal $I((CNH_4)^{\oplus})$. This dependence can be modeled by an additional assumption for the transformation probability $P(M \rightarrow (M + H)^+)$, which factorizes into a desorption probability P_D and an intrinsic ionization probability α_I. The transformation probability depends on total surface coverage like the transformation probability $P(M \rightarrow (CNH_4)^{\oplus})$ of the characteristic photoion, and α_I considers the catalytic effect of the substrate atoms on the intrinsic ionization of the sputtered β-alanine molecules. A simple formulation of α_I is given by the sum of a constant part $\alpha_{I,multi}$ and a part proportional to the surface density of the substrate atoms

CONCLUSION

Due to the decoupling of desorption and ionization by using laser postionization for the analysis of sputtered neutral molecules, a classification of the matrix effects into a sputter-induced matrix effect and an ionization matrix effect was possible. Whereas the sputter-induced matrix effect is found for the secondary particle emission of adenine and β-alanine, only the secondary ion emission of β-alanine is affected by the ionization matrix effect.

REFERENCES

1. Schnieders, A., Möllers, R., and Benninghoven, A., *Surf. Sci.* (accepted).
2. Möllers, R., *Flugzeitmassenspektrometrische Untersuchungen zur Zerstäubung und Photo-ionisierung organischer Moleküle*, PhD Thesis Westfälische Wilhelms-Universität Münster, 1996.
3. Schnieders, A., Schröder, M., Stapel, D., Arlinghaus, H. F., and Benninghoven, A., "Molecular Secondary Particle Emission from Molecular Overlayers under SF_5^+-Bombardment" in *Secondary Ion Mass Spectrometry SIMS XII*, edited by A. Benninhoven et al., Elsevier Science Publishers, Amsterdam, 2000, pp. 263 - 266.

The detection of Sr sputtered from metallic and biological matrices by double-resonant photoionization mass spectrometry

Erno Vandeweert, Jeroen Bastiaansen, Vicky Philipsen,
Peter Lievens, Roger E. Silverans

*Laboratorium voor Vaste-Stoffysica en Magnetisme, K.U. Leuven,
Celestijnenlaan 200D, B-3001 Leuven, Belgium.*

Helmut H. Telle

*Department of Physics, University of Wales Swansea,
Singleton Park, Swansea SA2 8PP, UK.*

Abstract. Resonance ionization mass spectrometry (RIMS) was used to obtain isotope and state selective information on Sr sputtered from metallic and biological matrices. In exploratory experiments Sr atoms were sputtered from bulk metal upon impact of 15 keV Ar^+ ions, and probed by stepwise resonant ionization using two-color schemes. Efficient ionization schemes were selected to excite ground-state originating atoms into autoionizing states. Cross sections for photoionization were found to be up to the order of 10^{-15} cm^2. The Sr content in bone fragments was probed utilizing these schemes. Even with minimal sample preparation, a detection limit of $\pm$ 50ppm Sr in the hydroxiapatite-matrix of the bone was demonstrated, with isotope specificity. While this is inferior to detection limits normally associated with RIMS, these preliminary experiments were carried out for sputtering from untreated, non-conducting matrix materials.

INTRODUCTION

Over the years, resonance ionization mass spectrometry (RIMS) combined with ion-beam sputtering has proven to be able to detect with unmatched selectivity and sensitivity trace amounts of atoms and molecules for a variety of substances in different matrices. The impact of an energetic ion on the sample results in the release of particles from the outermost layers of the surface. Neutral atoms form the dominant contribution in such a plume of sputtered material. Liquid-metal ion guns are nowadays capable to generate short pulses of ions and focus them down to spot sizes of the order of 50 nm. Surface-sensitive analysis with a previously unseen lateral and depth resolution in combination with minimal sample consumption is thus within reach provided that also efficient detection schemes are employed. Although secondary ion mass spectrometry (SIMS) is a widely used and well established surface microprobe technique, the normally very low yields of directly formed ions together with the severe variations as a function of matrix composition, make a reliable quantifiable SIMS analysis in many cases extremely difficult or even altogether impossible. Efficient laser-ionization of sputtered neutrals often is found to be a viable if not the only alternative (1).

CP584, *Resonance Ionization Spectroscopy 2000: 10th Int'l. Symp.*, edited by J. E. Parks and J. P. Young
© 2001 American Institute of Physics 0-7354-0024-5/01/$18.00

In this contribution we report on the feasibility of trace detection of strontium in metallic and organic matrices by selective laser-ionization of abundantly sputtered ground-state atoms. The interest in this element is fueled by the fact that - due to its chemical similarity - Sr atoms are known to replace calcium atoms, and hence accumulate in, e.g., human bone tissue. The spatially resolved quantitative analysis of the natural and anthropogenic variations of Sr isotope ratios in biomedical specimen is known to be highly interesting in a number of medical (2), biological (3) and environmental (4) applications.

EXPERIMENTAL SETUP

The experimental setup and procedures are described in detail elsewhere (5). In short, the RIMS apparatus for the sputtering studies reported here consists of an UHV chamber in which an ion gun directs Ar^+ ions with a kinetic energy of 15 keV onto a centrally located target at 45° incidence. Cleaning of the sample surfaces was obtained by a prolonged continuous ion-bombardment prior to each experiment. During the actual measurements, the ion gun was operated in a long pulse mode (~6 µs) and the plume of sputtered particles was intersected, parallel to the sample surface foil, by two overlapping laser beams from a pulsed optical parametric oscillator and a pulsed dye laser system. These tunable laser systems generate linearly polarized laser light with laser pulses of ~6 ns duration the wavelength range from 225 to 1600 nm and pulse energies from 4 mJ in the UV up to 50 mJ in the visible range were available. The atoms sputtered in a polar angle interval of ~10° around the surface normal were photoionized and subsequently detected in a time-of-flight (TOF) mass spectrometer. TOF signals were preamplified, recorded with a digital oscilloscope and stored on a PC.

RESULTS AND DISCUSSION

The envisaged studies depend crucially on the use of highly efficient schemes used for the ionization of atoms subsequent to their release from the target to discriminate a small number of Sr atoms against a high background of other species. Detailed calculations on the population dynamics were therefore performed to select and evaluate ionization schemes, accessible using the laser systems available in the experimental setups of our laboratories (2). This theoretical study was followed by an experimental investigation on the efficiency of two-color two-step double-resonant ionization schemes (6). Such ionization schemes are characterized by the inclusion of a resonant ionization step into an autoionizing state. Fig. 1 shows a spectral scan over the

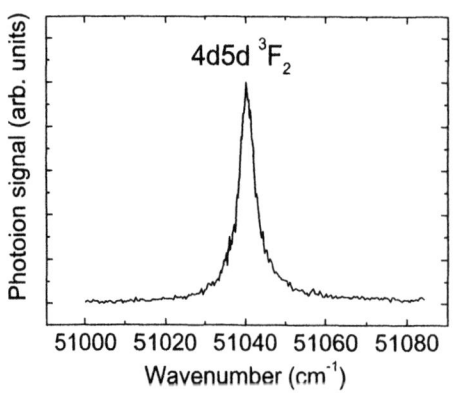

Figure 1. Ionization spectrum showing the even parity $4d\,5d\,^3F_2$ autoionizing state.

$4d\,5d\ ^{3}F_{2}$ autoionizing state (7). This spectrum was obtained by tuning the excitation-laser wavelength to the ground-state originating transition $5s^{2}\ ^{1}S_{0} \to 4d\,5p\ ^{3}D_{1}^{o}$. From the saturation behavior of the ionization step, cross sections for photoionization σ_I were deduced. Values for σ_I up to $5.\times10^{-15}$ cm^{2} were found, an increase of over 3 orders of magnitude compared to the cross sections for photoionization for schemes with a non-resonant ionization step (Fig. 2). It was thus

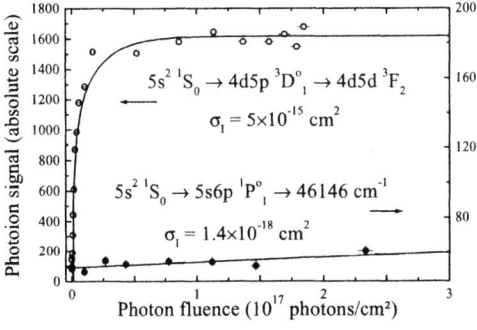

$5s^{2}\ ^{1}S_{0} \to 4d5p\ ^{3}D_{1}^{o} \to 4d5d\ ^{3}F_{2}$
$\sigma_I = 5\times10^{-15}$ cm^{2}

$5s^{2}\ ^{1}S_{0} \to 5s6p\ ^{1}P_{1}^{o} \to 46146$ cm^{-1}
$\sigma_I = 1.4\times10^{-18}$ cm^{2}

Photon fluence (10^{17} photons/cm^2)

Figure 2. Comparison of the saturation behavior of the ionization step in a two-color two-step ground-state originating scheme including an autoionizing state (open symbols, left scale) and into the structureless continuum (closed symbols, right scale).

shown that the use of an autoionizing state allows to saturate the overall ionization even at moderate laser pulse energies. Although a second tunable laser system is required, these ionization schemes are highly favored since the highest achievable ionization efficiency (by ionizing every atom of interest present in the ionization volume during the laser pulse) is combined with a high degree of isotope selectivity.

Following these exploratory experiments, the Sr content in bio-organic matrices (bone tissue) was probed by two-color RIMS. Three different samples were prepared : a pure bone fragment and bone fragments soaked in resp. a 2000 *parts per million* (ppm) and a 5000 ppm SrCl$_2$ in water. During the soaking process, Ca in the bone matrix is replaced by the chemically resembling Sr. The actual Sr content in the bone fragments was later assessed using laser-induced breakdown spectroscopy (LIBS) and comparing the measured values with a calibration curve obtained from a series of reference pellets of CaCO$_3$ containing known amounts of Sr (8). Note that the CaCO$_3$ matrix closely resembles bone hydroxiapatite - Ca$_{10}$(PO$_4$)$_6$(OH)$_2$ - in its properties.

During the RIMS measurements, care was taken to avoid charging of the non-conductive sample surface by placing a conducting mesh in front of the sample and keeping the ion pulse duration sufficiently short and the repetition rate low. In Fig. 3 a comparison between mass spectra obtained using two-color two-step ionization of a non-treated bone fragment

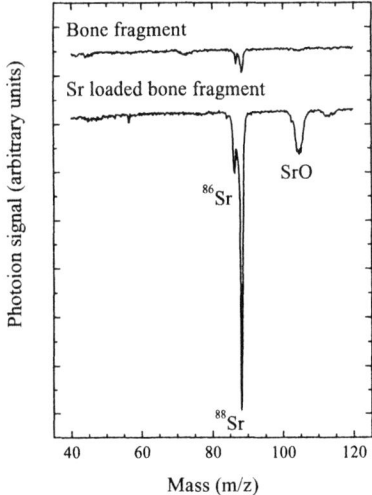

Bone fragment

Sr loaded bone fragment

SrO

^{86}Sr

^{88}Sr

Mass (m/z)

Figure 3. Comparison between mass spectra obtained on a non-treated bone fragment (upper trace) and a bone fragment kept in a water solution containing 5000 ppm SrCl$_2$ (lower trace).

(containing a few hundred ppm of Sr from natural sources (8)) and a bone fragment kept in a water solution containing 5000 ppm $SrCl_2$ is shown. From these experiments the instrumental detection limit could be estimated to be ~50 ppm Sr in a Ca matrix. While this is far inferior to detection limits normally associated with RIS it has to be kept in mind that these preliminary experiments were carried out by sputtering from untreated, non-conducting matrix materials with minimal sample preparation. It should be noted that at this stage only very limited efforts were put into avoiding the notorious space-charge effects encountered for such specimen and which hamper the full exploitation of the selectivity of RIS.

In conclusion, we have shown results from a feasibility study to detect trace amounts of Sr in metallic and biological matrices. For this, efficient two-color two-step photoionization schemes were devised and evaluated. Future experiments will include the use of focused beams from Ar^+ and liquid-metal ion guns to allow for fully spatially resolved detection of Sr traces in biomedical matrices. Ultimately, this know-how could help to quantitatively monitor the build-up of radioactive Sr isotopes, and their spatial distribution, in bone tissue of humans and animals near radioactively contaminated sites.

ACKNOWLEDGMENTS

This work is financially supported by the Fund for Scientific Research - Flanders (Belgium) (F.W.O.), by the Flemish Concerted Action (G.O.A.) Research Program and by the Interuniversity Poles of Attraction Program (I.U.A.P.) - Belgian State, Prime Minister's Office - Federal Office for Scientific, Technical and Cultural Affairs. P.L. and E.V. are postdoctoral fellows of the F.W.O. Financial support for part of this work within the framework of the British-Flemish ARC programme (Project V7.002.97 N5) is gratefully acknowledged.

REFERENCES

1. K.F. Willey, H.F. Arlinghaus, and T.J. Whitaker, Routine analysis at sub-micron resolution through the use of sputter initiated resonance ionization spectroscopy, Appl. Surf. Science **144-145** (1999) 36.
2. R.O. Jones, R.M. Perks, and H.H. Telle, Isotope-specific resonance ionization mass spectrometry using continuous-wave diode and pulsed dye lasers for trace detection and enrichment of elements of environmental and medical importance, exemplified for Strontium (Sr) – Part I: Theoretical modelling, Rap. Comm. in Mass Spectrom. **10** (1996) 1725-1738.
3. R.H. Wasserman, Strontium as a tracer for calcium in biological and clinical research, Clinical Chemistry **44** (1998) 437-439.
4. B.A. Bushaw and B.D. Cannon, Diode laser based resonance ionization mass spectrometric measurement of strontium-90, Spectrochim. Acta Part B **52** (1997) 1839-1854.
5. P. Lievens, E. Vandeweert, P. Thoen, and R.E. Silverans, Resonant photoionization of Ni I into autoionizing states, Phys. Rev. A **54** (1996) 2253.
6. V. Philipsen, J. Bastiaansen, G. Verschoren, P. Lievens, E. Vandeweert, R.E. Silverans, and H.H. Telle, Double-resonant photoionization spectroscopy of Sr I, Spectrochim. Acta B **55** (2000) 1539.
7. M. Aymar, E. Luc-Koenig, and S. Watanabe, R-matrix calculation of eigenchannel multichannel quantum defect parameters for strontium, J. Phys. B **20** (1987) 4325-4345.
8. H.A. McKenzie and L.E. Smythe (eds.), Quantitative Trace Analysis of Biological Materials (Elsevier Amsterdam, 1988).
9. J.O. Cáceres, B. Béscos del Castillo, O. Samek, H.H. Telle, and A. González Ureña, Quantitative compositional analysis of mineralized tissue, *1st Int. Conf. Laser-Induced Breakdown Spectroscopy and Applications*, Pisa, Italy (October 2000).

Two-color two-step laser ionization spectroscopy of uranium sputtered from thin metallic films

E. Vandeweert, V. Philipsen, P. Lievens, R.E. Silverans

Laboratorium voor Vaste-Stoffysica en Magnetisme, K.U. Leuven,
Celestijnenlaan 200D, B- 3001 Leuven, Belgium

C. Grüning

Institut für Kernchemie, Universität Mainz, D-55099 Mainz, Germany

F. Miserque, M. Betti, T. Gouder, and N. Erdmann

European Commission Joint Research Centre,
Institute for Transuranium Elements, Postfach 2340,
D-76125 Karlsruhe, Germany

Abstract. Resonance ionization mass spectrometry (RIMS) was used to obtain state selective information on uranium atoms released from thin films by keV ion beam bombardment. First experiments were conducted to evaluate double-resonant laser ionization schemes. The high selectivity of such ionization schemes is further illustrated by comparing the flight-time distributions of the sputtered particles obtained both by resonant and non-resonant multiphoton (355 nm) laser post-ionization.

INTRODUCTION

The detection of uranium using resonance ionization spectroscopy (RIS) has drawn quite some attention over the years. In most studies, however, researchers focused on elaborate sample preparation (1) often combined with sophisticated three-color ionization schemes exploiting transitions with favorable isotope shifts (2). Such schemes can be optimized to ensure a high isotope selectivity needed for, e.g., laser-ionization based isotope separation. In view of the development of sensitive techniques for routine analysis of individual radioactive particles obtained after environmental sampling (3), we focus here our attention to more simple two-step laser ionization spectroscopy in combination with ion-beam sputtering. The impact of an energetic ion on the sample results in the release of particles from the outermost layers of the surface. Neutral atoms form the dominant contribution in such a plume of sputtered material. Liquid-metal ion guns are nowadays capable to generate short pulses of ions and focus them down to spot sizes of the order of 50 nm. Surface-sensitive analysis with a previously unseen lateral and depth resolution in combination with minimal sample consumption is thus within reach provided that also efficient detection schemes are employed. Although secondary ion mass spectrometry (SIMS) is a widely used and well established surface microprobe technique, it suffers from several drawbacks. The low yields of directly formed ions together with the severe variations as function of matrix composition, e.g., from oxidized surfaces, make a reliable quantifiable SIMS analysis

CP584, *Resonance Ionization Spectroscopy 2000: 10th Int'l. Symp.*, edited by J. E. Parks and J. P. Young
© 2001 American Institute of Physics 0-7354-0024-5/01/$18.00

extremely difficult or even altogether impossible. Moreover, isobaric interferences (such as $^{238}U/^{238}Pu$ or $^{241}Pu/^{241}Am$) cannot be resolved by SIMS analysis.

EXPERIMENTAL SETUP

The experimental setup and procedures are described in detail elsewhere (4). The RIMS apparatus for the sputtering studies reported here consists of an UHV chamber in which an ion gun directs Ar^+ ions with a kinetic energy of 15 keV onto a centrally located target foil at $45°$ incidence in a long pulse mode (~1 μs) for the spectroscopy measurements. For the flight-time measurements short pulses of the order of 200 ns are used. Taking into account the sputter yield for U upon 15 keV Ar^+ bombardment, this implies that per detection cycle the mass equivalent of only ~5×10^6 atoms is released from the surface. This plume of sputtered particles is intersected, parallel to the foil, by two overlapping, loosely focused laser beams from a pulsed optical parametric oscillator and a pulsed dye laser system. These tunable laser systems generate linearly polarized laser light with laser pulses of ~6 ns in the wavelength range from 225 to 1600 nm and pulse energies from 4 mJ in the UV up to 50 mJ in the visible range. The atoms sputtered in a polar angle interval of ~$10°$ around the surface normal are photoionized and subsequently detected in a time-of-flight mass spectrometer. Time-of-flight spectra were preamplified, recorded with a digital oscilloscope and finally stored on a PC.

The samples consisted of 0.5 μm thick layers of depleted ($^{235}U/^{238}U = 0.3\%$) uranium sputter-deposited onto 5×5 mm^2 Ta foils. Although samples were stored and transported under vacuum conditions, oxidation of the upper layers of the metallic films could not be avoided. A typical mass spectrum obtained from such an oxidized metal surface contains mass peaks that correspond to U, UO, and UO_2. When the sample surface is subjected to a continuous Ar^+ bombardment for a short period (30 s) the upper oxide layers are removed, resulting in a strong increase of the U signal. This procedure was repeated periodically during the course of the experiment.

RESULTS AND DISCUSSION

The preliminary experiments presented in this study were aimed to evaluate the efficiency and the specificity of resonant and non-resonant ionization schemes. We therefore explored the continuum structure of uranium near the first ionization limit using two-color two-step photoionization spectroscopy. In such experiments, the photon energy of one laser is fixed on a resonant transition. A scan of the excitation laser wavelength revealed several strong transitions originating from different metastable states which are substantially populated during the sputter event. Dedicated experiments have shown that the population on metastable states during ion-beam sputtering of metals add up to a fraction that can be larger than the fraction of ground-state atoms. The processes governing the population on these states are not fully understood (5). By scanning the wavelength of the ionization laser, a large number of autoionizing states were observed. Fig. 1 shows part of the ionization spectrum obtained with the excitation laser wavelength fixed to a transition form the ground state $^5L_6^o$ to an intermediate state at $E = 30795.4$ cm^{-1} $(J = 5)$. The efficiency of ionization schemes including an autoionizing state was compared to that with a non-resonant ionization step. In Fig. 2, the saturation behavior of the ionization steps of both ionization schemes are shown. Incorporating an autoionizing state allows to saturate the overall ionization process at only a fraction of the laser energy needed for a scheme employing a non-resonant ionization step. Although this implies the use of a second tunable laser system, such double-resonant ionization schemes allow to increase the selectivity of the RIS process by suppression of the isobaric interferences from non-resonantly ionized species.

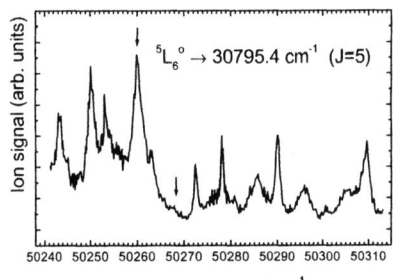

Figure 1. Ionization spectrum showing a large number of odd parity autoionizing states in the continuum near the first ionization threshold of U I. Arrows indicate the total energies in the two-color two-step ionization schemes with and without an autoionizing state which were compared with respect to their overall ionization efficiency.

The selectivity of the double-resonant two-color two-step ionization schemes can further be demonstrated by comparing the flight-time distributions of the sputtered particles obtained by resonant and non-resonant multiphoton (355 nm) laser post-ionization upon sputtering of an oxidized sample. Flight-time distributions can be obtained by measuring the photoion signal obtained from these species as function of the delay time between the impact of a short ion pulse on the sample and the ionizing laser pulses. When the sample-to-ionization distance is known, such experiments allow to obtain information on the kinetic energy distribution with which the particles are ejected during the sputter event. The normalized flight-time distributions of U and UO obtained using non-resonant (355 nm, ~5 J/cm^2) laser-ionization are within the accuracy of the experiment indistinguishable (upper panel of Fig. 3). This proves that a large fraction of the detected atom signal originates from photodissociated molecules by interaction with the intense non-resonant photon field in the ionization volume. In the lower panel of Fig. 3 flight-time distributions of the same species are shown but now a double-resonant ionization scheme was employed to ionize ground-state atoms with a maximal intensity of only 50 mJ/cm^2. Even such low laser energy is enough to ionize a sizable number of UO molecules. From the clear differences in both flight-time distributions we conclude that in this case the laser intensity is too low to induce photodissociation of the oxide and that the largest fraction of photoionized uranium stems from two-step laser ionization of directly sputtered atoms.

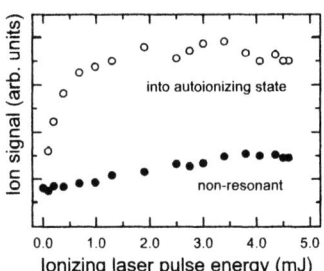

Figure 2. Comparison of the saturation behavior of the ionization step in two-color two-step schemes including a transition into an autoionizing state (open symbols) and a non-resonant ionization step (closed symbols).

In conclusion, we demonstrated that two-color two-step ionization schemes that include an autoionizing state can be successfully applied to detect extremely small amounts (below 10^7 atoms per detection cycle) of ground-state uranium atoms sputtered from surfaces involving minimal sample treatment. An abundant population on a large number of metastable states combined with the high state-selectivity of RIS implies that the reservoir of detectable (i.e., ground-state) atoms is reduced to less than 50 % of the sputtered particles compared to non-resonant photoionization. However, the overall ionization efficiency of such double-resonant ionization schemes is so high that the laser intensities can be attenuated with several orders of magnitude to obtain the same ion signal intensities as in a non-resonant multiphoton experiment. Isobaric interferences can thus be largely avoided and photofragmentation of molecular species can be strongly suppressed. The large number of intermediate states available for

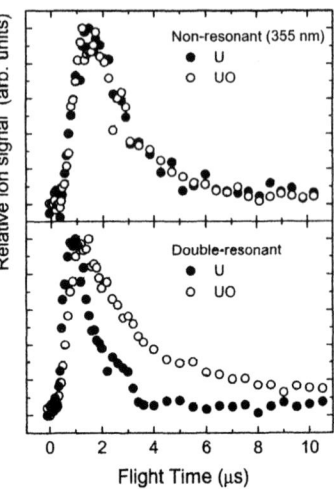

Figure 3. Flight-time distributions of sputtered U and UO particles obtained by non-resonant multiphoton (355 nm) (upper panel) and double-resonant laser-ionization (lower panel).

U I and autoionizing states observed so far should encourage research into the design of one-color two-step schemes in which the ionization transition is in near-resonance with an autoionizing state (4). Such schemes allow to design a conceptual much simpler (and thus more economical) experimental setup incorporating only one tunable laser system, without trading in on the detection efficiency or selectivity.

ACKNOWLEDGMENTS

This work is financially supported by the Fund for Scientific Research - Flanders (Belgium) (F.W.O.), by the Flemish Concerted Action (G.O.A.) Research Program and by the Interuniversity Poles of Attraction Program (I.U.A.P.) - Belgian State, Prime Minister's Office - Federal Office for Scientific, Technical and Cultural Affairs. P.L. and E.V. are postdoctoral fellows of the F.W.O.

REFERENCES

1. A.W. McMahon, J.D. Gilmour, M.B. Hernandez, and M. Rateitzak, A RIS-TOF instrument for the measurement of ultratrace quantities of uranium and plutonium, AIP Conf. Proc. **454** (1998) 269.
2. A. Coste, R. Avril, P. Blancard, J. Chatelet, D. Lambert, J. Legre, S. Liberman, and J. Pinard, New spectroscopic data on high-lying excited levels of atomic uranium, J. Opt. Soc. Am. **72** (1982) 103.
3. N. Erdmann, A. Benninghoven, M. Betti, T. Gouder, C. Grüning, F. Kollmer, P. Lievens, F. Miserque, V. Philipsen, R.E. Silverans, E. Vandeweert, Evaluation of a system for trace and particle analysis based on resonance ionization of sputtered neutrals, contribution to this Conference, Proceedings RIS-2000.
4. P. Lievens, E. Vandeweert, P. Thoen, and R.E. Silverans, Resonant photoionization of Ni I into autoionizing states, Physical Review A **54** (1996) 2253.
5. B.J. Garrison, N. Winograd, R. Chatterjee, Z. Postawa, A. Wucher, E. Vandeweert, P. Lievens, V. Philipsen, and R.E. Silverans, Sputtering of atoms in fine structure states: a probe of excitation and de-excitation events, Rap. Comm. in Mass Spectrom. **12** (1998) 1266.